ORDINARY DIFFERENTIAL EQUATIONS
AND APPLICATIONS

Essential Textbooks in Mathematics

ISSN: 2059-7657

The *Essential Textbooks in Mathematics* series explores the most important topics that undergraduate students in Pure and Applied Mathematics are expected to be familiar with.

Written by senior academics as well lecturers recognised for their teaching skills, they offer, in around 200 to 400 pages, a precise, introductory approach to advanced mathematical theories and concepts in pure and applied subjects (e.g. Probability Theory, Statistics, Computational Methods, etc.).

Their lively style, focused scope, and pedagogical material make them ideal learning tools at a very affordable price.

Published:

Ordinary Differential Equations and Applications
 by Enrique Fernández-Cara (University of Seville, Spain)

A First Course in Algebraic Geometry and Algebraic Varieties
 by Flaminio Flamini (University of Rome "Tor Vergata", Italy)

Analysis in Euclidean Space
 by Joaquim Bruna (Universitat Autònoma de Barcelona, Spain &
 Barcelona Graduate School of Mathematics, Spain)

Introduction to Number Theory
 by Richard Michael Hill (University College London, UK)

A Friendly Approach to Functional Analysis
 by Amol Sasane (London School of Economics, UK)

A Sequential Introduction to Real Analysis
 by J M Speight (University of Leeds, UK)

Essential Textbooks in Mathematics

ORDINARY DIFFERENTIAL EQUATIONS AND APPLICATIONS

Enrique Fernández-Cara

University of Seville, Spain

W☺ World Scientific

NEW JERSEY • LONDON • SINGAPORE • BEIJING • SHANGHAI • HONG KONG • TAIPEI • CHENNAI • TOKYO

Published by

World Scientific Publishing Europe Ltd.

57 Shelton Street, Covent Garden, London WC2H 9HE

Head office: 5 Toh Tuck Link, Singapore 596224

USA office: 27 Warren Street, Suite 401-402, Hackensack, NJ 07601

Library of Congress Cataloging-in-Publication Data

Names: Fernández-Cara, Enrique, author.

Title: Ordinary differential equations and applications /
   Enrique Fernández-Cara, University of Seville, Spain.

Description: New Jersey : World Scientific, [2024] | Series: Essential textbooks in mathematics,
   2059-7657 | Includes bibliographical references.

Identifiers: LCCN 2023003677 | ISBN 9781800613935 (hardcover) |
   ISBN 9781800613966 (paperback) | ISBN 9781800613942 (ebook for institutions) |
   ISBN 9781800613959 (ebook for individuals)

Subjects: LCSH: Differential equations--Textbooks.

Classification: LCC QA372 .F38 2024 | DDC 515/.352--dc23/eng20230415

LC record available at https://lccn.loc.gov/2023003677

British Library Cataloguing-in-Publication Data

A catalogue record for this book is available from the British Library.

For any available supplementary material, please visit
https://www.worldscientific.com/worldscibooks/10.1142/Q0410#t=suppl

Desk Editors: Logeshwaran Arumugam/Adam Binnie/Shi Ying Koe

Typeset by Stallion Press
Email: enquiries@stallionpress.com

This book is dedicated to Rosa Echevarría
(for one million reasons that I have in mind)

Preface

Differential equations can be viewed as a tool well suited to bring mathematics closer to real life and describe phenomena with origin in physics, chemistry, biology, economics, etc. Of course, this "invites" scientists and engineers to use them.

On the other hand, differential equations are at the starting point of a lot of purely mathematical activity. Thus, a large part of functional analysis has been motivated by the need to solve questions coming from the analysis of differential systems. The same can be said of numerical analysis. Also, we must remember the relevant role played by differential equations in the formulation and resolution of problems in harmonic analysis, differential geometry, probability calculus, etc.

Consequently, the interest in differential equations is justified from a double viewpoint: as significative and inspiring examples of questions and results arising in many areas in mathematics and also as a machinery that must be well understood in order to apply mathematics to real-world problems.

The main goals in this book have been to provide a rigorous introduction to the theoretical study of this important tool and, also, to demonstrate its utility with applications coming from many fields.

The book has its origin in several "classical" or "standard" courses given by the author in the University of Seville for decades. It has been an objective to make compatible the rigor of the underlying mathematical theory and the richness of applicability. Thus, together with existence, uniqueness, regularity, continuous dependence on data and parameters, permanent interpretation of the laws, and the role of the data, the behavior of the solutions and other elements is offered. Although briefly, the connections to control and optimization have also been indicated.

The author hopes that this volume works as a fruitful introduction to the subject.

About the Author

Enrique Fernández-Cara received a Grade in Mathematics from the University of Seville (Spain) and a Doctorate in Mathematics from this university and also University Pierre and Marie Curie (Paris 6, France). He has worked in Seville, Paris, and Clermont-Ferrand and, also, as visiting professor, in several Brazilian universities: UNICAMP, UFF in Nitéroi, UFPE in Recife, and UFPB in Joao Pessoa. He has been the advisor of 30 PhD theses and has participated in more than 20 research projects in Spain, France, and Brazil. He has published more than 180 papers in peer-reviewed international journals and has attended more than 50 international conferences, most of them related to control theory and PDEs. He has attained the maximal qualification of six 6-year research periods by the Spanish agency CNEAI (the last one in 2015). His main areas of interest are the analysis of PDEs from both the theoretical and numerical viewpoints, with special emphasis on nonlinear equations and systems from physics, biology, etc., and the study of related control and optimization problems. In particular, he has considered problems for semilinear and nonlinear elliptic, parabolic, and hyperbolic systems, homogeneous and non-homogeneous Navier–Stokes equations, Oldroyd viscoelastic systems and variants, other nonlinear nonscalar systems from tumor growth modeling, etc. He has contributed to regularity results, new numerical schemes (some of them based on parallelization), new control solution techniques, and related applications. More information can be found at https://personal.us.es/cara/.

Acknowledgments

I would like to thank my colleagues Blanca Climent, Manuel Delgado, and Rosa Echevarría for their help in the elaboration of a previous version of this book. Especially, I am indebted to Rosa Echevarría . . .

The Notation and Preliminary Results

ODE, ODS: ordinary differential equation, ordinary differential system.

PDE: partial differential equation.

CP: Cauchy problem.

$\mathbb{R}$: the field of real numbers.

$\mathbb{R}_+$: the set of nonnegative real numbers.

$\mathbb{R}_+^N$: the set $\mathbb{R}^{N-1} \times \mathbb{R}_+$.

I, J: intervals in $\mathbb{R}$ (open or not, bounded or not).

$\Omega, G, U, \mathcal{O}$: open sets in $\mathbb{R}^N$.

$\partial\Omega$: the boundary of Ω.

δ_{ij}: *Kronecker's symbol;* $\delta_{ij} = 1$ if $i = j$, $\delta_{ij} = 0$ otherwise.

$C^0(I), C^0(U)$: the vector space of continuous functions $\varphi : I \mapsto \mathbb{R}$ or $\varphi : U \mapsto \mathbb{R}$.

$C^k(I), C^k(U)$: the subspace of $C^0(I)$ or $C^0(U)$ formed by the functions that possess continuous derivatives up to the kth order.

$C^\infty(I), C^\infty(U)$: the intersection of all $C^k(I)$ or $C^k(U)$ with $k \geq 1$.

We will denote by $|\cdot|$ the Euclidean norm in $\mathbb{R}^N$; $B(x_0; r)$ (resp. $\overline{B}(x_0; r)$) will stand for the open (resp. closed) ball of center x_0 and radius r.

In order to get maximal profit when reading this book, it is preferable to be familiar with basic results for real-valued functions of one and several real variables (continuity, differentiability, integrability).

In particular, it is adequate to understand and be able to apply the following results, which will be frequently used:

- The *Chain Rule* (differentiation of composed mappings) for functions defined in a set in $\mathbb{R}^n$ with values in $\mathbb{R}^m$.
- Continuity and differentiability criteria for functions defined from integrals depending on parameters (*Leibniz's Rule*).

The background needed in this book is contained and very well explained in the references (Rudin, 1976; Smith, 1998; Spivak, 1970).

There are a huge number of texts devoted to the theoretical study of ODEs. At the end of the book, the reader can find a collection of significative samples, where we can find "classical" texts (Brauer and Nohel, 1969; Coddington and Levinson, 1955; Eastham, 1965; Hartman, 2002; Petrovskii, 1966).

Other well-known references are Cordunneanu (1971), Hochstadt (1975), Simmons (1977), Miller and Michel (1982), Piccinini *et al.* (1984), Amann (1990), and Braun (1993).

Finally, the interested reader can take a look at Robinson (2004), Hsu (2006), Chicone (2006), Arnold (2006), Ding (2007), Cronin (2008), Barbu (2016), Nandakumaran *et al.* (2017), Graef *et al.* (2018a, 2018b), and Rao (1981), for other more recent presentations.

Contents

Introduction

The origin of this book is a collection of Notes written by the author to give several classical courses on differential equations in the University of Seville. They were titled "Mathematical Analysis III," "Ordinary Differential Equations," "Mathematical Modeling," etc.

The main aim is to provide an introduction to the theory of ordinary differential equations.

In view of their ubiquity, it is in general very difficult to identify the origin of many of the arguments and results. Like many other people, I have assumed that they belong to the scientific community. Consequently, it is appropriate to thank all the colleagues that contributed somehow to the material in this book. It is also adequate to apologize to anyone who thinks that a particular result or argument that follows belongs to him or her and I have forgotten to mention the authorship.

In general terms, an *ordinary differential equation* is an equality in a function space where there is an unknown (a function of one real variable) that appears together with one or several of its derivatives. The unknown is usually denoted the *dependent variable*. It is assumed to be a function of the so-called *independent variable*. By definition, the *order* of the equation is the maximal order of a derivative found in the identity.

Analogously, we can speak of *ordinary differential systems*. This is a collection of equalities of the same kind where there are not only one but several dependent variables (and their derivatives). We will use ODE and ODS as acronyms for ordinary differential equation and ordinary differential system, respectively.

Almost all the time (but not always), we will consider equations and systems with a *normal structure*. This means that the derivatives of the highest order are isolated and can be written alone, at one side of the identity.

Our main goals are to provide a rigorous introduction to the theoretical study of these important tools and, also, to put in evidence their usefulness as models or representations of reality by introducing and describing applications coming from many fields.

The contents of the book are the following.

Chapter 1 is devoted to presenting some fundamental concepts arising in ODE theory and motivations in physics, engineering, etc. We will formulate the *Cauchy problem* for equations and systems of any order. We will also describe the most relevant *elementary integration methods*, devoted to solving explicitly some specific equations: separable ODEs, linear ODEs of the first and second-order, homogeneous ODEs, Bernoulli ODEs, etc.

In many cases, it will be shown that these equations have a motivation coming not necessarily from inside mathematics. Thus, examples with origin in mechanics, nuclear chemistry, population dynamics, cell biology, finance, etc., will be given. I hope this will convince the reader that, for many purposes in science and engineering, it is convenient and fruitful to acquire some knowledge of the theory.

Here, it must be observed that we can speak of elementary methods only in a very small number of cases. Actually, this is a motivation for the theoretical analysis achieved in Chapters 2–4: we accept that the solutions cannot be explicitly found and we try however to identify and explain their properties in detail. Let us mention that this also justifies the need and relevance of the related numerical analysis and methods that have been the objective of a lot of work, especially along the last 50 or 60 years.

From the examples considered in this chapter, it becomes clear that an ODE possesses not one but an infinity of solutions. Obviously, the same can be said for an ODS. Accordingly, it is completely reasonable to "complete" or "complement" the equation or system with some conditions and formulate and analyze the resulting problem.

This is the inspiring principle of the Cauchy problem, also called the *initial-value problem*. For example, in the case of a first-order ODE, what we do is to add one initial-value condition, that is, information on the value of the unknown at a fixed value of the independent variable. This way, we expect to identify one single function (a solution to the equation satisfying this requirement).

The arguments and results in this and the other chapters have been complemented with applications, with the aim to illustrate the theory and justify its interest. This has been a permanent concern during the writing of the book.

The objective of Chapter 2 is to present classical local results for the Cauchy problem for a first-order "normal" ODS. We start with some background on basic mathematical analysis: the definition and properties of spaces of continuous functions, the concept of contraction in a metric space, the Banach Fixed-Point Theorem, etc.

The central result in this chapter is *Picard's Theorem*, which furnishes the existence and uniqueness of a local solution to the Cauchy problem. We also present *Peano's Theorem*. Under less restrictive conditions, this result guarantees the existence of at least one local solution.

We end the chapter with a brief mention of *implicit ODEs* (i.e. those that do not have a normal structure) and the associated Cauchy problems.

Chapter 3 is devoted to analyzing uniqueness. Here, the main tool is *Gronwall's Lemma,* a well-known result that has considerable interest by itself and can be used in many contexts. In the main result in this chapter, we prove that under the assumptions in Picard's Theorem, two solutions to the same Cauchy problem defined in the same interval must coincide.

In Chapter 4, we analyze the extension (or prolongation) of local solutions and the existence and uniqueness of a maximal (i.e. non-extendable) solution to the Cauchy problem. We present some results that allow to characterize maximal solutions in terms of their behavior near the endpoints of their interval of definition. Thus, we examine with detail "blow-up" phenomena and we try to interpret their occurrence in physical terms, in view of their relevance in many applications.

In Chapter 5, we consider *linear* ODEs and ODSs. This is an important class of equations and systems where a lot of things can be said. For instance, for a linear ODE of order n, the family of solutions possesses a vector structure; more precisely, it is a *linear manifold* of dimension n and, consequently, if we are able to compute $n + 1$ solutions, then we know all of them. On the other hand, linear systems are the first to be considered to model real-world phenomena. Indeed, when we try to make explicit a law coming (for example) from physics, our first attempt will naturally be to use a polynomial formula of the first order.

We start the study with a close look at *homogeneous linear* systems, where we find some basic concepts: fundamental matrices, fundamental systems, Wronskians, etc. In a second step, we use the so-called *Lagrange method* to solve general non-homogeneous systems. The arguments show that, if we are able to compute a fundamental matrix, then all the solutions to the system are known.

Then, we consider constant coefficients systems of the linear kind. For them, the explicit computation of a fundamental matrix is achievable. Here, the key idea relies on power series in the space of square matrices. In particular, we can introduce the exponential of a matrix and, after this, the semigroup solution to a homogeneous linear system with constant coefficients.

Chapter 6 concerns boundary-value problems for linear ODSs. They appear when we complement an ODS with *boundary conditions*, that is, information on the behavior of the unknown (and its derivatives) at the endpoints of the interval where we try to find the solution. The main result in this chapter is the so-called *Alternative Theorem*. As will be shown, it permits to find out whether or not a solution to the problem exists (and how many there are) by studying a linear algebraic system.

Chapter 7 is devoted to deducing regularity results. First, we see that the smoothness of the solutions "improves" as we impose stronger regularity properties on the model functions. This provides *local regularity* results. Then, we adopt a global viewpoint and we investigate the way the solutions depend on the data and possible parameters (or additional functions) arising in the problem. In particular, we prove that, under the hypotheses in Picard's Theorem, the maximal solution to a Cauchy problem depends continuously on the initial data. Under somewhat stronger assumptions, it is also possible to prove differentiability and even identify the corresponding derivatives.

It must be emphasized that these results are of major relevance from the viewpoint of applications. Indeed, in practice we are always subject to work with "erroneous" data. If they come to us as the result of a measurement, it is clear that they are only approximate. Even more, if our interest is to use the data for computational purposes, we will unavoidably truncate the values in our computer. Consequently, only a continuous dependence result can ensure that our conclusions (and our computations) are significant. Without them, it will not be clear in any way whether ODEs are actually useful to model real phenomena.

Chapter 8 is an introduction to *stability theory*. The stability of a solution of an ODS is a crucial property. In this respect, a lot of work has been carried out in the last decades, especially in connection with high dimension nonlinear systems. For reasons of space, we will only present in this chapter a few results, but we will try to motivate further analysis in some interesting and challenging directions.

A typical stability question is the following: consider an ODE such that the function identically equal to zero is a solution (the so-called *trivial* solution); then, is it true or not that any other solution sufficiently close to zero at a fixed point keeps close everywhere?

The goal of Chapter 9 is to present the *method of characteristics* for partial differential equations of the first order. A partial differential equation is an expression where we find an unknown function that depends on several real variables together with some of its partial derivatives. As before, we say that the order of the equation is the maximal order of the derivatives of the unknown (we use the acronym PDE in the sequel). Again, these equations possess a lot of motivations in science and engineering and their analysis and exact and approximate resolution are of major interest from many viewpoints.

The study of PDEs is a much more complex subject and there is no way to give a significative set of results in a reduced amount of chapters. There is, however, a connection of first-order PDEs and ODSs. Thus, we can associate to any given PDE a particular ordinary system (the *characteristic* system), whose solutions can be used to construct the solutions to the former. This key fact allows to formulate and solve, at least locally, the Cauchy problem for a PDE.

It will be seen in this chapter that these ideas are natural when we interpret the PDE as a law governing the behavior of a physical phenomenon in several different situations.

Chapter 10 contains a brief introduction to control theory. More precisely, we consider systems where some data are to be fixed in accordance with appropriate criteria and we exhibit a theory where that provides means to make a good choice. For example, if in the ODS we find a parameter α that we are free to choose and we want to get a solution close to a desired prescribed function y_d, it can be appropriate to solve an extremal problem where the function to minimize is the "distance" of the solution to y_d.

This is the kind of problem considered here. We will present existence and uniqueness results and also we will give *characterization* results that allow to identify the "good" or even the "best" choice.

Along these chapters, I have included a considerable amount of exercises. They have been conceived and designed to help to understand the theoretical contents and also, to reinforce the connection of differential equations and problems coming from anywhere.

In many cases, useful "indications" and "hints" have been given. On the other hand, the difficulty level of some sections labeled with an asterisk * is higher and they can be omitted on first reading.

Finally, Chapter 11 contains a set of additional notes and remarks that serve to illustrate and complete the previous material. They are of many kinds: historical, purely technical, application-directed, etc. Possibly, many of them, suitably rewritten and extended, will appear in a second part of this book.

Chapter 1

Basic Concepts

In this chapter, we will give basic definitions, we will present the problems
to solve, and we will anticipate some important results.

We will also introduce and describe some elementary methods of
integration of first- and second-order ODEs. They will be illustrated with
a collection of applications.

1.1 The Fundamental Definitions

Let us begin by indicating what is an ordinary differential equation and
what is an ordinary differential system.

Definition 1.1. Let $N \geq 1$ be an integer and let $\mathcal{O} \subset \mathbb{R}^{2N+1}$ be a non-
empty open set. An ordinary differential system (ODS) of the first order
and dimension N is a collection of equalities of the form

$$\begin{cases} F_1(t, y_1, \ldots, y_N, y_1', \ldots, y_N') = 0, \\ \quad \cdots \\ F_N(t, y_1, \ldots, y_N, y_1', \ldots, y_N') = 0, \end{cases} \tag{1.1}$$

where the $F_i : \mathcal{O} \subset \mathbb{R}^{2N+1} \mapsto \mathbb{R}$ are continuous functions. Usually, we will
abridge the notation and write it as

$$F(t, y, y') = 0, \tag{1.2}$$

grouping all the y_i (resp. the y_i' and F_i) in y (resp. y' and F). $\qquad\square$

1

If $\mathcal{O}$ is of the form $\mathcal{O} = \Omega \times \mathbb{R}^N$ where $\Omega \subset \mathbb{R}^{N+1}$ is an open set and

$$F(t, y, p) = p - f(t, y) \qquad \forall (t, y, p) \in \mathcal{O} \tag{1.3}$$

for some $f : \Omega \mapsto \mathbb{R}^N$, it is said that (1.2) *possesses normal structure*; sometimes, it is said that (1.2) *is written in normal form*. Then, instead of (1.2), we write

$$y' = f(t, y). \tag{1.4}$$

In (1.2) and (1.4), t is the *independent variable* and $y_1, \ldots, y_N$ are the *dependent variables*, that is, the *unknowns*.

When $N = 1$, it is said that (1.2) (resp. (1.4)) is an *ordinary differential equation* (ODE) (resp. an ordinary differential equation with normal structure).

Let us present some examples of first-order ODEs:

$$\sqrt{1 + (y')^2} - 1 - 2ty = 0, \quad 2tyy' + t - y^2 = 0, \quad y' = (1 - 2t)y - y^2 + 2t.$$

The last one has normal structure.

On the other hand, examples of non-scalar ODSs are the following:

$$\begin{cases} y_1' + (y_2')^2 + 2y_2 y_2' + 4t^2 - 1 = 0 \\ (y_1')^2 + y_2' + 2y_1 y_1' - 4t = 0 \end{cases} \qquad \begin{cases} y_1' = 4y_1 - y_2 + \dfrac{1 - t}{1 + t^2} \\ y_2' = -y_1 + 4y_2 - y_3 + \sqrt{1 + t^2} \\ y_3' = -y_2 + 4y_3 + \dfrac{1 + t}{1 + t^2}. \end{cases}$$

Very frequently, we will use a slightly different notation.

Thus, it will be usual to choose the symbol x instead of t to appoint the independent variable; sometimes, the dependent variables will be denoted u, v, etc.

All this will make full sense if we want to insist in the meaning and interpretations given to the equations or systems.

Now, we will present some very simple examples, together with their motivations in geometry and physics.

Example 1.1.

(a) The ODE

$$y' = y + h \tag{1.5}$$

(where h is a prescribed constant) can be used to identify the curves in the plane $y = y(x)$ whose slope at each point is equal to the sum of the ordinate and the fixed quantity h (Figure 1.1).

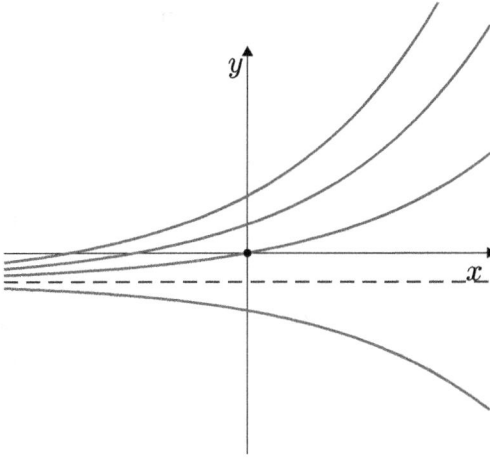

Fig. 1.1   The curves determined by (1.5) for $h = 1$.

(b) The two-dimensional first-order ODS

$$x' = k, \qquad y' = x + y \qquad (1.6)$$

(where k is a constant, we have changed the notation and the unknowns are denoted x and y), serves to identify the trajectories followed by the points in the plane (x, y) moving with speed equal to k in the x direction and equal to $x + y$ in the y direction. □

Definition 1.2. Let $n \geq 1$ be an integer and let $\mathcal{O} \subset \mathbb{R}^{n+2}$ be a non-empty open set. An ordinary differential equation of order n is an expression of the form

$$F(t, y, y', \ldots, y^{(n)}) = 0, \qquad (1.7)$$

where $F : \mathcal{O} \subset \mathbb{R}^{n+2} \mapsto \mathbb{R}$ is a continuous function. □

If $\mathcal{O}$ is of the form $\mathcal{O} = \Omega \times \mathbb{R}$, where $\Omega \subset \mathbb{R}^{n+1}$ is an open set and there exists a continuous function $f : \Omega \mapsto \mathbb{R}$ such that

$$F(t, y, p_1, \ldots, p_n) = p_n - f(t, y, p_1, \ldots, p_{n-1}) \quad \forall (t, y, p_1, \ldots, p_{n-1}) \in \mathcal{O}, \qquad (1.8)$$

we say that (1.7) *possesses normal structure*. In such a case, instead of (1.7), we write the equation in the form

$$y^{(n)} = f(t, y, y', \ldots, y^{(n-1)}). \qquad (1.9)$$

Again, in (1.7) and (1.9), t (resp. y) is the independent variable (resp. the dependent variable or unknown).

Obviously, it is also easy to give the definition of an ODS of order $n \geq 2$ and dimension $N \geq 2$.

In the sequel, we will present results concerning the ODSs of the first order and ODEs of order $n \geq 1$. As shown in what follows, it will not be difficult to extend most of them to systems of order and dimension greater than 1.

The ordinary differential equations and systems have participated in a fundamental way in the development of mathematics since the origins of *Differential Calculus*, in the 17th century; for a brief historical review, see Section 11.1.1.

Example 1.2. The second-order ODEs

$$y'' = -g \quad \text{and} \quad y'' = -g - \frac{k}{m}y' \tag{1.10}$$

(where g, k, and m are positive constants) allow to identify the trajectories followed by bodies that fall vertically.

In the first case, the body is subject to the action of the Gravity force, exclusively.

In the second case, we are assuming that an additional vertical force acts, in a sense opposite to the motion of the body; this can be interpreted as a resistance force opposed by the surrounding medium. $\square$

In the sequel, we will speak almost all the time of ODEs and ODSs written in normal forms.

Unless otherwise specified, we will assume that Ω is a non-empty open set in $\mathbb{R}^{N+1}$ or $\mathbb{R}^{n+1}$ and f is a continuous function on Ω that takes values in $\mathbb{R}^N$ or $\mathbb{R}$.

Let us now indicate what is a solution to (1.4):

Definition 1.3. Let $I \subset \mathbb{R}$ be a non-degenerate interval and let $\varphi : I \mapsto \mathbb{R}^N$ be a given function. It will be said that φ is a solution in I to (1.4) if

(1) φ is continuously differentiable in I.
(2) $(t, \varphi(t)) \in \Omega \quad \forall t \in I$.
(3) $\varphi'(t) = f(t, \varphi(t)) \quad \forall t \in I$. $\square$

We can give a similar definition of solution to (1.9):

Definition 1.4. Let $I \subset \mathbb{R}$ be a non-degenerate interval and let $\varphi : I \mapsto \mathbb{R}$ be a given function. It will be said that φ is a solution in I to (1.9) if

(1) φ is n times continuously differentiable in I.
(2) $(t, \varphi(t), \varphi'(t), \ldots, \varphi^{(n-1)}(t)) \in \Omega \quad \forall t \in I$.
(3) $\varphi^{(n)}(t) = f(t, \varphi(t), \varphi'(t), \ldots, \varphi^{(n-1)}(t)) \quad \forall t \in I$. $\qquad\square$

It is convenient to explain with precision exactly what "continuously differentiable in I" means (or n times continuously differentiable in I).

If I is an open interval, the meaning is well known: φ is continuously differentiable in I if it is differentiable at each $t \in I$ with derivative $\varphi'(t)$ and the function $t \mapsto \varphi'(t)$ is continuous.

Henceforth, in order to fix ideas and avoid ambiguity, we will say that φ is continuously differentiable in I (I being an open interval or not) if it coincides with the restriction to I of a function $\tilde{\varphi}$ defined and continuously differentiable in an open interval containing I.

By induction, we can extend this concept and speak of functions m times continuously differentiable in I for any integer $m \geq 1$.

For example, the function $\varphi_1 : [0, 1] \mapsto \mathbb{R}$, given by

$$\varphi_1(t) = \begin{cases} \sqrt{t} & \text{if } 0 < t \leq 1 \\ 0 & \text{if } t = 0 \end{cases}$$

is continuous but not continuously differentiable in $[0, 1]$; accordingly, we will put $\varphi_1 \in C^0([0, 1])$ and $\varphi_1 \notin C^1([0, 1])$.

Contrarily, the function $\varphi_2 : [0, 1] \mapsto \mathbb{R}$, where

$$\varphi_2(t) = \begin{cases} \dfrac{1}{t} \sin t & \text{if } 0 < t \leq 1 \\ 1 & \text{if } t = 0 \end{cases}$$

is m times continuously differentiable in $[0, 1]$ for all $m \geq 1$; thus, we write $\varphi_2 \in C^m([0, 1])$ for all m or, equivalently, $\varphi_2 \in C^\infty([0, 1])$.

Example 1.3.

(a) It is immediate that, for every $C \in \mathbb{R}$, the function $\varphi(\cdot\,; C)$, given by

$$\varphi(x; C) = Ce^x - h \quad \forall x \in \mathbb{R}$$

is a solution in $\mathbb{R}$ to (1.5); see Figure 1.1.

(b) Also, for any $C_1, C_2 \in \mathbb{R}$, the function $\varphi(\cdot\,; C_1, C_2)$, given by

$$\varphi(t; C_1, C_2) = (kt + C_1, (C_2 - k)e^t - k(1 + t) - C_1) \quad \forall t \in \mathbb{R}$$

is a solution in $\mathbb{R}$ to (1.6).

(c) Finally, for any $C_1, C_2 \in \mathbb{R}$, the function $\psi(\cdot\,; C_1, C_2)$, with

$$\psi(t; C_1, C_2) = -\frac{1}{2}gt^2 + C_1 t + C_2 \quad \forall t \in \mathbb{R}$$

is a solution in $\mathbb{R}$ to the first equation in (1.10).   □

The ODEs can be regarded as a very useful tool for the description, analysis, and resolution of a great variety of problems in science.

For example, it makes sense to look for the *ODE of a family of functions.* Thus, assume that we know of the existence of a one-parameter family of regular functions $\varphi(\cdot\,; C)$ (depending on C), characterized by the identities

$$\Phi(x, \varphi(x; C), C) \equiv 0, \tag{1.11}$$

where Φ is a regular function defined in an appropriate open set in $\mathbb{R}^3$. We can interpret (1.11) as an implicit definition of the family.

The functions $y = \varphi(\cdot\,; C)$ must satisfy the equalities

$$\Phi(x, y, C) = 0 \tag{1.12}$$

and

$$\Phi_x(x, y, C) + \Phi_y(x, y, C)y' = 0, \tag{1.13}$$

where Φ_x and Φ_y stand for the partial derivatives of Φ.

Frequently, it is possible to eliminate C using (1.12) and (1.13). This leads to a relation that "links" the variables x, y, and y', that is, an ODE (not necessarily with a normal structure).

It is then clear by construction that the starting functions $\varphi(\cdot\,; C)$ are solutions to this ODE.

Example 1.4. Consider the family of functions $\varphi(\cdot\,; C)$, with

$$\varphi(x; C) = Cx^2 + \frac{1}{C} \quad \forall x \in \mathbb{R}, \tag{1.14}$$

for all $C > 0$. Arguing as before, we get the identities

$$y - Cx^2 - \frac{1}{C} = 0, \quad y' = 2Cx,$$

that make it possible to eliminate C and get the ODE

$$x(y')^2 - 2yy' + 4x = 0.$$

It is easy to see that the functions (1.14) are solutions in $\mathbb{R}$; indeed, they are all continuously differentiable in $\mathbb{R}$ and satisfy the identity (Figure 1.2).

$$x(\varphi'(x; C))^2 - 2\varphi(x; C)\varphi'(x; C) + 4x = 0 \quad \forall x \in \mathbb{R}. \quad\quad □$$

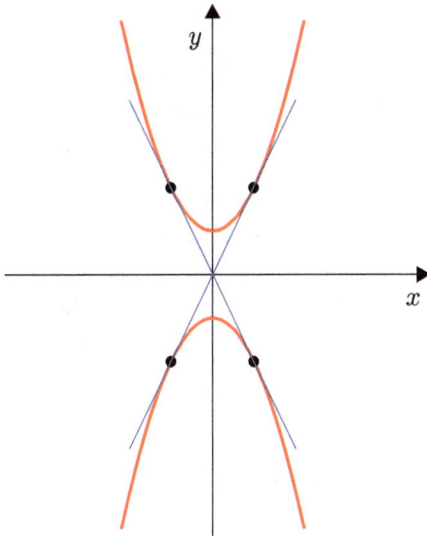

Fig. 1.2   Two functions of the family (1.14) (for $C = 1$ and $C = -1$).

The ODEs can also help to state and solve many geometrical problems. Thus, they serve in many cases to identify the curves characterized by a particular property. It suffices to get a precise statement of the property.

Example 1.5. Let us try to determine the curves in the plane that satisfy the following property:

Given a point (x, y) on the curve, the associated normal at (x, y) passes through the origin.

Let $y = \varphi(x)$ be one of these curves. The previous property indicates that, in the plane (X, Y), any straight line

$$Y = \varphi(x) - \frac{1}{\varphi'(x)}(X - x)$$

touches $(0, 0)$. Therefore, we must have $\varphi(x) + x/\varphi'(x) = 0$, that is, the ODE

$$yy' + x = 0.$$

The elementary integration methods given in Section 1.3 can be used here to show that the solutions to this ODE are the functions defined implicitly by the relations

$$x^2 + y^2 = C,$$

where C is a positive constant. Consequently, the curves we are searching for are arcs of circumferences centered at the origin. $\square$

That the ODEs are useful tools becomes clear when we invoke them to model phenomena with origin in physics, biology, engineering, etc.

Indeed, many problems with origin in these fields can be "written" in terms of ODEs or ODSs. Their formulation helps to understand the underlying phenomena and, in many cases, serves as a first step to obtain useful information.

At this moment, we will try to illustrate these assertions with one example coming from *population dynamics*. In the following sections and chapters, many more examples will be given.

Example 1.6. Let us assume that we want to specify with details the evolution of an isolated population, where many individuals live together and share unlimited resources.

Let us denote by $p(t)$ the relative number of alive individuals at time t, that is, $p(t) = P(t)/M$, where $P(t)$ is the true number of individuals and M is a large quantity, for instance of the order of the maximal size of $P(t)$. It is then usual to accept that the function $t \mapsto p(t)$ is continuous and even differentiable.

With unlimited means, it is reasonable to assume that, at each t, the population grows or decreases with a rate that is proportional to the population itself. This leads to the ODE

$$\dot{p} = ap, \tag{1.15}$$

proposed by Malthus in 1798,[1] where $a \in \mathbb{R}$ (Figure 1.3).

The constant a must be interpreted as the difference of births and deaths per time unit.

Obviously, a positive a is a sign of health: there are more births than deaths and the population grows. Contrarily, if $a < 0$, something goes wrong, maybe a disease (or the mean age is large) and the population is expected to diminish and possibly become extinct.

---

[1] Here, we denote by $\dot{p}$ the time derivative of p. We will use this notation when we want to emphasize the meaning of the "time derivative" of a function, that is, its instantaneous variation rate with respect to time.

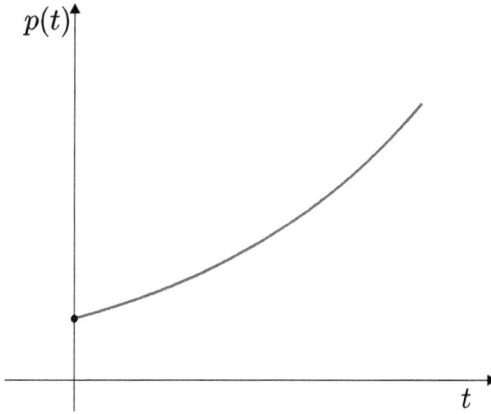

Fig. 1.3   A solution to (1.15) with $a > 0$ and $p(0) > 0$.

It is immediate that any function of the form

$$p_C(t) = Ce^{at} \quad \forall t \in \mathbb{R} \tag{1.16}$$

(where C is an arbitrary constant) is a solution in $\mathbb{R}$ to (1.15). On the other hand, it will be seen very soon that the unique positive solutions to (1.15) are precisely these functions.

Consequently, Malthus's hypothesis (or law) leads to populations that increase or decrease exponentially.

To fix ideas, let us assume from now on that $a > 0$ in (1.15).

From empirical results, it is known that this model is realistic only for small values of t (when the population is still small with respect to the resources at hand).

For greater times, it is much more appropriate to assume that the individuals *compete* and it is more realistic to change (1.15) by the so-called *logistic equation*

$$\dot{p} = ap - bp^2, \tag{1.17}$$

where $a, b > 0$.[2] The solutions to (1.17) possess a more complex structure. In fact, we will see later that the positive solutions are the functions

$$\varphi_C(t) = \frac{a}{b + Ce^{-at}} \quad \forall t \in \mathbb{R} \tag{1.18}$$

---

[2]For a brief historical review on this and other similar ODEs and some additional information, see Note 11.1.2.

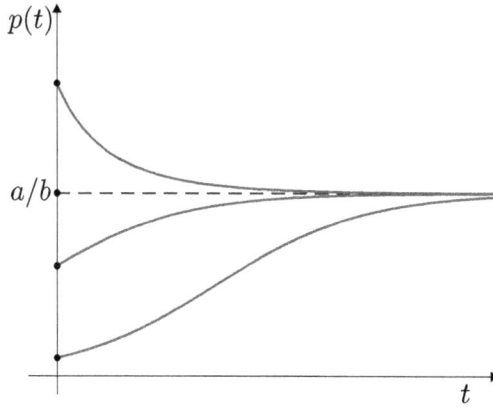

Fig. 1.4   The solutions to (1.17) with $p(0) > 0$.

and

$$\psi_C(t) = \frac{a}{b - Ce^{-at}} \quad \forall t \in \left(\frac{1}{a} \log \frac{C}{b}, +\infty\right), \tag{1.19}$$

where we, respectively, have $C \geq 0$ and $C > 0$; see Figure 1.4.

In particular, we note that, for $C = 0$, the solution is defined for all $t \in \mathbb{R}$ and takes the constant value a/b. Another constant solution to (1.17) is the function $\zeta(t) \equiv 0$ (the so-called trivial solution).

Note that the structure and properties of all these functions are clearly different from those observed in (1.16). We see that all the positive solutions converge to a/b as $t \to +\infty$ and also that, even if they are small at $t = 0$, they separate from 0 as t grows.

This behavior will be explained in what follows.   □

Let us now see that any ODE of the nth order can be equivalently rewritten as a first-order ODS of dimension n.

For simplicity, we will consider Eq. (1.9), although a similar argument is valid for any ODE that is not written in the normal form.

Let us introduce the "new variables"

$$y_1 = y, \ y_2 = y', \dots, \ y_n = y^{(n-1)}. \tag{1.20}$$

In short, we will write $Y = (y_1, \dots, y_n)$. Then, the nth order ODE (1.9) can be rewritten in the form

$$Y' = F(t, Y), \quad \text{where} \ F(t, Y) := (y_2, \dots, y_n, f(t, Y)) \quad \forall (t, Y) \in \Omega. \tag{1.21}$$

It is immediate that (1.9) and (1.21) are *equivalent* in the following sense:

- If $\varphi : I \mapsto \mathbb{R}$ is a solution in I to (1.9) and $\Phi : I \mapsto \mathbb{R}^n$ is given by
$$\Phi(t) = (\varphi(t), \varphi'(t), \ldots, \varphi^{(n-1)}(t)) \quad \forall t \in I,$$
  then Φ is a solution in I to (1.21).
- Conversely, if $\Phi : I \mapsto \mathbb{R}^n$ solves (1.26) in I and we set
$$\varphi(t) := \Phi_1(t) \quad \forall t \in I,$$
  then φ is a solution in I to (1.9).

This correspondence will be used very frequently in the sequel.

Specifically, many results proved in the following chapters for (1.4) (local and global existence of solutions, uniqueness, regularity, etc.) will lead via (1.20) to similar results for (1.9).

1.2 The Cauchy or Initial-Value Problem

In view of the previous examples, it is reasonable to expect that, if the interval I is appropriately fixed, a first-order ODE possesses (at least) an infinite one-parameter family of solutions in I.

This can be written symbolically in the form
$$y = \varphi(t; C), \quad \text{with } C \in \mathbb{R}.$$
It is also reasonable to expect that an ODS of the first order and dimension N possesses a family of solutions
$$y = \varphi(t; C_1, \ldots, C_N), \quad \text{with } C_1, \ldots, C_N \in \mathbb{R}$$
and an ODE of the nth order admits a family of solutions
$$y = \psi(t; C_1, \ldots, C_n), \quad \text{with } C_1, \ldots, C_n \in \mathbb{R}.$$

Consequently, it is completely justified that we consider in the sequel problems containing ODEs or ODSs together with additional conditions that allow to select one solution within the family of solving functions.

The first of them will be the *initial-value problem* (or Cauchy problem). It is described as follows:

Definition 1.5. An initial-value problem (or Cauchy problem) for (1.4) is a system of the form

(CP)
$$\begin{cases} y' = f(t, y) \\ y(t_0) = y_0, \end{cases}$$

where $(t_0, y_0) \in \Omega$. $\qquad \square$

The second equality in (CP) is the *initial condition*; (t_0, y_0) is the Cauchy datum.[3]

Let I be an interval containing t_0 and let $\varphi : I \mapsto \mathbb{R}^N$ be given. It will be said that φ is a solution to (CP) in I if it solves (1.4) in I and, furthermore,

$$\varphi(t_0) = y_0.$$

Definition 1.6. An initial-value problem for (1.9) is a system of the form

$$(CP)_n \quad \begin{cases} y^{(n)} = f(t, y, y', \dots, y^{(n-1)}) \\ y(t_0) = y_0^{(0)}, \ y'(t_0) = y_0^{(1)}, \dots, \ y^{(n-1)}(t_0) = y_0^{(n-1)}, \end{cases}$$

where $(t_0, y_0^{(0)}, \dots, y_0^{(n-1)}) \in \Omega$ (Figure 1.5). $\qquad\square$

The last n equalities are called *initial conditions* or *Cauchy conditions*; this time, the Cauchy datum is $(t_0, y_0^{(0)}, \dots, y_0^{(n-1)})$.

Of course, if I is an interval containing t_0 and $\varphi : I \mapsto \mathbb{R}$, we say that φ is a solution in I to $(CP)_n$ if it solves in I the nth order ODE (1.9) and satisfies

$$\varphi(t_0) = y_0^{(0)}, \ \varphi'(t_0) = y_0^{(1)}, \dots, \varphi^{(n-1)}(t_0) = y_0^{(n-1)}.$$

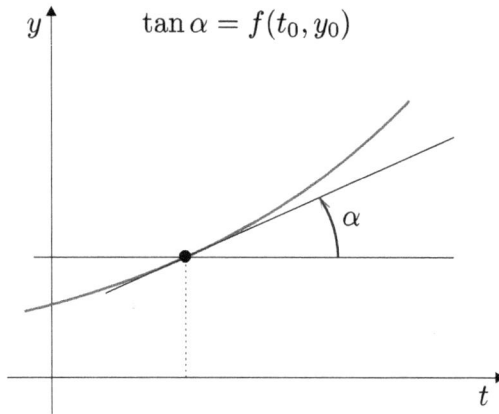

Fig. 1.5   The solution to (CP) in a neighborhood of t_0.

[3]It seems that initial-value problems for differential equations have been stated (and solved) for the first time by Agustin Louis Cauchy (1789–1857); for more details, see Note 11.1.3.

In a large part of this book, our main interest will be to study Cauchy problems for ODEs and ODSs with a normal structure. The data will be the precise equation or system (that is, the function f) and the initial values of the independent and dependent variables (that is, the Cauchy data).

With this information, we will try to "solve" the corresponding Cauchy problem.

If we cannot compute the solution explicitly (and this will be the most frequent case), at least we will try to prove that it exists and is unique, we will try to determine the largest definition interval, analyze its regularity, etc.[4]

Let us give some examples.

Example 1.7.

(a) A CP for (1.5) is the following:

$$\begin{cases} y' = y + h \\ y(0) = 0. \end{cases} \tag{1.22}$$

Obviously, the Cauchy datum is here $(x_0, y_0) = (0, 0)$. Among all the solutions to (1.5) in $\mathbb{R}$ considered in Example 1.3(a), we only find one solution to (1.22): the function corresponding to $C = h$.

The results in the following chapters show that this is in fact the unique solution to (1.22).

(b) A CP for the ODS (1.6) is

$$\begin{cases} x' = k, \quad y' = x + y \\ x(0) = 0, \quad y(0) = 0. \end{cases} \tag{1.23}$$

Looking at the set of solutions in $\mathbb{R}$ shown in Example 1.3(b), we see that only one of them, the function corresponding to $C_1 = 0$ and $C_2 = 2k$, solves (1.23).

It will be seen in what follows that this is the unique solution to (1.23).

---

[4]From the viewpoint of physics, the formulation of a CP is completely natural. Indeed, physical laws can frequently be written in terms of differential equations. On the other hand, *to fix the Cauchy data* means to indicate exactly what is the status of the system at a prescribed time instant $t = t_0$. It is generally accepted that this suffices to determine the evolution of the system, that is to say, the values of the solution in a whole interval containing t_0. Similar considerations hold for problems with origins in biology, chemistry, economics, etc.

(c) Now, let us present a CP for the first second-order ODE in (1.10). It is the following:

$$\begin{cases} y'' = -g \\ y(0) = y_0, \quad y'(0) = v_0, \end{cases} \tag{1.24}$$

where $y_0, v_0 \in \mathbb{R}$.

Again, in the family of functions $\psi(\cdot; C_1, C_2)$ indicated in Example 1.3(c), we find exactly one solution to (1.24): the function corresponding to $C_1 = v_0$ and $C_2 = y_0$.

It will be clear later that this is the unique solution.

(d) It makes perfect sense to consider Cauchy problems in the context of population dynamics and, in particular, for the Malthus and logistic models.

For example, let us take $t_0 = 0$. Given $p_0 > 0$, the problem

$$\begin{cases} \dot{p} = ap \\ p(0) = p_0 \end{cases} \tag{1.25}$$

permits to identify the isolated Malthus population whose size at $t = 0$ is given by p_0.

Among all the functions in (1.16), we find one and only one solution to (1.25): the function corresponding to $C = p_0$:

$$p(t) = p_0 \, e^{at} \quad \forall t \in \mathbb{R}.$$

On the other hand, for each $p_0 > 0$, in the family of functions (1.18) and (1.19), we find only one solution to

$$\begin{cases} \dot{p} = ap - bp^2 \\ p(0) = p_0. \end{cases}$$

It is given by (1.18) with $C = a/p_0 - b$ if $0 < p_0 \leq a/b$ and (1.19) with $C = b - a/p_0$ if $p_0 > a/b$.

In other words,

$$p(t) = \frac{a}{b + \left(\dfrac{a}{p_0} - b \right) e^{-at}} \quad \forall t \in (t^*, +\infty),$$

where $t^* = -\infty$ if $0 < p_0 \leq a/b$ and $t^* = \dfrac{1}{a} \log \left(1 - \dfrac{a}{bp_0} \right)$ if $p_0 > a/b$.

This is interpreted as follows:

(a) If $p_0 < a/b$, the initial population size is small and it grows with time, converging exponentially to this value.

(b) In the exceptional case $p_0 = a/b$, the population remains constant for all times.

(c) Finally, if for some reason we have a too large population at time $t = 0$ (that is, $p_0 > a/b$), then it decreases and, again, it converges to a/b as $t \to +\infty$.

Accordingly, it is usual to interpret a/b as the "natural" threshold of the considered population. $\qquad\square$

At the end of the previous section, we saw that any ODE of order n can be equivalently rewritten as a first-order ODS of dimension n, in the sense that we can establish a one-to-one correspondence from the solutions to the former onto the solutions of the latter.

Clearly, this correspondence can also be established between solutions to associated CPs.

More precisely, let us consider the initial-value problem $(\text{CP})_n$ and let us perform again the change of variables (1.20). Then, $(\text{CP})_n$ can be rewritten in the form

$$\begin{cases} Y' = F(t, Y) \\ Y(t_0) = Y_0, \end{cases} \tag{1.26}$$

where F is as in (1.21),

$$F(t, Y) := (y_2, \dots, y_n, f(t, Y)) \quad \forall (t, Y) \in \Omega \text{ and } Y_0 := (y_0^{(0)}, \dots, y_0^{(n-1)}). \tag{1.27}$$

We have that $(\text{CP})_n$ and (1.26) are equivalent. For the sake of clarity, let us specify:

- If $\varphi : I \mapsto \mathbb{R}$ is a solution in I to $(\text{CP})_n$ and $\Phi : I \mapsto \mathbb{R}^n$ is given by

$$\Phi(t) := (\varphi(t), \varphi'(t), \dots, \varphi^{(n-1)}(t)) \quad \forall t \in I,$$

  then Φ is a solution in I to (1.26).
- Conversely, if $\Phi : I \mapsto \mathbb{R}^n$ is a solution in I to (1.26) and

$$\varphi(t) := \Phi_1(t) \quad \forall t \in I,$$

  then φ solves $(\text{CP})_n$ in I.

To end this section, let us see that the ODEs and ODSs can be completed with conditions of other kinds, not necessarily related to the values of the solution and its derivatives at a fixed value of the independent variable.

For example, this happens when we consider *boundary-value problems*.

An example of a boundary-value problem for an ODE of the second order is the following:

$$\begin{cases} y'' = f(x, y, y'), & x \in [\alpha, \beta] \\ y(\alpha) = y_\alpha, & y(\beta) = y_\beta. \end{cases} \tag{1.28}$$

Here, it is assumed that (for instance) $f : \mathbb{R}^3 \mapsto \mathbb{R}$ is a continuous function, $\alpha, \beta \in \mathbb{R}$, $\alpha < \beta$ and $y_\alpha, y_\beta \in \mathbb{R}$. Observe that the independent variable has been denoted here as x; this will be justified in the following, see Chapter 6.

The last two equalities in (1.28) are the so-called *boundary conditions*. By definition, a solution to (1.28) is a function $\varphi : [\alpha, \beta] \mapsto \mathbb{R}$ that solves in $[\alpha, \beta]$ the ODE in (1.28) and, furthermore, satisfies

$$\varphi(\alpha) = y_\alpha, \quad \varphi(\beta) = y_\beta.$$

Boundary-value problems of this kind can serve (for instance) to describe the stationary state of a mechanical system subject to appropriate efforts whose endpoints are at prescribed positions.

1.3　Some Elementary Integration Methods

In this part of the chapter, we are going to present some elementary methods for the solution (or integration) of simple ODEs of the first- and second-order.

In general terms, it can be said that the *integration* of (1.4) is the process that allows to find an explicit family of solutions $\varphi(\cdot\,; C)$ (depending on C, that may be viewed as a parameter).

It is usual to say that this family is a *general solution* (although, as we will see, not always are all the solutions to the ODE in the family).

Sometimes, all we can do is to describe the functions $\varphi(\cdot\,; C)$ through a formula of the form

$$\Phi(t, y; C) = 0.$$

In these cases, we will say that the general solution is given implicitly.

It is known that the computation of a general solution can only be achieved for a very reduced amount of particular equations; see Dou (1969) and Note 11.1.4.

However, it is interesting to know what can be done for some simple ODEs; for more information on this subject, see de Guzmán *et al.* (1978) and Kiseliov *et al.* (1973).

In the remainder of the chapter, we will be rather informal. The methods and results will be presented with a "quick" analysis; in particular, we will not pay much attention to the intervals of definition of the functions. The goal is to find the solutions $\varphi(\cdot\,;C)$.

Consequently, it makes sense to advance in the computations, find a possible expression, investigate which is the maximal interval of definition, and, then, check that we have found what we were looking for.

1.3.1   *The most simple cases*

The most simple situation appears when the dependent variable is not present in the right-hand side:

$$y' = f(t). \tag{1.29}$$

The integration of (1.29) is immediate:

$$\varphi(t; C) = F(t) + C,$$

where F is a primitive of f (arbitrarily chosen, but fixed).

It is said that an ODE is *separable* if it has the following structure:

$$y' = \frac{a(t)}{b(y)}. \tag{1.30}$$

Let A and B be primitives of a and b, respectively. If we introduce the symbols $\mathrm{d}y$ and $\mathrm{d}t$ and put

$$y' = \frac{\mathrm{d}y}{\mathrm{d}t},$$

it is tempting to rewrite (1.30) in the form

$$b(y)\,\mathrm{d}y = a(t)\,\mathrm{d}t.$$

After integration in the left with respect to the variable y and in the right with respect to the variable t, we obtain an implicit expression of the general solution:

$$B(y) = A(t) + C. \tag{1.31}$$

Then, what we have to do is to "solve" for y in (1.31) and get something as follows:

$$\varphi(t; C) = B^{-1}(A(t) + C).$$

Note that, if B^{-1} is well defined, for every C, a function $\varphi(\cdot\,;C)$ defined in this way solves (1.30).

In classical terms, we can say that the integration of (1.29) (resp. (1.30)) needs one (resp. two) quadrature(s).

Example 1.8.

(a) Consider the separable ODE

$$y' = 1 + y^2.$$

Following the previous steps, we find

$$\frac{dy}{1 + y^2} = dt,$$

whence

$$\arctan y = t + C.$$

For every C, this identity can be locally inverted. This leads to the following solutions:

$$\varphi(t; C) = \tan(t + C).$$

It is easy to check that the $\varphi(\cdot\,; C)$ are solutions to the ODE (each of them defined in an appropriate interval).

We observe that the functions $\varphi(\cdot\,; C)$ "go to infinity" at some finite values of t.

Thus, there are solutions to this ODE defined only on a bounded interval and not on the whole $\mathbb{R}$, and although the equation is very simple, it has a normal structure and corresponds to a very regular function $f = f(t, y)$, perfectly defined in $\mathbb{R}^2$.

This will be rigorously proved in the following, in Chapter 4, for ODEs for which f grows (for instance) quadratically as $|y| \to +\infty$.

(b) Now, let us consider the CP

$$\begin{cases} 2yy' + (1 + y^2)\sin t = 0 \\ y(\pi/2) = 1. \end{cases} \quad (1.32)$$

Here, the ODE can be equivalently rewritten as a separable equation (Figure 1.6):

$$\frac{2yy'}{1 + y^2} = -\sin t.$$

With some computations like those above, we are led to the following implicit expression:

$$1 + y^2 = \exp(\cos t + C).$$

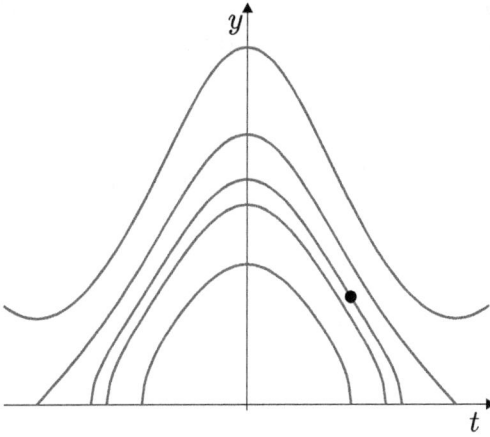

Fig. 1.6   The solutions to the ODE in (1.32) for various C and the initial datum $(\pi/2, 1)$.

Consequently, it seems reasonable to test solutions of the form

$$\varphi(t; C) = \pm\sqrt{-1 + \exp(\cos t + C)}.$$

It is again clear that these functions, defined in suitable intervals, solve the ODE (check, please).

If we impose the initial condition in (1.32), we deduce at once that the sign in the previous square root must be positive and C must be equal to $\log 2$.

Obviously, the solution (denoted by φ) must be defined in an interval that contains 0. The maximal one is $(-T_*, T_*)$, where T_* is defined by

$$T_* \in (0, \pi), \quad \cos T_* = -\log 2.$$

This way, we have found the following solution to the CP:

$$\varphi(t) = \sqrt{-1 + 2\exp(\cos t)} \quad \forall t \in (-T_*, T_*).$$

Note that the positive solutions φ to the ODE in (1.32) can be computed more quickly by introducing $z := y^2$, rewriting the equation in the form $z' + (1 + z)\sin t = 0$, and arguing in a similar way as before. $\square$

1.3.2   *Linear equations*

A *linear* ODE of the first order takes the form

$$y' = a(t)y + b(t), \tag{1.33}$$

where a and b are continuous functions in the interval $I \subset \mathbb{R}$. For a motivation and a justification of the relevance of linear ODEs, see Note 11.1.5.

In this section, when we speak of solutions to (1.33), we implicitly mean *solutions in the whole interval I*.[5]

If $b \equiv 0$, it is said that (1.33) is a *homogeneous linear* equation. In this case, we are dealing with the ODE

$$y' = a(t)y, \tag{1.34}$$

that can be reduced to being separable. Arguing as before, we get at once a general solution $\varphi_h(\cdot\,;C)$:

$$\varphi_h(t;C) = Ce^{A(t)},$$

where A is a primitive of a.[6]

Example 1.9. Let us consider again Malthus's Law (1.15). Now, it becomes clear that the solutions to this ODE are the functions (1.16). $\square$

If $b \not\equiv 0$, we observe that the whole family of solutions to (1.33) can be obtained by adding to a fixed solution the solutions to (1.34). In other words, the general solution to (1.33) is $\varphi(\cdot\,;C)$, where

$$\varphi(t;C) = \varphi_p(t) + \varphi_h(t;C)$$

and φ_p is a particular solution to (1.33).[7]

In order to compute a particular solution to (1.33), it suffices to apply the so-called *Lagrange constant variation method*.

Specifically, what we have to do is to search for a solution φ_p of the form $\varphi_p(t) = \varphi_h(t;C_p(t))$, where the function C_p is determined by imposing (1.33).

---

[5]In Chapter 5, it will be seen that, if the equation is linear and the coefficients are continuous in I, then all the solutions are defined in I.

[6]In particular, we see that the family of solutions to (1.34) is a vector space of dimension 1 for the usual operation laws. This property characterizes the first-order homogeneous linear ODEs. Indeed, we can associate to any family of functions of the form $y = Cm(t)$ the homogeneous linear ODE $y' = (m'(t)/m(t))\,y$.

[7]This means that the family of solutions to (1.33) is a linear manifold whose support space coincides with the family of solutions to (1.34). It is traditional to say that

The general solution to a non-homogeneous linear ODE is the sum of a particular solution and the general solution to the associated homogeneous linear equation.

Thus, we put

$$\varphi_p(t) = C_p(t)e^{A(t)}.$$

If φ_p satisfies (1.33), we necessarily have

$$C'_p(t) = e^{-A(t)}b(t),$$

whence a valid choice of C_p is

$$C_p(t) = \int_{t_0}^t e^{-A(s)}b(s)\,ds$$

(where $t_0 \in I$). This leads to the general solution to (1.33):

$$\varphi(t;C) = e^{-A(t)}\int_{t_0}^t e^{-A(s)}b(s)\,ds + Ce^{A(t)}.$$

Example 1.10. Consider the ODE

$$y' + y\cos t = \sin t\cos t. \qquad (1.35)$$

The associated homogeneous linear ODE is $y' = -y\cos t$. Its general solution is given by $\varphi_h(t;C) = Ce^{-\sin t}$.

In order to compute a particular solution to (1.35), we write

$$\varphi_p(t) = C_p(t)e^{-\sin t}$$

and we impose (1.35), which yields $C'_p(t) = e^{\sin t}\sin t\cos t$. Therefore, we get $C_p(t) = e^{\sin t}(\sin t - 1)$ and $\varphi_p(t) = \sin t - 1$.

Finally, the general solution to (1.35) is

$$\varphi(t;C) = Ce^{-\sin t} + \sin t - 1. \qquad \square$$

1.3.3 *Bernoulli equations*

A Bernoulli ODE has the form

$$y' = a(t)y + b(t)y^r, \quad \text{where } r \neq 0,1 \qquad (1.36)$$

and a and b are continuous in I.[8]

If $r > 0$, we always have that the function

$$\varphi_0(t) \equiv 0$$

is a solution (this is called the *trivial* solution).

---

[8]If $r = 0$ (resp. $r = 1$), the ODE is linear (resp. linear and homogeneous).

Let us indicate how the other (nontrivial) solutions to (1.36) can be found.

We can use two methods. The first of them relies on the change of variable $z = y^{1-r}$, that leads to the following reformulation of (1.36) as a linear ODE:

$$z' = (1-r)a(t)z + (1-r)b(t).$$

For the second method, we put $y = u(t)v(t)$. After replacing in (1.36), we get

$$u'v + uv' = a(t)uv + b(t)u^r v^r.$$

Consequently, in order to solve (1.36), it suffices to find a function $u \neq 0$ such that $u' = a(t)u$ and then all the functions v satisfying $v' = b(t)u^{r-1}v^r$.

In other words, what we will have to do is to compute a nontrivial solution to a homogeneous linear ODE and, then, solve a second ODE for v that can be reduced to being separable.

Both methods allow to find the general solution easily.

Example 1.11. Consider again the logistic equation (1.17). This is of the Bernoulli kind (with $r = 2$). Accordingly, we introduce the new variable $q = p^{-1}$ and rewrite the equation in terms of q:

$$\dot{q} = -aq + b,$$

whence we find that $q(t) = Ce^{-at} + b/a$ and the following general solution to (1.17) is given by

$$p(t) = \frac{a}{b + Ce^{-at}}.$$

This confirms that the positive solutions to (1.17) are the functions φ_C and ψ_C that we see in (1.18) and (1.19).

Recall that we also have the trivial solution $p_0(t) \equiv 0$. □

Example 1.12. Consider the ODE

$$y' + \frac{1}{t}y = \frac{\log t}{t}y^2. \tag{1.37}$$

Again, we have the trivial solution

$$\varphi_0(t) \equiv 0.$$

Let us apply the first method described above. The change of variable is $z = y^{-1}$ and the resulting ODE for z is

$$z' - \frac{1}{t}z = -\frac{\log t}{t}.$$

The general solution to this linear ODE is $\zeta(t; C) = Ct + \log t + 1$. Thus, the general solution to (1.37) is given by

$$\varphi(t; C) = \frac{1}{Ct + \log t + 1}. \tag{1.38}$$

If we apply the second method, with $y = uv$, the functions u and v must, respectively, satisfy $u' = -u/t$ and $v' = (\log t) \, v^2/t^2$. Consequently, a good choice of u is $u = 1/t$, whence we get $v = t/(Ct + \log t + 1)$ and the general solution is again (1.38). $\qquad\square$

1.3.4 *Riccati equations*

We deal now with equations of the form

$$y' = a(t) + b(t)y + c(t)y^2, \tag{1.39}$$

where a, b, and c are continuous in I and $a \not\equiv 0$ (if $a \equiv 0$, we get again a Bernoulli equation).

We see that this equation is not too complicated. In spite of this, it cannot be solved explicitly in general; see Note 11.1.4 to this purpose.

However, if a particular solution φ_p is known, the general solution can be found. Indeed, after the change of variable

$$y = \varphi_p + \frac{1}{u},$$

the ODE is transformed into $u'u^{-2} = -(b(t) + 2c(t)\varphi_p(t)) \, u^{-1} - c(t)u^{-2}$ and, after multiplication by u^2, we arrive at the linear ODE

$$u' = -(b(t) + 2c(t)\varphi_p(t))u - c(t).$$

Example 1.13. Consider the Cauchy problem

$$\begin{cases} y' = 2t + (1 - 2t)y - y^2 \\ y(0) = 0. \end{cases} \tag{1.40}$$

It is immediate that the function $\varphi_p(t) \equiv 1$ is a solution to the ODE (Figure 1.7). Thus, we put

$$y = 1 + \frac{1}{u}. \tag{1.41}$$

After replacing in the equation and simplifying the terms, we find the linear ODE

$$u' = (1 + 2t)u + 1.$$

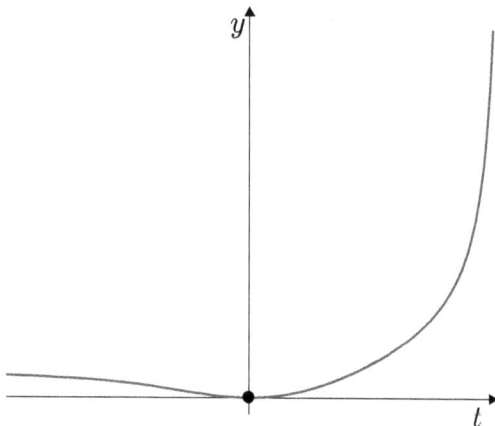

Fig. 1.7   The solution to (1.40).

Taking into account (1.41), we must find a solution to this ODE satisfying the initial condition

$$u(0) = \frac{1}{y(0) - 1} = -1.$$

Arguing as in Section 1.3.2 and coming back to the variable y, the solution to (1.40) is easily found as follows:

$$\varphi(t) = 1 + \frac{e^{-(t^2 + t)}}{\displaystyle\int_0^t e^{-(s^2 + s)}\, \mathrm{d}s - 1}.$$

It is not difficult to check that φ is well defined and solves (1.40) in the whole $\mathbb{R}$.   □

1.3.5   Homogeneous equations

It will be said that the ODE (1.4) is *homogeneous* if f is a *homogeneous function of order zero*, that is, the following property is satisfied:

If $(t, y) \in \Omega$, then $(\lambda t, \lambda y) \in \Omega$ and $f(\lambda t, \lambda y) = f(t, y)$ $\forall \lambda > 0$.

Obviously, we then have

$$f(t, y) = f(1, y/t) \quad \text{for all } (t, y) \in \Omega \text{ with } t > 0.$$

Homogeneous ODEs can be solved with the change of variable $y = tu$. Indeed, this allows to go from (1.4) to the new ODE

$$t\,u' + u = f(1, u), \tag{1.42}$$

that can be reduced to a separable equation.

Example 1.14. Consider the ODE

$$y' = \frac{y + \sqrt{t^2 - y^2}}{t}. \tag{1.43}$$

Clearly, it is homogeneous. With $y = t\,u$, the ODE becomes $t\,u' + u = u \pm \sqrt{1 - u^2}$, whence

$$t\,u' = \pm\sqrt{1 - u^2}. \tag{1.44}$$

Let us assume for the moment that $\sqrt{1 - u^2} \neq 0$. Then (1.44) can be reduced to a separable ODE that can be easily integrated:

$$\frac{du}{\sqrt{1 - u^2}} = \pm\frac{dt}{t},$$

whence $\arcsin u = \pm \log |Ct|$ and we obtain the following general solution:

$$\varphi(t; C) = \pm t\,\sin(\log |Ct|).$$

To this family of functions we must add two particular solutions that were "lost" when we divided by $\sqrt{1 - u^2}$. In terms of u, they correspond to the following constant functions: $u = 1$ and $u = -1$. Coming back to the y variable, we get

$$\varphi_1(t) = t \quad \text{and} \quad \varphi_{-1}(t) = -t. \qquad \square$$

The following ODEs can also be solved easily:

$$y' = f\left(\frac{at + by + r}{ct + dy + s}\right), \tag{1.45}$$

with $a, b, c, d, r, s \in \mathbb{R}$.

Indeed, if the straight lines $at + by + r = 0$ and $ct + dy + s = 0$ are parallel, we can assume that (1.45) has the form

$$y' = f\left(\frac{\mu(ct + dy) + r}{ct + dy + s}\right).$$

Then, the change of variable $z = ct + dy$ leads to a separable ODE.

Contrarily, if these straight lines are not parallel, there must exist a point (t_0, y_0) where they meet. By introducing the change of independent and dependent variables $T = t - t_0$, $Y = y - y_0$ and taking into account that

$$\frac{dy}{dt} = \frac{dY}{dT}\frac{dT}{dt} = \frac{dY}{dT},$$

we get the equivalent ODE

$$Y' = f\left(\frac{aT + bY}{cT + dY}\right),$$

that is homogeneous.

Example 1.15. We consider now the ODE

$$y' = \frac{t + y + 2}{t - y - 2}.$$

The straight lines $t+y+2 = 0$ and $t-y-2 = 0$ meet at $(0, -2)$. Accordingly, let us introduce the new variables T and Y, with $T = t$, $Y = y + 2$. This leads to the homogeneous equation

$$Y' = \frac{T + Y}{T - Y}.$$

After appropriate computations, we find the following implicit expression for the general solution:

$$y + 2 - t \tan\left(\log|C\sqrt{t^2 + (y + 2)^2}|\right) = 0. \qquad \square$$

1.3.6 *Exact equations**

Let P and Q be continuous functions in the open set $\Omega \subset \mathbb{R}^2$, with $Q \neq 0$ in Ω. Consider the following ODE:

$$P(t, y) + Q(t, y)\, y' = 0. \tag{1.46}$$

It is said that (1.46) is an *exact* ODE if there exists a function Φ, defined and continuously differentiable in Ω, such that

$$\Phi_t = P \quad \text{and} \quad \Phi_y = Q \quad \text{in } \Omega. \tag{1.47}$$

Recall that the subindices stand for partial derivatives with respect to the corresponding variables.

If (1.46) is exact and Φ satisfies (1.47), any solution φ satisfies

$$\frac{d}{dt}\Phi(t, \varphi(t)) = \Phi_t(t, \varphi(t)) + \Phi_y(t, \varphi(t))\varphi'(t)$$

$$= P(t, \varphi(t)) + Q(t, \varphi(t))\varphi'(t) \equiv 0.$$

Consequently, we can view the following identity as an "implicit" definition of the general solution:

$$\Phi(t, y) = C.$$

Let us assume that P and Q are continuously differentiable in Ω. The following assertions are well known (see for example, Cartan, 1967):

- If the ODE (1.46) is exact, then we necessarily have

$$P_y = Q_t \quad \text{in } \Omega. \tag{1.48}$$

- Conversely, if Ω is *simply connected*[9] and we have (1.48), then (1.46) is exact.

For example, if Ω is a rectangle, P and Q are continuously differentiable in Ω and (1.48) holds, it is not difficult to construct a function Φ (unique up to an additive constant) satisfying (1.47): it suffices to choose $(t_0, y_0) \in \Omega$ and set

$$\Phi(t, y) := \int_{t_0}^{t} P(\tau, y_0) \, d\tau + \int_{y_0}^{y} Q(t, \eta) \, d\eta \quad \forall (x, y) \in \Omega. \tag{1.49}$$

In the general case of a simply connected open set Ω, we can do similarly by previously fixing a polygonal path from (t_0, y_0) to any other point (t, y) formed by segments parallel to the axes and integrating along them alternatively with respect to τ and η. For instance, if

$$\Omega = \{(t, y) \in \mathbb{R}^2 : 0 < y - t < 2\}$$

and we take $(t_0, y_0) = (0, 1)$ and $(t, y) = (3/2, 3)$, we introduce the points

$$a_0 = (0, 1), \quad a_1 = (1/2, 1), \quad a_2 = (1/2, 2), \quad a_3 = (3/2, 2), \quad a_4 = (3/2, 3)$$

and the closed segments $S_i := [a_i, a_{i+1}]$ for $0 \le i \le 3$ and we set

$$\Phi(t, y) := \int_{S_0} P(\tau, y_0) \, d\tau + \int_{S_1} Q(1/2, \eta) \, d\eta$$
$$+ \int_{S_2} P(\tau, 2) \, d\tau + \int_{S_3} Q(3/2, \eta) \, d\eta.$$

It can be checked that this value is in fact independent of the chosen polygonal path. The function Φ is thus well defined and continuously differentiable in the whole set Ω and moreover satisfies (1.47).

---

[9]This means that Ω has no hole. More precisely, the complementary set of Ω is connected or, equivalently, any *circle* in Ω is homological to zero.

In our context, in order to find a general solution to (1.46), we will first integrate (for instance) P with respect to the variable t and write

$$\Phi(t,y) = \int P(t,y)\,dt + \psi(y).$$

Then, we will determine ψ (up to an additive constant) by imposing

$$\frac{\partial}{\partial y} \int P(t,y)\,dt + \psi'(y) = Q(t,y). \tag{1.50}$$

Example 1.16. Consider the ODE

$$y' = -\frac{t^3 + ty^2}{t^2 y + y^3}.$$

We have $P(t,y) - t^3 + ty^2$ and $Q(t,y) - t^2 y + y^3$. Both functions are polynomial in $\mathbb{R}^2$ and the ODE is exact ($P_y \equiv Q_t$). In order to find Φ, we write

$$\Phi(t,y) = \int P(t,y)\,dt + \psi(y) = \frac{t^4}{4} + \frac{t^2 y^2}{2} + \psi(y).$$

By imposing (1.50), we get $t^2 y + \psi'(y) = t^2 y + y^3$. Therefore, we can take

$$\psi(y) = \frac{y^4}{4} \quad \text{and} \quad \Phi(t,y) = \frac{1}{4}(t^2 + y^2)^2.$$

Finally, the general solution is given by the identity

$$t^2 + y^2 = C.$$

Thus, we see that the solutions to the ODE can be identified with arcs of circumferences in the plane centered at $(0,0)$.

Note that the equation is homogeneous. Consequently, the general solution can also be found following a different method.

This task is left (and recommended) to the reader. □

1.3.7 *Reducible to exact equations, integrating factors**

Let us consider again an equation of the form (1.46), with $\Omega \subset \mathbb{R}^2$ a simply connected open set and P and Q continuously differentiable in Ω.

Assume that (1.46) is not exact. It will be said that the function $\mu = \mu(t,y)$ is an *integrating factor* in Ω for this ODE if it is defined, continuously differentiable, and nonzero in Ω and the ODE

$$\mu(t,y)P(t,y) + \mu(t,y)Q(t,y)\,y' = 0 \tag{1.51}$$

is exact.

The ODEs (1.46) and (1.51) are equivalent, in the sense that they possess the same solutions. Consequently, if P and Q are given, it can be interesting to know how we can find an associated integrating factor.

Since Ω is simply connected, the goal is to find μ such that

$$(\mu(t,y)P(t,y))_y = (\mu(t,y)Q(t,y))_t \quad \text{in } \Omega.$$

In many cases, it is possible to find such a function μ by solving an auxiliary ODE. This can be achieved if we look for a μ independent of t, independent of y or, more generally, of the form

$$\mu(t,y) = \mu_0(X(t,y)),$$

where μ_0 is a function of one real variable and $X(t,y)$ is an appropriately well-chosen expression.

Example 1.17. Let us consider the ODE

$$t(1-y) + (t^2 + y)y' = 0.$$

Let us try to compute an integrating factor. We must find a function μ satisfying

$$(\mu(t,y)t(1-y))_y = (\mu(t,y)(t^2+y))_t,$$

that is to say,

$$t(1-y)\mu_y - (t^2+y)\mu_t = 3t\mu.$$

Let us look for a function μ of the form $\mu(t,y) = \mu_0(t^2 + y^2)$. To this end, μ_0 must satisfy

$$\big(2ty(1-y) - 2t(t^2+y)\big)\,\mu_0'(t^2+y^2) = 3t\,\mu_0(t^2+y^2).$$

After simplification, with $z = t^2 + y^2$, we see that

$$\frac{d\mu_0}{dz}(z) = -\frac{3}{2z}\mu_0(z),$$

which is a homogeneous linear ODE for μ_0. Therefore, a valid integrating factor is

$$\mu(t,y) = (t^2 + y^2)^{-3/2}.$$

Now, taking into account that the corresponding ODE (1.51) is exact, we can compute a function Φ such that $\Phi_t = \mu P$ and $\Phi_y = \mu Q$. Proceeding as in Section 1.3.6, the following is easily found:

$$\Phi(t,y) = (y-1)(t^2+y^2)^{-1/2}$$

and, finally, we get the general solution

$$(y-1)(t^2+y^2)^{-1/2} = C.$$

We leave to the reader the task to find the solutions to the CPs corresponding to the initial data $(0,1)$ and $(0,2)$. □

In the following paragraphs, we will consider some second-order ODEs with normal structure:

$$y'' = f(t,y,y'). \tag{1.52}$$

As expected, an associated general solution will be a family of functions $\varphi(\cdot\,;C_1,C_2)$ (parameterized by two arbitrary constants C_1 and C_2) that solve (1.52).

1.3.8   *Second-order incomplete equations*

First, let us consider an ODE of the form

$$y'' = f(t).$$

In this case, we get immediately the general solution

$$\varphi(t;C_1,C_2) = \int F(t)\,\mathrm{d}t + C_1 t + C_2,$$

where F is a primitive of f.

For an equation of the form

$$y'' = f(t,y'),$$

the change of variable $z = y'$ leads to a first-order ODE:

$$z' = f(t,z).$$

On the other hand, any ODE

$$y'' = f(y,y')$$

can also be reduced to the first order. It suffices to introduce the new variable p, with $p(y(t)) = y'(t)$. This gives

$$y'' = \frac{\mathrm{d}}{\mathrm{d}t}p(y(t)) = \frac{\mathrm{d}p}{\mathrm{d}y}(y(t))\,y'(t) = \frac{\mathrm{d}p}{\mathrm{d}y}(y(t))\,p(y(t))$$

and thus

$$\frac{\mathrm{d}p}{\mathrm{d}y}p = f(y,p).$$

After computing a general solution, the task is reduced to the solution of an additional first-order ODE.

Example 1.18. Let us consider the ODE

$$y'' = \frac{(2y + 3y')y'}{y}. \tag{1.53}$$

As before, let us introduce $p = p(y)$, with $p(y(t)) = y'(t)$. This leads to the ODE

$$\frac{dp}{dy} = \frac{3}{y}p + 2, \tag{1.54}$$

that is linear and nonhomogeneous. Now, y (resp. p) is the independent (resp. dependent) variable.

With the techniques in Section 1.3.2, we find at once the general solution to (1.54):

$$p = Ky^3 - y, \tag{1.55}$$

where K is an arbitrary constant.

We must interpret (1.55) as a second first-order ODE. More precisely, coming back to the definition of p, we see from (1.55) that

$$y' = Ky^3 - y$$

and, consequently, y must solve a Bernoulli equation. Finally, arguing as in Section 1.3.3, we find the general solution to (1.53):

$$\varphi(t; C_1, C_2) = \left(C_1 + C_2 e^{2t}\right)^{-1/2},$$

where C_1 and C_2 are arbitrary constants. □

1.3.9   Second-order linear equations

A *linear* ODE of the second order takes the form

$$y'' = a_1(t)y' + a_0(t)y + b(t), \tag{1.56}$$

where a_0, a_1, and b are continuous functions in the interval $I \subset \mathbb{R}$.

Again, as in Section 1.3.7, we will refer to solutions defined in the whole interval I.

Many of the arguments that follow will be generalized in what follows, when we speak of linear ODSs of the first-order and linear ODEs of order n.

Let us first assume that $b \equiv 0$:

$$y'' = a_0(t)y + a_1(t)y'. \tag{1.57}$$

In this case, we say that the equation is *linear homogeneous* and, again, we see that the set of solutions is a linear space for the usual laws.

Let us accept that the dimension of this space is 2 (this will be proved later, see Chapter 5).

Then, in order to solve (1.57), it will suffice to find two linearly independent solutions φ_1 and φ_2. The general solution $\varphi_h(\cdot\,; C_1, C_2)$ will be given by

$$\varphi_h(t; C_1, C_2) = C_1\varphi_1(t) + C_2\varphi_2(t). \tag{1.58}$$

Example 1.19. Consider the linear homogeneous ODE

$$y'' = a(t)y'. \tag{1.59}$$

With the change of variable $z = y'$, (1.59) can be rewritten as a first-order ODE and we can compute the general solution

$$z = C_1 e^{A(t)},$$

where C_1 is arbitrary and A is a primitive of a. Therefore, the general solution to (1.59) is

$$y = C_1 \int e^{A(t)}\,\mathrm{d}t + C_2.$$

In this particular case, the linearly independent solutions are φ_1 and φ_2, with

$$\varphi_1(t) = \int e^{A(t)}\,\mathrm{d}t \quad \text{(a primitive of } e^A\text{)}, \quad \varphi_2(t) = 1.$$

□

If $b \not\equiv 0$, the family of solutions to (1.56) is a linear manifold supported by the space of solutions to (1.57).

Consequently, the general solution is the sum of a particular (arbitrary but fixed) solution φ_p and the general solution to the associated homogeneous ODE:

$$\varphi(t; C_1, C_2) = \varphi_p(t) + \varphi_h(t; C_1, C_2) = \varphi_p(t) + C_1\varphi_1(t) + C_2\varphi_2(t).$$

As a consequence, in order to find all the solutions to (1.56), we must compute

(1) Two linearly independent solutions φ_1 and φ_2 to (1.57).
(2) One particular solution φ_p to (1.56).

Let us now see that it is always possible to compute φ_p if φ_1 and φ_2 are known.

To this purpose, we will apply a new version of Lagrange's method of variation of constants (see Section 1.3.2).

The fundamental idea is, as before, to replace the constants in (1.58) by functions of t and write

$$\varphi_p(t) = C_1(t)\varphi_1(t) + C_2(t)\varphi_2(t). \tag{1.60}$$

Then, what we do is to ask φ_p to satisfy (1.56).

Specifically, we impose the following:

$$\begin{cases} C_1'(t)\varphi_1(t) + C_2'(t)\varphi_2(t) = 0 \\ C_1'(t)\varphi_1'(t) + C_2'(t)\varphi_2'(t) = b(t). \end{cases} \tag{1.61}$$

Under these conditions, it is clear that φ_p must satisfy

$$\varphi_p'(t) = C_1(t)\varphi_1'(t) + C_2(t)\varphi_2'(t)$$

and

$$\varphi_p''(t) = C_1(t)\varphi_1''(t) + C_2(t)\varphi_2''(t) + b(t)$$

and, taking into account that φ_1 and φ_2 solve (1.57), we immediately deduce that φ_p solves (1.56).[10]

Example 1.20. Consider the ODE

$$y'' = y' + 6y + 5t. \tag{1.62}$$

The functions $\varphi_1(t) = e^{3t}$ and $\varphi_2(t) = e^{-2t}$ are solutions to the associated linear homogeneous equation. Obviously, they are linearly independent. Let us compute a particular solution to the nonhomogeneous ODE (1.62).

Let us put

$$\varphi_p(t) = C_1(t)e^{3t} + C_2(t)e^{-2t},$$

with C_1 and C_2 satisfying

$$\begin{cases} C_1'(t)e^{3t} + C_2'(t)e^{-2t} = 0 \\ 3C_1'(t)e^{3t} - 2C_2'(t)e^{-2t} = 5t. \end{cases}$$

---

[10]This process is completely general, applicable to any second-order linear ODE. It will be seen in Chapter 5 that it can be simplified in some particular situations.

Then

$$C_1'(t) = te^{-3t}, \quad C_2'(t) = -te^{2t}$$

and possible choices of $C_1(t)$ and $C_2(t)$ are

$$C_1(t) = -\frac{1}{9}(3t+1)e^{-3t}, \quad C_2(t) = -\frac{1}{4}(2t-1)e^{2t}.$$

This furnishes the particular solution

$$\varphi_p(t) = \frac{5}{36}(1-6t).$$

Finally, the general solution to (1.62) is

$$\varphi(t; C_1, C_2) = \frac{5}{36}(1-6t) + C_1 e^{3t} + C_2 e^{-2t}. \qquad \Box$$

The explicit computation of two independent solutions to (1.57) is only possible in some particular cases. We will indicate now two of them.

1.3.9.1 *The case of constant coefficients*

Here, we have $I = \mathbb{R}$ and the ODE is the following:

$$y'' = a_0 y + a_1 y', \tag{1.63}$$

where $a_0, a_1 \in \mathbb{R}$.

In order to find a couple of independent solutions, we consider the so-called *characteristic polynomial* of (1.63), given by

$$p(\lambda) := \lambda^2 - a_1\lambda - a_0$$

and the *characteristic equation* $p(\lambda) = 0$. Let λ_1 and λ_2 be the roots of $p(\lambda)$ and let us assign to these numbers the functions φ_1 and φ_2 as follows:

- If λ_1 and λ_2 are real and distinct, we take

$$\varphi_i(t) = e^{\lambda_i t}, \quad i = 1, 2.$$

- If $\lambda_1 = \lambda_2$ (a double real root), we set

$$\varphi_1(t) = e^{\lambda_1 t}, \quad \varphi_2(t) = te^{\lambda_1 t}.$$

- Finally, if λ_1 and λ_2 are conjugate complex roots, that is,

$$\lambda_1 = \alpha + i\beta, \quad \lambda_2 = \alpha - i\beta$$

and for some α and β, we take

$$\varphi_1(t) = e^{\alpha t}\cos(\beta t), \quad \varphi_2(t) = e^{\alpha t}\sin(\beta t).$$

In all cases, it is not difficult to check that φ_1 and φ_2 are linearly independent and solve (1.57).

Example 1.21. Consider the ODE

$$y'' = -2y' - 5y + \cos(2t) - 4\sin(2t). \qquad (1.64)$$

The associated linear homogeneous equation is

$$y'' = -2y' - 5y.$$

The characteristic equation possesses two conjugate complex solutions: $\lambda_1 = -1 + 2i$ and $\lambda_2 = -1 - 2i$. Thus, we find the following linearly independent solutions:

$$\varphi_1(t) = e^{-t}\cos(2t), \quad \varphi_2(t) = e^{-t}\sin(2t).$$

After applying Lagrange's method, we easily find the particular solution $\varphi_p(t) = \cos(2t)$. Consequently, the general solution to (1.64) is

$$\varphi(t; C_1, C_2) = \cos(2t) + C_1 e^{-t}\cos(2t) + C_2 e^{-t}\sin(2t),$$

where, once more, C_1 and C_2 are arbitrary constants. $\qquad \square$

Example 1.22. Let us consider again the ODEs (1.10), where g, k, and m are positive constants. We already know that the general solution of the first of these ODEs is

$$\psi(t; C_1, C_2) = -g\frac{t^2}{2} + C_1 t + C_2.$$

In particular, the CP (1.24) possesses the following solution:

$$\phi(t) = y_0 + v_0 t - g\frac{t^2}{2}. \qquad (1.65)$$

Let us compute the general solution to the second ODE. For simplicity, let us set $\kappa := k/m$. The associated homogeneous ODE is

$$y'' + \kappa y' = 0.$$

Its general solution is given by

$$\psi_h(t; C_1, C_2) = C_1 + C_2 e^{-\kappa t}.$$

On the other hand, a particular solution to the non-homogeneous equation is $\psi_p(t) = -gt/\kappa$, so we get the following general solution:

$$\psi(t; C_1, C_2) = -\frac{g}{\kappa}t + C_1 + C_2 e^{-\kappa t} = -\frac{mg}{k}t + C_1 + C_2 e^{-kt/m}.$$

This shows that the CP

$$\begin{cases} y'' = -g - \dfrac{k}{m} y' \\ y(0) = y_0 \,, \; y'(0) = v_0 \end{cases} \tag{1.66}$$

(where y_0 and v_0 are given in $\mathbb{R}$) possesses the following solution:

$$\psi(t) = y_0 - \frac{mg}{k} t + \frac{m}{k} \left(v_0 + \frac{mg}{k} \right) \left(1 - e^{-kt/m} \right)$$

$$= y_0 + v_0 t - \left(g + \frac{k}{m} v_0 \right) \left(\frac{t^2}{2!} - \left(\frac{k}{m} \right) \frac{t^3}{3!} + \left(\frac{k}{m} \right)^2 \frac{t^4}{4!} + \cdots \right). \tag{1.67}$$

By comparing (1.65) and (1.67), we can appreciate the effect and relevance of the friction term $-\frac{k}{m} y'$ in the second ODE in (1.10) (Figure 1.8).   □

1.3.9.2 *Second-order Euler equations*

Here, we deal with equations of the form

$$t^2 y'' = a_1 t y' + a_0 y, \tag{1.68}$$

where $a_0, a_1 \in \mathbb{R}$. We will try to find solutions in $I = (0, +\infty)$.

In order to solve (1.68), it suffices to perform the change of variable

$$\tau = \log t \quad (\text{that is, } t = e^\tau).$$

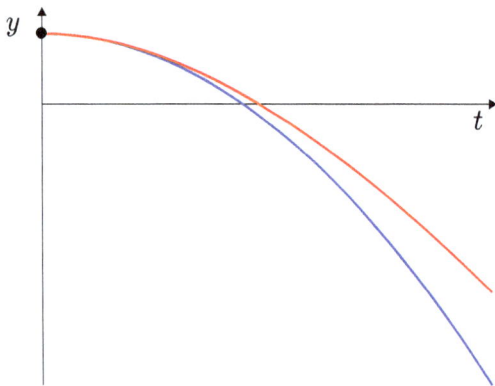

Fig. 1.8   The solutions to (1.66) for $k = 0$ (blue) and $k > 0$ (red).

Denoting by $\dot{y}$ and $\ddot{y}$ the first and second derivative of y with respect to the new variable τ, one has the following:

$$y' = \frac{dy}{dt} = \frac{dy}{d\tau}\frac{d\tau}{dt} = \frac{1}{t}\dot{y},$$

$$y'' = \frac{d}{dt}\left(\frac{1}{t}\dot{y}\right) = -\frac{1}{t^2}\dot{y} + \frac{1}{t^2}\ddot{y}.$$

Therefore, the ODE can be rewritten as follows:

$$\ddot{y} = (a_1 + 1)\dot{y} + a_0 y,$$

that is linear homogeneous of constant coefficients.

Example 1.23. Let the Euler equation

$$t^2 y'' = ty' - y \tag{1.69}$$

be given. With $t = e^\tau$, using the previous notation, we see that

$$\ddot{y} = 2\dot{y} - y. \tag{1.70}$$

The general solution of (1.70) is $\psi(\tau; C_1, C_2) = C_1 e^\tau + C_2 \tau e^\tau$ and this indicates that the general solution to (1.69) is

$$\varphi(t; C_1, C_2) = C_1 t + C_2 t \log t. \qquad \square$$

Exercises

E.1.1 Find the general solutions to the following linear ODEs:

(1) $(1 + x^2)y' + xy = (1 + x^2)^{5/2}$.

   Sol.: $y = \left(x + \frac{1}{5}x^5 + \frac{2}{3}x^3 + C\right)(1 + x^2)^{-1/2}$.

(2) $y' + \frac{xy}{1 + x^2} = 1 - \frac{x^3 y}{1 + x^4}$.

   Sol.: $y = \left(C + \int (1 + x^2)^{1/2}(1 + x^4)^{1/4}\,dx\right)(1 + x^2)^{-1/2}$ $(1 + x^4)^{-1/4}$.

(3) $y' = \frac{2xy}{3x^2 - y^2}$. **Sol.:** $y^2 - x^2 = Cy^3$, $y = 0$.

   Hint: Exchange the roles of the variables x and y. In other words, instead of $y = y(x)$, try to find its inverse $x = x(y)$; use that $\frac{dx}{dy} = \left(\frac{dy}{dx}\right)^{-1}$ to get a "new" form of the equation.

E.1.2 Solve the following Cauchy problems:

(1) $y' - 2ty = t$, $y(0) = 1$. **Sol.:** $y = \dfrac{1}{2}\left(3e^{t^2} - 1\right)$.

(2) $yy' + (1+y^2)\sin t = 0$, $y(0) = 1$. **Sol.:** $y = \left(2e^{2(\cos t - 1)} - 1\right)^{1/2}$.

(3) $(1+t^2)y' + 4ty = t$, $y(1) = \dfrac{1}{4}$. **Sol.:** $y = \dfrac{1}{4}$.

(4) $(1+e^t)yy' = e^t$, $y(0) = 1$. **Sol.:** $y = \left(1 + \log\dfrac{(1+e^t)^2}{4}\right)^{1/2}$.

E.1.3 Find the general solutions to the following ODEs:

(1) $(2x+2y-4)y' + (4x+4y-2) = 0$. **Sol.:** $(x+y+1)^3 = Ce^{2x+y}$.

(2) $y' + \dfrac{1}{x}y = x^3 y^3$. **Sol.:** $y = \pm\dfrac{1}{x\sqrt{C-x^2}}$, $y - 0$.

(3) $4xy^2 + (3x^2y - 1)y' = 0$. **Sol.:** $\alpha = -2$, $y(x^2y-1)^2 = C$.

     Hint: Introduce the new variable z, with $y = z^\alpha$ for an adequate α.

(4) $(3x-7y-3)y' = 3y - 7x + 7$. **Sol.:** $(y+x-1)^5(y-x+1)^2 = C$.

E.1.4 Find the general solutions to the following ODEs:

(1) $y' = a^{t+y}$, $a > 0$. **Sol.:** $a^t + a^{-y} = C$.

(2) $y' + t^2 y = 1$. **Sol.:** $y = e^{-t^3/3}\left(C + \displaystyle\int e^{t^3/3}\,dt\right)$.

(3) $y' = \dfrac{3t^2}{t^3 + y + 1}$. **Sol.:** $t^3 = Ce^y - (y+2)$.

(4) $y'\sin t - y\cos t = -\dfrac{\sin^2 t}{t^2}$; among all the solutions, find those satisfying $\lim\limits_{t\to\infty} y(t) = 0$. **Sol.:** $y = \dfrac{\sin t}{t}$.

E.1.5 Find the curves in the plane satisfying the following properties:

(1) For each x, the area $S(x)$ of the region bounded by the curve, the OX axis and the straight lines $X = 0$ and $X = x$ is given by $S(x) = a^2 \log\dfrac{y(x)}{a}$, where a is a positive constant. **Sol.:** $y = \dfrac{a^2}{a-x}$.

(2) At any point, the slope of the tangent is n times the slope of the straight line joining the point to the origin. **Sol.:** $y = Cx^n$.

(3) The segment of a tangent to the curve between the axes is always divided in half at the point of contact. **Sol.:** $y = C/x$.

(4) The distance from the origin to the tangent at a point of the curve is equal to the abscissa of the point.
Sol.: $y = \pm\sqrt{x(C-x)}$.

(5) At any point, the area of the triangle formed by the tangent, the normal, and the OX axis is equal to the slope of the normal.

(6) The length of the segment of a tangent to the curve between the axes is always equal to 1.

(7) The midpoint of the segment determined in the OY axis by the tangent and the normal at any point is always $(0, -1/2)$.

E.1.6 Find the general solutions to the following ODEs:

(1) $(y^4 - 3t^2)y' = -ty$. **Sol.:** $t = \pm y^2\sqrt{Cy^2 + 1}$.

(2) $t^2 y' + 2t^3 y = y^2(1 + 2t^2)$.

Sol.: $y = e^{-t^2}\left(C - \int \left(\dfrac{1 + 2t^2}{t^2}\right)e^{-t^2}\,dt\right)^{-1}$.

(3) $(t+y)(y'+1) = y'$. **Sol.:** $y = (1-t) \pm (C - 2t)^{1/2}$.

E.1.7 Find the general solutions to the following ODEs:

(1) $y' = \dfrac{1}{x\sin y + 2\sin 2y}$. **Sol.:** $x = 4(1 - \cos y) + Ce^{-\cos y}$.

(2) $(x - 2y - 1) = (6y - 3x - 2)y'$.
Sol.: $3(x - 2y) + 2\log|x - 2y| = 5x + C$.

(3) $y' + x\sin 2y = xe^{-x^2}\cos^2 y$.

Sol.: $y = \arctan\left(\left(\dfrac{x^2}{2} + C\right)e^{-x^2}\right)$.

Hint: Perform the change of variable $u = \tan y$.

E.1.8 Rewrite the search of φ such that $\int_0^1 \varphi(\alpha x)\,d\alpha = k\varphi(x)$ for all $x \in (0, +\infty)$ with $\varphi \in C^1(0, +\infty)$ as an ODE and compute the corresponding general solution. **Sol.:** $\varphi(x) = Cx^{(1-k)/k}$.

E.1.9 Find the general solutions to the following ODEs:

(1) $(x - y + 3) + (3x + y + 1)y' = 0$.
Sol.: $y + x - 1 = C\exp\left(\dfrac{2(x+1)}{y + x - 1}\right)$.

(2) $(1 + e^{x/y}) + e^{x/y}(1 - \frac{x}{y})y' = 0$, $y(1) = 1$. **Sol.:** $x + ye^{x/y} = 1 + e$.

(3) $(x - y^2) + 2xyy' = 0$. **Sol.:** $y^2 = x\left(C + \log\dfrac{1}{|x|}\right)$.

E.1.10 Find the general solutions to the following ODEs:

(1) $(ty + t^2 y^3)y' = 1$. **Sol.:** $t = \dfrac{1}{2 - y^2 + Ce^{-y^2/2}}$.

(2) $y' - y\tan t = \sec t$, $y(0) = 0$. **Sol.:** $y = t\sec t$.

(3) $3t^2 - 2t - y = (t - 2y - 3y^2)y'$. **Sol.:** $t^3 + y^3 - t^2 + y^2 - ty = C$.

E.1.11 Find the curves in the plane satisfying the following properties:

(1) The tangent at any point P touches the OX axis at a point Q such that the triangle OPQ has an area equal to b^2.
Sol.: $x = Cy \pm b^2/y$.

(2) $(0,1)$ is a point on the curve and, moreover, the tangent at any point P touches the OX axis at a point Q such that the midpoint of the segment PQ describes the parabola $y^2 = 2ax$ as P moves on the curve. **Sol.:** $x = -(y^2 \log|y|)/(4a)$.

E.1.12 Find the curves in the plane that are orthogonal to the curves of the following families:

(1) $y = ax^2$. **Sol.:** $x^2 + 2y^2 = C$.
(2) $y^2 + 3x^2 - 2ax = 0$. **Sol.:** $x^2 = y^2(Cy + 1)$.
(3) $x^2 - y^2 = a^2$. **Sol.:** $y = C/x$.
(4) $y^2 + 2ax = a^2$, $a > 0$. **Sol.:** $y^2 - 2Cx = C^2$, $C > 0$.
(5) $y^2 = 4(x - a)$. **Sol.:** $y = Ce^{-x/2}$.

Hint: In each case, find first the ODE $y' = f(x,y)$ of the family; then, use the new ODE $z' = -1/f(x,z)$ to compute the orthogonal curves.

E.1.13 The "half-life" of a radioactive substance is the time taken by an initial amount c_0 to be reduced to $c_0/2$. Disintegration is governed by the linear equation $c' = -\lambda c$, where $\lambda > 0$ is the so-called disintegration constant.

(1) Prove that the half-life does not depend on the considered initial quantity and is given by

$$T = \frac{\log 2}{\lambda}.$$

(2) Assume that a radioactive substance keeps the 95% of its initial mass after 50 h. Determine the corresponding half-life (this is the usual experimental way that the half-life or equivalently the disintegration constant λ can be determined).
Sol.: $T = 50 \log 2 / \log(20/19) \sim 675'67$ h.

E.1.14 *Newton's Cooling Law* asserts that the rate of change of the temperature of a body is proportional to the difference between its temperature and the temperature of the surrounding air. Assume that a room is kept at constant temperature, equal to $70°F$. Assume that a body initially at $350°F$ passes to $150°F$ at $t = 45$ min. At which time will the body be at $80°F$?

E.1.15 Let a spherical piece of naphthalene have a radius of 3 cm. Assume that, after 1 month, the radius has diminished 1 cm. Taking into account that the evaporation speed is proportional to the surface, compute the radius as a function of time. At which time will the piece disappear completely?

E.1.16 Consider the second-order ODE

$$y'' = -ky + f(t),$$

where $k = \omega_0^2 > 0$ and $f(t) \equiv a\sin\omega t$ for some positive constants a, ω_0, and ω. This ODE describes the motion of a body of mass $m = 1$ submitted to the action of elastic forces (that are proportional to the distance to the origin) and additional external forces (that are periodic, of period ω and intensity a).

(1) Solve this ODE for any positive ω_0, ω, and a.
(2) Prove that this ODE can have unbounded solutions if and only if $\omega = \omega_0$.

This explains the occurrence of *resonance*, a phenomenon that can be the origin of great disasters; for more details, see Braun (1993).

E.1.17 Solve the following ODEs:

(1) $y'' = x(y')^3$. **Sol.:** $y = \arcsin\dfrac{x}{C} + k$, $y = C$.
(2) $xy'' + x(y')^2 = y'$. **Sol.:** $y = \log|x^2 + C_1| + C_2$.
(3) $y'' + yy' = 0$. **Sol.:** $y = C_1 - \dfrac{2C_1}{e^{C_1(x+C_2)} + 1}$.
(4) $yy'' = (y')^2 + yy'$. **Sol.:** $y = C_1 e^{C_2 e^x}$.

E.1.18 Solve the ODE $xy'' + (x-2)y' - 3y = 0$ using that $y_1 = x^3$ is a particular solution. **Sol.:** $y = C_1 x^3 \int \dfrac{e^{-x}}{x^4}\,dx + C_2 x^3)$.

Hint: Introduce the new unknown u, with $y = y_1 u$.

Chapter 2

The Cauchy Problem: Local Analysis

This chapter is devoted to the study of initial-value problems for a first-order ODS and for a higher-order ODE. We will prove results concerning the existence and uniqueness of a local solution (that is, a solution in principle only defined in a "small" interval around a specified value of the independent variable).

To this purpose, we will first have to recall several basic properties of Banach spaces. In particular, we will speak of the space of continuous functions of one variable in a compact interval.

2.1   Integral Formulation

We will consider the Cauchy problem

(CP) $$\begin{cases} y' = f(t, y) \\ y(t_0) = y_0, \end{cases}$$

where $f : \Omega \mapsto \mathbb{R}^N$ is continuous and $(t_0, y_0) \in \Omega$. As in Chapter 1, $\Omega \subset \mathbb{R}^{N+1}$ is a non-empty open set.

In the sequel, if $I \subset \mathbb{R}$ is an interval, $C^0(I; \mathbb{R}^N)$ denotes the set of continuous functions $\varphi : I \mapsto \mathbb{R}^N$. On the other hand, $C^1(I; \mathbb{R}^N)$ will stand for the set of continuously differentiable functions on I. When $N = 1$, we will simply put $C^0(I)$ or $C^1(I)$.

Recall that, in order to clarify the situation as much as possible, we have adopted the following convention (see Chapter 1):

- If I is open, it is understood that $\varphi : I \mapsto \mathbb{R}^N$ is continuously differentiable in I if it satisfies the usual property: there exists the derivative $\varphi'(t)$ at every $t \in I$ and, moreover, the function $t \mapsto \varphi'(t)$ is continuous in I.
- Contrarily, if I contains at least one endpoint, it will be said that φ is continuously differentiable in I if there exists a continuously differentiable extension $\tilde{\varphi}$ of φ to an open interval $J \supset I$.

Both $C^0(I; \mathbb{R}^N)$ and $C^1(I; \mathbb{R}^N)$ are (infinite dimensional) linear spaces for the usual laws. For simplicity, when we write $C^0(I; \mathbb{R}^N)$, we will refer not only to the set of continuous functions, but to the corresponding linear space.

The main result in this section is the following:

Lemma 2.1. *Let I be an interval containing t_0 and let $\varphi : I \mapsto \mathbb{R}^N$ be a given function. The following assertions are equivalent:*

(1) *φ is a solution to* (CP) *in I.*
(2) *$\varphi \in C^0(I; \mathbb{R}^N)$, $(t, \varphi(t)) \in \Omega$ for all $t \in I$ and*

$$\varphi(t) = y_0 + \int_{t_0}^{t} f(s, \varphi(s)) \, \mathrm{d}s \quad \forall t \in I. \tag{2.1}$$

Proof. First, let us assume that φ solves (CP) in I. Then $\varphi \in C^1(I; \mathbb{R}^N)$ and $(t, \varphi(t)) \in \Omega$ for all $t \in I$. Furthermore, for all $t \in I$ one has

$$\varphi(t) = \varphi(t_0) + \int_{t_0}^{t} \varphi'(s) \, \mathrm{d}s = y_0 + \int_{t_0}^{t} f(s, \varphi(s)) \, \mathrm{d}s.$$

Consequently, one has (2.1).

Conversely, let us assume that $\varphi \in C^0(I; \mathbb{R}^N)$, $(t, \varphi(t)) \in \Omega$ in I and (2.1) holds. Then $\varphi(t_0) = y_0$; moreover, φ coincides in I with a function in $C^1(I; \mathbb{R}^N)$ whose derivative at each t is equal to $f(t, \varphi(t))$. This proves that φ is a solution in I to (CP). $\qquad\square$

This is a very useful lemma. Indeed, it makes it possible to replace the differential problem (CP) by an integral equation.

In principle, it can be expected that solving this integral equation is more within our possibilities than solving (CP) directly: the search of a solution is carried out in a larger set of functions and, also, to integrate is in some sense "easier" than to differentiate.

Remark 2.1. If we interpret the variables y and t as the position of a physical particle in the Euclidean space $\mathbb{R}^N$ and the time passed since it started moving, we see that the formulation of (CP) relies on:

- Fixing an initial position at time $t = t_0$.
- Setting up a law that indicates the way the velocity y' of the particle at time t depends on t and the position y.

In (2.1), this is rephrased.
Specifically, we say there (again) which is the position at $t = t_0$.
On the other hand, we can interpret the integral identity asserting that, at any time t, the position $\varphi(t)$ can be obtained adding to the initial position y_0 the "sum" of all "small displacements" $f(s, \varphi(s))\,ds$ produced between t_0 and t. □

In view of the previous lemma, if we want to solve (CP), we have to look for a function in $C^0(I; \mathbb{R}^N)$ satisfying (2.1).
Consequently, the more we can say on $C^0(I; \mathbb{R}^N)$, the more appropriate will be the situation to deduce an existence result.
In the reminder of the chapter, it will be assumed that I is a non-degenerate compact interval:

$$I = [a, b], \quad \text{with} \quad -\infty < a < b < +\infty.$$

As a consequence, the functions $\varphi \in C^0(I; \mathbb{R}^N)$ are *uniformly continuous*, that is, they satisfy the following property: for any $\varepsilon > 0$, there exists $\delta > 0$ such that

$$t, s \in I, \quad |t - s| \leq \delta \Rightarrow |\varphi(t) - \varphi(s)| \leq \varepsilon.$$

Note that, in the vector space $C^0(I; \mathbb{R}^N)$, the mapping $\varphi \mapsto \|\varphi\|$, where

$$\|\varphi\| := \sup_{t \in I} |\varphi(t)| \quad \forall \varphi \in C^0(I; \mathbb{R}^N)$$

is a norm. It is frequently called the norm of the supremum or *norm of the uniform convergence*.
Of course, this is not the unique possible norm in $C^0(I; \mathbb{R}^N)$. However, unless otherwise stated, with the notation $C^0(I; \mathbb{R}^N)$ we will refer to this normed space.

Recall that, as any other norm, $\|\cdot\|$ induces a *distance* in $C^0(I;\mathbb{R}^N)$. It is given by

$$\text{dist}\,(\varphi_1,\varphi_2) := \|\varphi_1 - \varphi_2\| \quad \forall \varphi_1, \varphi_2 \in C^0(I;\mathbb{R}^N).$$

Consequently, we can also speak of the *metric space* $C^0(I;\mathbb{R}^N)$ and it is completely meaningful to consider open, closed, compact, connected, etc., sets in $C^0(I;\mathbb{R}^N)$ and also Cauchy and convergent sequences in this space.

For example, a non-empty set $G \subset C^0(I;\mathbb{R}^N)$ is *open* if, for any $\varphi \in G$ there exists $\varepsilon > 0$ such that the ball $B(\varphi;\varepsilon)$ (centered at φ and of radius ε) is contained in G.

A non-empty set $K \subset C^0(I;\mathbb{R}^N)$ is *compact* if any open covering of K possesses a finite subcovering; equivalently, K is compact if and only if any sequence in K possesses a subsequence that converges to a function $\varphi \in K$. And so on . . . .

A crucial property of $C^0(I;\mathbb{R}^N)$ is given in the following result:

Theorem 2.1. $C^0(I;\mathbb{R}^N)$ *is a Banach space, that is, a complete normed space. In other words, any Cauchy sequence in $C^0(I;\mathbb{R}^N)$ is convergent.*

Proof. Let $\{\varphi_n\}$ be a Cauchy sequence in the normed space $C^0(I;\mathbb{R}^N)$. Let us see that there exists $\varphi \in C^0(I;\mathbb{R}^N)$ such that

$$\lim_{n\to\infty} \|\varphi_n - \varphi\| = 0. \tag{2.2}$$

To this end, we will use that I is compact and, also, that the Euclidean space $\mathbb{R}^N$ is complete.

We know that, for each $\varepsilon > 0$, there exists $n_0(\varepsilon)$ such that, if $m, n \geq n_0(\varepsilon)$, then

$$\|\varphi_n - \varphi_m\| \leq \varepsilon.$$

Let $\bar{t} \in I$ be given. Obviously, $\{\varphi_n(\bar{t})\}$ is a Cauchy sequence in $\mathbb{R}^N$; indeed, for each $\varepsilon > 0$ there exists $n_0(\varepsilon)$ such that, if $m, n \geq n_0(\varepsilon)$, then

$$|\varphi_n(\bar{t}) - \varphi_m(\bar{t})| \leq \varepsilon.$$

Consequently, $\{\varphi_n(\bar{t})\}$ converges in the complete normed space $\mathbb{R}^N$ to some point.

Let us put

$$\varphi(\bar{t}) := \lim_{n\to\infty} \varphi_n(\bar{t}).$$

Since $\bar{t}$ is arbitrary in I, this argument allows to introduce a well-defined function $\varphi : I \mapsto \mathbb{R}^N$. We will see now that $\varphi \in C^0(I;\mathbb{R}^N)$ and (2.2) holds.

Let $\varepsilon > 0$ be given. If $m, n \geq n_0(\varepsilon)$, we have

$$|\varphi_n(t) - \varphi_m(t)| \leq \varepsilon \quad \forall t \in I.$$

Taking limits as $m \to \infty$, we deduce that, if $n \geq n_0(\varepsilon)$, then

$$|\varphi_n(t) - \varphi(t)| \leq \varepsilon \quad \forall t \in I,$$

that is,

$$\sup_{t \in I} |\varphi_n(t) - \varphi(t)| \leq \varepsilon.$$

Since ε is arbitrarily small, we have proved that

$$\sup_{t \in I} |\varphi_n(t) - \varphi(t)| \to 0 \quad \text{as} \quad n \to +\infty.$$

Thus, in order to achieve the proof, it suffices to check that φ is continuous. This can be seen as follows.

Again, let $\varepsilon > 0$ be given and let us set $q = n_0(\varepsilon/3)$. Since φ_q is uniformly continuous in I, there exists $\delta > 0$ such that

$$t, s \in I \ |t - s| \leq \delta \Rightarrow |\varphi_q(t) - \varphi_q(s)| \leq \varepsilon/3.$$

Therefore, if $t, s \in I$ and $|t - s| \leq \delta$, we get

$$|\varphi(t) - \varphi(s)| \leq |\varphi(t) - \varphi_q(t)| + |\varphi_q(t) - \varphi_q(s)| + |\varphi_q(s) - \varphi(s)| \leq \varepsilon.$$

Since $\varepsilon > 0$ is arbitrarily small, we have $\varphi \in C^0(I; \mathbb{R}^N)$, as desired. $\qquad \square$

Remark 2.2.

(a) As already said, in this proof we have used in an essential way that I is compact and $\mathbb{R}^N$ is complete.

In a very similar way, it is possible to introduce the normed space $C^0(K; Y)$, where (K, d) is a compact metric space and Y is (for example) a normed space with norm $\| \cdot \|_Y$. The norm in $C^0(K; Y)$ is now

$$\|\varphi\|_{C^0(K;Y)} := \sup_{k \in K} \|\varphi(k)\|_Y \quad \forall \varphi \in C^0(K; Y).$$

Then, if Y is a complete normed space for the norm $\| \cdot \|_Y$, so is $C^0(K; Y)$ for the norm $\| \cdot \|_{C^0(K;Y)}$.

(b) $C^0(I; \mathbb{R}^N)$ is an infinite-dimensional Banach space.

In other words, for each $m \geq 1$ it is possible to find a set of m linearly independent functions in $C^0(I; \mathbb{R}^N)$ (consider for example the polynomial functions on I of degree $\leq m - 1$).

In fact, $C^0(I; \mathbb{R}^N)$ is probably the most significant example of Banach space of infinite dimension (for more details, see Section 2.4). □

Remark 2.3. The norm of the uniform convergence $\|\cdot\|$ is, essentially, the unique norm in the linear space $C^0(I; \mathbb{R}^N)$ for which we get completeness. In other words, for any other norm not equivalent to $\|\cdot\|$, completeness is lost.

In order to illustrate this assertion, let us assume for simplicity that $N = 1$, let us consider the norm $\|\cdot\|_1$, with

$$\|\varphi\|_1 := \int_I |\varphi(t)|\,dt \quad \forall \varphi \in C^0(I)$$

and let us see that there exist (many) Cauchy sequences in the normed space $(C^0(I), \|\cdot\|_1)$ that do not converge.

This is not difficult to check: it suffices to choose $c \in \overset{\circ}{I}$ and take

$$\varphi_n(t) = \begin{cases} 0 & \text{if } t \leq c \\ n(t-c) & \text{if } c < t < c + 1/n \\ 1 & \text{if } t \geq c + 1/n \end{cases}$$

for all $n \geq 1$; then, if we had $\|\varphi_n - \varphi\|_1 \to 0$ as $n \to +\infty$ for some φ, this function would necessary satisfy

$$\varphi(t) = \begin{cases} 0 & \text{if } t \leq c \\ 1 & \text{if } t > c, \end{cases}$$

which is obviously an absurdity.

More details on the lack of completeness for certain norms in $C^0(I; \mathbb{R}^N)$ can be found in Exercises E.2.4 and E.2.5. □

2.2 Contractions and Fixed Points

Let us come back to the CP. Let us assume for the moment that $N = 1$ and $\Omega = \mathbb{R}^2$.

In the sequel, for any $\delta > 0$, the following notation will be used: $I_\delta := [t_0 - \delta, t_0 + \delta]$. According to Lemma 2.1, $\varphi : I_\delta \mapsto \mathbb{R}$ solves (CP) in I_δ if and only if $\varphi \in C^0(I_\delta)$ and

$$\varphi(t) = y_0 + \int_{t_0}^t f(s, \varphi(s))\,ds \quad \forall t \in I_\delta. \tag{2.3}$$

Let us introduce the mapping $\mathcal{T} : C^0(I_\delta) \mapsto C^0(I_\delta)$, with

$$\mathcal{T}(\varphi)(t) := y_0 + \int_{t_0}^t f(s, \varphi(s)) \, ds \quad \forall t \in I_\delta, \;\; \forall \varphi \in C^0(I_\delta). \tag{2.4}$$

It is not difficult to prove that this mapping is well defined in $C^0(I_\delta)$ and takes values in the same space $C^0(I_\delta)$. Also, it is clear that φ is a solution to (CP) in I if and only if φ solves the *fixed-point equation*

$$\mathcal{T}(\varphi) = \varphi, \quad \varphi \in C^0(I_\delta). \tag{2.5}$$

Accordingly, we see that what we have to do in order to solve (CP) is to prove the existence (or, even better, the existence and uniqueness) of a solution to (2.5).

In the following lines, we will recall a related result for general equations of this kind. Then, the result will be applied in the context of (2.5).

For these purposes, we have to introduce the concept of *contraction*:

Definition 2.1. Let (X, d) be a metric space and let $\mathcal{T} : X \mapsto X$ be a given mapping. It will be said that $\mathcal{T}$ is a contraction if there exists α with $0 \leq \alpha < 1$ such that

$$d(\mathcal{T}(x), \mathcal{T}(y)) \leq \alpha \, d(x, y) \quad \forall x, y \in X.$$

$\square$

Obviously, any contraction $\mathcal{T} : X \mapsto X$ is uniformly continuous in X, that is, for each $\varepsilon > 0$ there exists $\delta > 0$ such that

$$x, y \in X, \;\; d(x, y) \leq \delta \;\Rightarrow\; d(\mathcal{T}(x), \mathcal{T}(y)) \leq \varepsilon.$$

This will be used in what follows.

The main result in this section is the so-called *Contraction Mapping Principle*, also known as *Banach's Fixed-Point Theorem*:

Theorem 2.2 (Banach). *Let (X, d) be a metric space and let $\mathcal{T} : X \mapsto X$ be a contraction. Then $\mathcal{T}$ possesses exactly one fixed-point, i.e. there exists a unique $\hat{x} \in X$ satisfying $\mathcal{T}(\hat{x}) = \hat{x}$.*

Proof. The proof is constructive. More precisely, we will introduce an appropriate sequence (the so-called successive approximations), we will check that it satisfies the Cauchy criterion and, finally, that it converges to the unique fixed point of $\mathcal{T}$.

The sequence of successive approximations is defined as follows:

- First, $x_0 \in X$ is given (arbitrary but fixed).
- Then, for any $n \geq 0$, we set

$$x_{n+1} = \mathcal{T}(x_n). \tag{2.6}$$

The identities (2.6) define without ambiguity a sequence $\{x_n\}$ in X.

Let us see that this is a Cauchy sequence (we will use here that $\mathcal{T}$ is a contraction).

If $m > n \geq 1$, one has

$$\begin{aligned} d(x_m, x_n) = d(\mathcal{T}(x_{m-1}), \mathcal{T}(x_{n-1})) &\leq \alpha \, d(x_{m-1}, x_{n-1}) \\ = \alpha \, d(\mathcal{T}(x_{m-2}), \mathcal{T}(x_{n-2})) &\leq \alpha^2 \, d(x_{m-2}, x_{n-2}) \\ &\leq \cdots \leq \alpha^n \, d(x_{m-n}, x_0). \end{aligned} \tag{2.7}$$

On the other hand, by using the triangular inequality $m - n - 1$ times, we see that

$$d(x_{m-n}, x_0) \leq d(x_{m-n}, x_{m-n-1}) + d(x_{m-n-1}, x_{m-n-2}) + \cdots + d(x_1, x_0).$$

Arguing as in (2.7), we get that

$$\begin{aligned} d(x_{m-n}, x_0) &\leq \alpha^{m-n-1} \, d(x_1, x_0) + \cdots + \alpha \, d(x_1, x_0) + d(x_1, x_0) \\ &= (1 + \alpha + \cdots + \alpha^{m-n-1}) \, d(x_1, x_0) \\ &\leq \frac{1}{1-\alpha} \, d(x_1, x_0). \end{aligned}$$

Consequently,

$$d(x_m, x_n) \leq \frac{\alpha^n}{1-\alpha} \, d(x_1, x_0) \quad \text{for } m > n \geq 1 \tag{2.8}$$

and, taking into account that $\alpha < 1$, this proves that $\{x_n\}$ is a Cauchy sequence in (X, d).

Let $\hat{x}$ be the limit of the sequence $\{x_n\}$ (we are using here that the considered metric space is complete). We have that $x_{n+1} = \mathcal{T}(x_n)$ converges at the same time to $\hat{x}$ and $\mathcal{T}(\hat{x})$ (we are using here that $\mathcal{T}$ is continuous).

Consequently, $\hat{x} = \mathcal{T}(\hat{x})$ and this proves in particular that there exists at least one fixed point of $\mathcal{T}$.

Now, let us assume that $\hat{x}_1$ and $\hat{x}_2$ are two different fixed points. Then

$$d(\hat{x}_1, \hat{x}_2) = d(\mathcal{T}(\hat{x}_1), \mathcal{T}(\hat{x}_2)) \leq \alpha \, d(\hat{x}_1, \hat{x}_2).$$

But, since $0 \leq \alpha < 1$, this is impossible unless one has $d(\hat{x}_1, \hat{x}_2) = 0$, i.e. $\hat{x}_1 = \hat{x}_2$.

Therefore, the fixed point of $\mathcal{T}$ is unique. □

This result deserves several comments:

• As already stated, the proof is constructive: we have got a sequence of approximations of the fixed point and also "error" estimates corresponding to the nth step.

It is worth highlighting that, as a consequence of (2.8), the following *a priori* estimates hold:

$$d(\hat{x}, x_n) \leq \frac{\alpha^n}{1 - \alpha} d(x_1, x_0) \quad \forall n \geq 1. \tag{2.9}$$

On the other hand, it is easy to check that we also have some *a posteriori* estimates as follows:

$$d(\hat{x}, x_n) \leq \alpha^n d(\hat{x}, x_0) \quad \forall n \geq 1. \tag{2.10}$$

Observe that x_0 is arbitrary but its choice is determinant for the "quality" of the nth approximation.

• Theorem 2.2 can be regarded as a starting point of a theory conceived to solve fixed-point equations in metric spaces and, in particular, nonlinear differential equations in Banach function spaces.

Under the conditions of this result, the iterates (2.6) give origin to the so-called *method of successive approximations*.

From the algorithmic viewpoint, the method possesses two remarkable characteristics:

(1) It is *robust*, in the sense that, independently of the starting point x_0, the sequence always converges.
(2) The convergence rate is *linear*, that is, the amount of steps needed to guarantee a given number of exact decimal digits is proportional to this number. This is clear from (2.8) and (2.9).

This second property can make the method too slow.

Indeed, once we are sufficiently close to the fixed point, it can be interesting to "accelerate" and replace the x_n by some x'_n that go faster to the limit as $n \to +\infty$.

The algorithmic and computational aspects of the method of successive approximations have been the subject of a lot of work; see for instance (Berinde, 2007).

• All assumptions in Theorem 2.2 are needed to assert existence and uniqueness.

Indeed, let us take for instance $X = [1, +\infty)$ with the Euclidean metric (this is a complete metric space) and let us set

$$\mathcal{T}(x) = x + \frac{1}{x} \quad \forall x \in X.$$

Note that $\mathcal{T}$ is not a contraction, although it maps X into itself and

$$|\mathcal{T}(x) - \mathcal{T}(y)| < |x - y| \quad \forall x, y \in X, \quad x \neq y.$$

On the other hand, the corresponding fixed-point equation is not solvable in X.

• Theorem 2.2 can be applied in the following particular situation: $X \subset Y$ is a non-empty closed set of a Banach space Y (for the norm $\|\cdot\|_Y$, $\mathcal{T} : X \mapsto Y$ is well defined, maps X into itself, and is a contraction in X, that is,

$$\|\mathcal{T}(x) - \mathcal{T}(y)\|_Y \leq \alpha \|x - y\|_Y \quad \forall x, y \in X, \quad 0 \leq \alpha < 1.$$

Since X can be viewed as a complete metric space for the distance induced by $\|\cdot\|_Y$, as a consequence of Theorem 2.2, $\mathcal{T}$ possesses a unique fixed point in X.[1]

• There are many other fixed-point theorems that do not rely on the contractivity of the mapping.

For more information on this, see Note 11.2.1.

A useful consequence of Theorem 2.2 is the following:

Corollary 2.1. *Let (X, d) be a complete metric space and let $\mathcal{T} : X \mapsto X$ be such that $\mathcal{T}^m$ is a contraction for some $m \geq 1$. Then $\mathcal{T}$ possesses exactly one fixed point.*

Proof. Let us see that $\hat{x}$ is a fixed point of $\mathcal{T}$ if and only if it is a fixed-point of $\mathcal{T}^m$. This will suffice.

If $\hat{x}$ is a fixed point of $\mathcal{T}$, it is clear that $\mathcal{T}^m(\hat{x}) = \hat{x}$. Conversely, if $\mathcal{T}^m(\hat{x}) = \hat{x}$, then $\mathcal{T}(\hat{x})$ is a fixed point of $\mathcal{T}^m$, since

$$\mathcal{T}^m(\mathcal{T}(\hat{x})) = \mathcal{T}(\mathcal{T}^m(\hat{x})) = \mathcal{T}(\hat{x}).$$

Since the fixed point of $\mathcal{T}^m$ is unique, we necessarily have $\mathcal{T}(\hat{x}) = \hat{x}$. □

Observe that, in the conditions of this corollary, the sequence of successive approximations associated to $\mathcal{T}$, constructed as in the proof of Theorem 2.2, converges again toward the unique fixed point.

---

[1] In fact, this will be the result used in the following to solve (CP).

Indeed, for any $x_0 \in X$, the m sequences

$$\{T^{nm}(x_0)\},\ \{T^{nm+1}(x_0)\}, \ldots,\ \{T^{nm+m-1}(x_0)\}$$

converge to $\hat{x}$. Actually, we can also deduce *a priori* and *a posteriori* error estimates as follows:

$$\mathrm{d}(x_k, \hat{x}) \le \alpha^{(k-\mathrm{mod}(k,m))/m} \max\{\mathrm{d}(x_0, \hat{x}), \ldots, \mathrm{d}(x_{m-1}, \hat{x})\}, \qquad (2.11)$$

$$\mathrm{d}(x_k, \hat{x}) \le \frac{\alpha^{(k-\mathrm{mod}(k,m))/m}}{1-\alpha} \max\{\mathrm{d}(x_0, x_1), \ldots, \mathrm{d}(x_{m-1}, x_m)\}, \qquad (2.12)$$

where $\mathrm{mod}(k, m)$ is the remainder of the division of k by m.

2.3   Picard's Theorem

Let us assume again that $\Omega \subset \mathbb{R}^{N+1}$ is a non-empty open set and let $f : \Omega \mapsto \mathbb{R}^N$ be a given function.

We will present in this section an existence and uniqueness result for (CP).

As announced, we will rewrite (CP) as a fixed-point equation for a contraction defined on an appropriate complete metric space, we will apply Theorem 2.2 and we will deduce that there exists a unique *local* solution.

As already stated, this means that the solution exists, in principle, only in a small neighborhood of t_0.

Obviously, from the viewpoint of applications, this can be insufficient: we should want to have global solutions, that is, solutions defined as far as possible from t_0. Hence, it will be convenient to analyze in a subsequent step whether or not the solution can be extended to a "large" interval.

To get existence and uniqueness, it will be indispensable to require an appropriate property of f.

Specifically, in addition to continuity, f will have to be *locally Lipschitz-continuous with respect to y*.

Definition 2.2. It will be said that f is (globally) *Lipschitz-continuous with respect to the variable y in Ω* if there exists $L > 0$ such that

$$|f(t, y^1) - f(t, y^2)| \le L|y^1 - y^2| \quad \forall (t, y^1), (t, y^2) \in \Omega. \qquad (2.13)$$

It will be said that f is *locally Lipschitz-continuous with respect to y in Ω* if it is Lipschitz-continuous with respect to y in any compact set $K \subset \Omega$; that is to say, if f satisfies the following property: for any compact set $K \subset \Omega$, there exists $L_K > 0$ such that

$$|f(t, y^1) - f(t, y^2)| \leq L_K |y^1 - y^2| \quad \forall (t, y^1), (t, y^2) \in K. \qquad (2.14)$$

$\square$

It is usual to say that L (resp. L_K) is a *Lipschitz constant* in Ω (resp. in K).

It will be seen in what follows that, if f is continuous and its components f_i possess continuous partial derivatives with respect to the y_j in Ω, then f is locally Lipschitz-continuous with respect to y in Ω (see Lemma 2.4).

We will also see that there exist (many) continuous functions $f : \Omega \mapsto \mathbb{R}^N$ that are not locally Lipschitz-continuous with respect to y.

Example 2.1.

(a) Consider the function $f : \mathbb{R}^2 \mapsto \mathbb{R}$, with

$$f(t, y) := |y|^a \quad \forall (t, y) \in \mathbb{R}^2,$$

where $0 < a < 1$. Then, there is no neighborhood of $(t_0, 0)$ where f is Lipschitz-continuous. Indeed, if there were $r > 0$ and $L > 0$ such that

$$|f(t_0 y^1) - f(t_0, y^2)| \leq L|y^1 - y^2| \quad \forall y^1, y^2 \in (-r, r),$$

we would have $|y|^a \leq L|y|$ for all small $y > 0$, which is clearly impossible.

(b) Now, take f as before, with $a \geq 1$. If $a = 1$, it is clear that f is globally Lipschitz-continuous in $\mathbb{R}$, with Lipschitz constant $L = 1$. Indeed, we have in this case

$$||y^1| - |y^2|| \leq |y^1 - y^2| \quad \forall y^1, y^2 \in \mathbb{R}.$$

On the other hand, if $a > 1$, it is clear that f is continuous and even continuously differentiable in the whole $\mathbb{R}$. Consequently, it is locally Lipschitz-continuous in $\mathbb{R}$.

Obviously, the same can be said for a similar function in $\mathbb{R}^N$ where we change the absolute values by Euclidean norms. $\square$

The expected local existence and uniqueness result is the following:

Theorem 2.3 (Picard). *Assume that $f : \Omega \subset \mathbb{R}^{N+1} \mapsto \mathbb{R}^N$ is continuous and locally Lipschitz-continuous with respect to y in Ω and $(t_0, y_0) \in \Omega$. There exist $\tau > 0$ and a unique function $\varphi : I_\tau \mapsto \mathbb{R}^N$ that solves (CP) in I_τ.*

Proof. Let the compact set $K \subset \Omega$ and the constants $M > 0$ and $\delta > 0$ be given as follows (Figure 2.1):

- The point (t_0, y_0) belongs to the open set Ω. Consequently, there exist $a_0, b_0 > 0$ such that

$$K := [t_0 - a_0, t_0 + a_0] \times \overline{B}(y_0; b_0) \subset \Omega.$$

Observe that the set K is compact and convex.
- Since f is continuous, there exists

$$M := \max_{(t,y) \in K} |f(t, y)|.$$

It can be assumed that $M > 0$. Indeed, if we had $M = 0$, we would be in a trivial situation, with (CP) possessing exactly one solution in the interval $I_{a_0} = [t_0 - a_0, t_0 + a_0]$: the function φ_0, with

$$\varphi_0(t) = y_0 \quad \forall t \in [t_0 - a_0, t_0 + a_0].$$

- Finally, we take

$$\delta := \min\left(a_0, \frac{b_0}{M}\right). \tag{2.15}$$

In what follows, we will choose appropriately τ in $(0, \delta]$. Suppose that this has already been done and set

$$A_\tau := \{\varphi \in C^0(I_\tau; \mathbb{R}^N) : |\varphi(t) - y_0| \leq b_0 \ \forall t \in I_\tau\}.$$

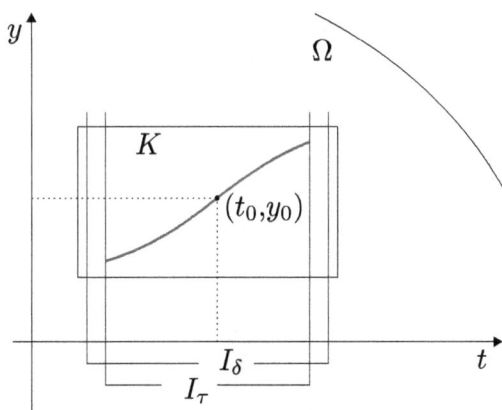

Fig. 2.1   Visualization of the proof of Theorem 2.3.

Let $d(\cdot,\cdot)$ be the distance induced by the norm in $C^0(I_\tau; \mathbb{R}^N)$. Then, (A_τ, d) is a complete metric space (because A_τ is a closed subset of the Banach space $C^0(I_\tau; \mathbb{R}^N)$).

For every $\varphi \in A_\tau$, one has $(s, \varphi(s)) \in K \subset \Omega$ for all $s \in I_\tau$. Consequently, the mapping $\mathcal{T}: A_\tau \mapsto C^0(I_\tau; \mathbb{R}^N)$ given by

$$\mathcal{T}(\varphi)(t) = y_0 + \int_{t_0}^t f(s, \varphi(s))\,ds \quad \forall t \in I_\tau, \quad \forall \varphi \in A_\tau, \tag{2.16}$$

is well defined.

We will need the following lemma, whose proof will be presented in what follows.

Lemma 2.2. *Under the previous conditions, a function φ is a solution to* (CP) *in I_τ if and only if one has*

$$\varphi \in A_\tau, \quad \varphi = \mathcal{T}(\varphi). \tag{2.17}$$

As a consequence, the proof will be achieved if we are able to choose τ in $(0, \delta]$ so that the following properties hold:

(1) $\mathcal{T}$ is A_τ-invariant, that is,

$$\mathcal{T}(\varphi) \in A_\tau \quad \forall \varphi \in A_\tau. \tag{2.18}$$

(2) $\mathcal{T}: A_\tau \mapsto A_\tau$ is a contraction, i.e. there exists $\alpha \in [0, 1)$ such that

$$\|\mathcal{T}(\varphi) - \mathcal{T}(\psi)\| \leq \alpha \|\varphi - \psi\| \quad \forall \varphi, \psi \in A_\tau. \tag{2.19}$$

In order to get (2.18), any $\tau \in (0, \delta]$ serves. This is because, with independence of the choice of τ, for any $\varphi \in A_\tau$ and any $t \in I_\tau$, one has

$$|\mathcal{T}(\varphi)(t) - y_0| = \left| \int_{t_0}^t f(s, \varphi(s))\,ds \right| \leq M\delta \leq b_0. \tag{2.20}$$

On the other hand, in order to find out what must be imposed to τ to get (2.19), let us fix $\varphi, \psi \in A_\tau$ and let us try to estimate $\|\mathcal{T}(\varphi) - \mathcal{T}(\psi)\|$ in terms of $\|\varphi - \psi\|$.

Let $t \in I_\tau$ be given and let us assume, for example, that $t \geq t_0$. We have

$$|\mathcal{T}(\varphi)(t) - \mathcal{T}(\psi)(t)| \leq \int_{t_0}^t |f(s, \varphi(s)) - f(s, \psi(s))|\,ds. \tag{2.21}$$

By assumption, there exists $L_K > 0$ such that

$$|f(t, y^1) - f(t, y^2)| \le L_K |y^1 - y^2| \quad \forall (t, y^1), (t, y^2) \in K.$$

On the other hand, in (2.21) the couples $(s, \varphi(s))$ and $(s, \psi(s))$ with s inside the interval $[t_0, t]$ belong to the compact set K. Consequently,

$$|\mathcal{T}(\varphi)(t) - \mathcal{T}(\psi)(t)| \le L_K \int_{t_0}^t |\varphi(s) - \psi(s)| \, ds \le L_K \tau \, \|\varphi - \psi\|$$

(and this must hold for any $t \in I_\tau$ with $t \ge t_0$).

In a very similar way, it can be proved that

$$|\mathcal{T}(\varphi)(t) - \mathcal{T}(\psi)(t)| \le L_K \int_t^{t_0} |\varphi(s) - \psi(s)| \, ds \le L_K \tau \, \|\varphi - \psi\|$$

when $t \in I_\tau$ and $t \le t_0$. Therefore,

$$|\mathcal{T}(\varphi)(t) - \mathcal{T}(\psi)(t)| \le L_K \tau \, \|\varphi - \psi\| \quad \forall t \in I_\tau.$$

We immediately deduce that

$$\|\mathcal{T}(\varphi) - \mathcal{T}(\psi)\| \le L_K \tau \|\varphi - \psi\| \tag{2.22}$$

(and this for all $\varphi, \psi \in A_\tau$).

Thus, whenever $0 < \tau < \min(\delta, 1/L_K)$, the requirements (2.18) and (2.19) are ensured.

This ends the proof. $\qquad\qquad\qquad\qquad\qquad\qquad\qquad\qquad\qquad \square$

Proof of Lemma 2.2. First, let us assume that φ is a solution in I_τ to the CP. If we prove that

$$|\varphi(t) - y_0| \le b_0 \quad \forall t \in I_\tau, \tag{2.23}$$

we will have $\varphi \in A_\tau$ and (2.17) will be fulfilled.

Suppose the contrary: there exists $\tilde{t} \in I_\tau$ such that $|\varphi(\tilde{t}) - y_0| > b_0$. For instance, we can assume that $\tilde{t} > t_0$. Then, it is clear that there must exist $\bar{t} \in (t_0, \tilde{t})$ such that

$$|\varphi(\bar{t}) - y_0| = b_0 \quad \text{and} \quad |\varphi(s) - y_0| < b_0 \quad \forall s \in [t_0, \bar{t}).$$

But this is impossible, since it leads to the following equalities and inequalities:

$$b_0 = |\varphi(\bar{t}) - y_0| = \left| \int_{t_0}^{\bar{t}} f(s, \varphi(s)) \, ds \right| \le M|\bar{t} - t_0| < M\delta \le b_0.$$

This proves (2.23) and, consequently, (2.17).

Conversely, if we have (2.17), it is an evident fact that φ solves (CP) in I_τ. □

Remark 2.4.

(a) Let us insist once more: this result is local in nature; there exists a "small" interval I_τ where the existence and uniqueness of the solution can be certified and it remains to know whether there exist solutions defined in "large" intervals.

In other words, we can now ask which is, in the conditions of Theorem 2.3, the largest interval where we can define a solution.

Questions of this kind will be analyzed later, in Chapter 4.

(b) Let δ be as in the proof of Theorem 2.3 and, again, set $I_\delta := [t_0 - \delta, t_0 + \delta]$. The inequalities (2.20) prove that $\mathcal{T}$ is well defined in

$$A_\delta := \{\varphi \in C^0(I_\delta; \mathbb{R}^N) : |\varphi(t) - y_0| \le b_0 \;\; \forall t \in I_\delta\}$$

and maps this set into itself.

Morcover, under the assumptions of Theorem 2.3, it can be proved that there exists $m \ge 1$ such that $\mathcal{T}^m : A_\delta \mapsto A_\delta$ is a contraction. Indeed, if $\varphi, \psi \in A_\delta$ and $m \ge 1$ are given, one has

$$|\mathcal{T}^m(\varphi)(t) - \mathcal{T}^m(\psi)(t)| \le \frac{(L_K|t - t_0|)^m}{m!}\|\varphi - \psi\| \;\; \forall t \in I_\delta,$$

whence we easily deduce the contractivity of $\mathcal{T}^m$ for m large enough (depending on δ and L_K).

Consequently, in view of Corollary 2.1, there exists a unique fixed-point $\mathcal{T}$ in A_δ.

This argument shows that the solution to (CP) furnished by Theorem 2.3 is defined not only in I_τ but in I_δ.

(c) It makes sense to ask if the assumptions in Theorem 2.3 are sharp or, contrarily, can be weakened.

In particular, the following question is in order: what can be said for (CP) if we only assume that f is continuous?

When f is (only) continuous, it can be proved that there exists $\rho > 0$ such that (CP) possesses *at least* one solution in I_ρ. This result is usually known as *Peano's Theorem*.

It will be presented in Section 2.5.

However, under these conditions, the solution is not unique in general. Thus, for instance, the Cauchy problem

$$\begin{cases} y' = y^{2/3} \\ y(0) = 0 \end{cases} \tag{2.24}$$

possesses infinitely many solutions in $[-\delta, \delta]$ for every $\delta > 0$.

Indeed, for any $\alpha, \beta \in (0, \delta)$, the associated function $\varphi_{\alpha,\beta}$, with

$$\varphi_{\alpha,\beta}(t) = \begin{cases} \left(\dfrac{t+\alpha}{3}\right)^3 & \text{if } -\delta \leq t \leq -\alpha \\ 0 & \text{if } -\alpha < t \leq \beta \\ \left(\dfrac{t-\beta}{3}\right)^3 & \text{if } \beta < t \leq \delta \end{cases}$$

is a solution in $[-\delta, \delta]$; see Figure 2.2. □

Let us now prove that the continuity of f and the partial derivatives of its components with respect to the y_j together imply the local Lipschitz-continuity with respect to y.

Note however that, as was shown in Example 2.1, for a function to be locally Lipschitz-continuous, differentiability is not needed.

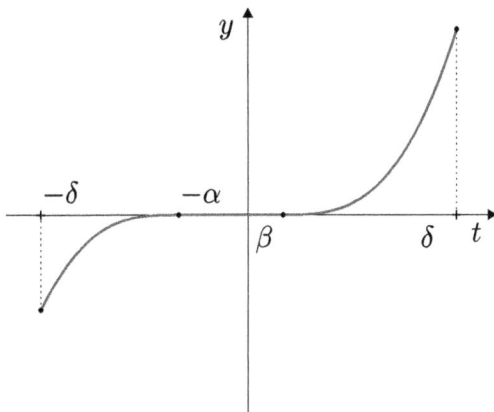

Fig. 2.2   A solution to the CP (2.24).

We will need the following two lemmas:

Lemma 2.3. *Let us assume that*

> *f is continuous and possesses continuous partial derivatives*   (2.25)
> *with respect to the y_j in $\Omega \subset \mathbb{R}^{N+1}$.*

Let $K \subset \Omega$ be a non-empty convex compact set. Then, there exists $L_K > 0$ with the following property:

$$|f(t, y^1) - f(t, y^2)| \leq L_K |y^1 - y^2| \quad \forall (t, y^1), (t, y^2) \in K. \qquad (2.26)$$

In general, L_K depends on K.

Proof. Let $(t, y^1), (t, y^2) \in K$ be given and let us fix i, with $1 \leq i \leq N$. Let us set

$$g(s) = f_i(t, y^2 + s(y^1 - y^2)) \quad \forall s \in [0, 1].$$

Since K is convex, $g : [0, 1] \mapsto \mathbb{R}$ is well defined. In view of (2.25), it is clear that $g \in C^1([0, 1])$. Consequently, there exists $\tilde{s} \in (0, 1)$ such that

$$f_i(t, y^1) - f_i(t, y^2) = g(1) - g(0) = g'(\tilde{s}).$$

From the *Chain Rule* (formula of differentiation of composed functions), the derivative of g can be expressed in terms of the partial derivatives of the f_i with respect to the y_j.

More precisely, we have

$$g'(\tilde{s}) = \sum_{j=1}^{N} \frac{\partial f_i}{\partial y_j}(t, y^2 + \tilde{s}(y^1 - y^2)) \, (y_j^1 - y_j^2).$$

Therefore,

$$
\begin{aligned}
|f_i(t, y^1) - f_i(t, y^2)| &= \left| \sum_{j=1}^{N} \frac{\partial f_i}{\partial y_j}(t, y^2 + \tilde{s}(y^1 - y^2)) \, (y_j^1 - y_j^2) \right| \\
&\leq \left(\sum_{j=1}^{N} \left| \frac{\partial f_i}{\partial y_j}(t, y^2 + \tilde{s}(y^1 - y^2)) \right|^2 \right)^{1/2} |y^1 - y^2| \\
&\leq L_{i,K} |y^1 - y^2|,
\end{aligned}
$$

where each $L_{i,K}$ is given by

$$L_{i,K} := \max_{(s,y) \in K} \left(\sum_{j=1}^{N} \left| \frac{\partial f_i}{\partial y_j}(s, y) \right|^2 \right)^{1/2}.$$

Here, we are using that K is compact and the partial derivatives of the f_i with respect to the y_j are continuous.

Taking into account that this argument works for all i, we easily obtain that

$$|f(t,y^1) - f(t,y^2)| = \left(\sum_{i=1}^{N} |f_i(t,y^1) - f_i(t,y^2)|^2\right)^{1/2}$$
$$\leq \left(\sum_{i=1}^{N} L_{i,K}^2\right)^{1/2} |y^1 - y^2|.$$

This proves the lemma, with

$$L_K = \left(\sum_{i=1}^{N} L_{i,K}^2\right)^{1/2}.$$ $\square$

Lemma 2.4. *Let us assume that $f : \Omega \mapsto \mathbb{R}^N$ is continuous. Then the following assertions are equivalent:*

(1) *f is locally Lipschitz-continuous with respect to y in Ω.*
(2) *For any $(t_*, y_*) \in \Omega$, there exist a neighborhood $\mathcal{N}(t_*, y_*) \subset \Omega$ and a positive constant $L(t_*, y_*)$ such that*

$$|f(t,y^1) - f(t,y^2)| \leq L(t_*,y_*)|y^1 - y^2| \quad \forall (t,y^1), (t,y^2) \in \mathcal{N}(t_*,y_*). \tag{2.27}$$

Proof. It is obvious that the first condition implies the second one. We have to prove the converse.

Thus, let $K \subset \Omega$ be a compact set and take

$$M := \max_{(t,y)\in K} |f(t,y)|.$$

Let $(t_*, y_*) \in K$ be given. By assumption, there exist $\varepsilon(t_*,y_*) > 0$ and $L(t_*,y_*) > 0$ such that

$$|f(t,y^1) - f(t,y^2)| \leq L(t_*,y_*)|y^1 - y^2| \quad \forall (t,y^1),(t,y^2) \in \overline{B}((t_*,y_*); \varepsilon(t_*,y_*)).$$

Since K is compact, we can find a finite collection of points $(t_*^1, y_*^1), \ldots, (t_*^p, y_*^p)$ in K such that

$$K \subset \bigcup_{i=1}^{p} B((t_*^i, y_*^i); \varepsilon(t_*^i, y_*^i)/2).$$

Let us set

$$r := \frac{1}{2} \min_{1 \le i \le p} \varepsilon(t_*^i, y_*^i), \quad L' = \max_{1 \le i \le p} L(t_*^i, y_*^i), \quad L_K = \max\left(L', \frac{2M}{r}\right).$$

Then the constant L_K is such that one has (2.14).

Indeed, if $(t, y^1), (t, y^2) \in K$ are given, either $|y^1 - y^2| \ge r$, which implies

$$\frac{|f(t, y^1) - f(t, y^2)|}{|y^1 - y^2|} \le \frac{2M}{r} \le L_K,$$

or $|y^1 - y^2| < r$. But, in this second case, the points (t, y^1) and (t, y^2) are in the same ball $B((t_*^i, y_*^i); \varepsilon(t_*^i, y_*^i))$, whence

$$|f(t, y^1) - f(t, y^2)| \le L'|y^1 - y^2| \le L_K|y^1 - y^2|. \qquad \square$$

An important consequence of the arguments in the proofs of the previous lemmas is the following:

Lemma 2.5. *Assume that $f : \Omega \mapsto \mathbb{R}^N$ is continuous and its components are continuously differentiable with respect to the y_j, with partial derivatives uniformly bounded in Ω. Then f is globally Lipschitz-continuous with respect to the variable y in Ω.*

In view of Lemmas 2.3 and 2.4, if f satisfies (2.25), it is locally Lipschitz-continuous with respect to the variable y in Ω.

Therefore, the following corollary of Theorem 2.3 holds:

Corollary 2.2. *Assume that one has (2.25) and $(t_0, y_0) \in \Omega$. Then, there exist $\tau > 0$ and a unique function $\varphi : I_\tau \mapsto \mathbb{R}^N$ that is a solution to (CP) in I_τ.*

In the last part of this section, we will consider CPs for ODEs of order $n \ge 1$.

We will adapt the previous existence and uniqueness results using the usual change of variables, already present in several arguments in Chapter 1.

Let $\Omega \subset \mathbb{R}^{n+1}$ be a non-empty open set and let $f : \Omega \mapsto \mathbb{R}$ be a given continuous function. The problem to study is

$$(CP)_n \quad \begin{cases} y^{(n)} = f(t, y, y', \ldots, y^{(n-1)}) \\ y(t_0) = y_0^{(0)}, \; y'(t_0) = y_0^{(1)}, \ldots, \; y^{(n-1)}(t_0) = y_0^{(n-1)}, \end{cases}$$

where $(t_0, y_0^{(0)}, \ldots, y_0^{(n-1)}) \in \Omega$.

For convenience, in problems of this kind, we will frequently use the abridged notation $Y := (y, y', \ldots, y^{(n-1)})$.

Definition 2.3. It will be said that f is (globally) Lipschitz-continuous with respect to the variable Y in Ω if there exists $L > 0$ such that

$$|f(t, Y^1) - f(t, Y^2)| \leq L|Y^1 - Y^2| \quad \forall (t, Y^1), (t, Y^2) \in \Omega. \qquad (2.28)$$

On the other hand, it will be said that f is locally Lipschitz-continuous with respect to the variable Y in Ω if, for every compact set $K \subset \Omega$, there exists $L_K > 0$ such that

$$|f(t, Y^1) - f(t, Y^2)| \leq L_K|Y^1 - Y^2| \quad \forall (t, Y^1), (t, Y^2) \in K. \qquad (2.29)$$

$\square$

Taking into account that $(\text{CP})_n$ is equivalent to the CP (1.26), where F and Y_0 are given by (1.27), thanks to Theorem 2.3 and Corollary 2.2, we immediately obtain the following existence and uniqueness results:

Theorem 2.4. *Assume that* $f : \Omega \subset \mathbb{R}^{n+1} \mapsto \mathbb{R}$ *is continuous and locally Lipschitz-continuous with respect to Y in Ω and* $(t_0, y_0^{(0)}, \ldots, y_0^{(n-1)}) \in \Omega$. *Then, there exist $\tau > 0$ and a unique function* $\varphi : I_\tau \mapsto \mathbb{R}$ *that solves* $(\text{CP})_n$ *in I_τ.*

Corollary 2.3. *Assume that*

$$f \text{ is continuous and possesses continuous partial derivatives} \atop \text{with respect to the variables } y, y', \ldots, y^{(n-1)} \text{ in } \Omega \qquad (2.30)$$

and $(t_0, y_0^{(0)}, \ldots, y_0^{(n-1)}) \in \Omega$. *Then the conclusion in Theorem 2.4 holds for* $(\text{CP})_n$.

2.4 Ascoli–Arzelà's Theorem*

For some reasons that will be discovered in the following, it is convenient to characterize the compact sets in $C^0(I; \mathbb{R}^N)$.

Recall that a set $\mathcal{F} \subset C^0(I; \mathbb{R}^N)$ is compact (resp. relatively compact) if any open covering of $\mathcal{F}$ (resp. $\overline{\mathcal{F}}$) possesses a finite subcovering.

In other words, every family $\{G_i\}_{i \in \mathcal{I}}$ of open sets satisfying $\mathcal{F} \subset \bigcup_{i \in \mathcal{I}} G_i$ (resp. $\overline{\mathcal{F}} \subset \bigcup_{i \in \mathcal{I}} G_i$) possesses a finite subfamily $\{G_{i_1}, \ldots, G_{i_p}\}$ such that

$$\mathcal{F} \subset \bigcup_{\ell=1}^{p} G_{i_\ell} \quad \left(\text{resp. } \overline{\mathcal{F}} \subset \bigcup_{\ell=1}^{p} G_{i_\ell} \right).$$

Since the topology we have defined in $C^0(I; \mathbb{R}^N)$ is induced by a distance, this happens if and only if from any sequence in $\mathcal{F}$ we can always extract a convergent subsequence whose limit belongs to $\mathcal{F}$ (resp. a convergent subsequence in $C^0(I; \mathbb{R}^N)$).

It is well known that, in any finite dimensional normed space, a set is compact (resp. relatively compact) if and only if it is closed and bounded (resp. if and only if it is bounded).

If the space is infinite dimensional, any compact set is necessarily closed and bounded and any relatively compact set is necessarily bounded.

Indeed, if X is a normed space for the norm $\| \cdot \|_X$ and $K \subset X$ is relatively compact, there exist finitely many points $u_1, \ldots, u_p$ in $\overline{K}$ such that $\overline{K} \subset \bigcup_{\ell=1}^p B(u_\ell; 1)$, whence any $u \in \overline{K}$ belongs to some $B(u_i; 1)$ and satisfies

$$\|u\|_X \leq \|u_i\|_X + \|u - u_i\|_X \leq C := \max_{1 \leq \ell \leq p} \|u_\ell\|_X + 1.$$

This shows that $\overline{K}$ is bounded and, obviously, the same happens to K.

On the other hand, any compact set $K \subset X$ is closed: if $u \in \overline{K}$ and $u \notin K$, then the set

$$F := \bigcap_{\varepsilon > 0} \overline{B}(u; \varepsilon)$$

is disjoint from K, whence $K \subset \bigcup_{\varepsilon > 0} \overline{B}(u; \varepsilon)^c$ and there must exist $\varepsilon_1, \ldots, \varepsilon_p$ such that

$$K \subset \bigcup_{\ell=1}^p \overline{B}(u; \varepsilon_\ell)^c;$$

but this means that, for $\varepsilon_0 = \min_{1 \leq \ell \leq p} \varepsilon_\ell$, one has $\overline{B}(u; \varepsilon_0) \cap K = \emptyset$, which is an absurdity.

Hence, in any infinite dimensional normed space, any compact set is closed and bounded. However, the converse of this assertion is not true.

In other words, compactness is a strictly stronger property than closedness and boundedness; for more details, see Exercise E.2.9, at the end of this chapter.[2]

Definition 2.4. Let $\mathcal{F} \subset C^0(I; \mathbb{R}^N)$ be given. It will be said that $\mathcal{F}$ is equi-continuous if, for any $\varepsilon > 0$, there exists $\delta > 0$ with the following property:

$$t, s \in I, |t - s| \leq \delta \quad \Rightarrow \quad |\varphi(t) - \varphi(s)| \leq \varepsilon \quad \forall \varphi \in \mathcal{F}. \qquad \square$$

---

[2]In fact, it can be proved that a normed space is finite dimensional if and only if any closed bounded set is compact; see, for example, Robertson and Robertson (1980).

The main result in this section is the following:

Theorem 2.5 (Ascoli–Arzelà). *Let $\mathcal{F} \subset C^0(I; \mathbb{R}^N)$ be given. Then, $\mathcal{F}$ is relatively compact if and only if it is bounded and equi-continuous.*

Proof. Let us first assume that $\mathcal{F}$ is relatively compact.

We know that it is bounded. Let us see that $\mathcal{F}$ is also equi-continuous.

Let $\varepsilon > 0$ and $\varphi \in \mathcal{F}$ be given, there exist $\psi_1, \dots, \psi_r \in \mathcal{F}$ with the following property:

$$\overline{\mathcal{F}} \subset \bigcup_{i=1}^{r} B(\psi_i; \varepsilon/3).$$

Moreover, for each i, there exists $\delta_i > 0$ such that

$$t, s \in I, \ |t - s| \leq \delta_i \quad \Rightarrow \quad |\psi_i(t) - \psi_i(s)| \leq \varepsilon/3.$$

Let us set $\delta = \min_{1 \leq i \leq r} \delta_i$ and let us assume that $t, s \in I$ and $|t - s| \leq \delta$. Then, one has $|\varphi(t) - \varphi(s)| \leq \varepsilon$.

Indeed, there exists j such that $\|\varphi - \psi_j\| \leq \varepsilon/3$. Consequently,

$$|\varphi(t) - \varphi(s)| \leq |\varphi(t) - \psi_j(t)| + |\psi_j(t) - \psi_j(s)| + |\psi_j(s) - \varphi(s)| \leq \varepsilon.$$

Since $\varepsilon > 0$ and $\varphi \in \mathcal{F}$ are arbitrary, this proves that $\mathcal{F}$ is equi-continuous.

Conversely, let us assume that $\mathcal{F}$ is bounded and equi-continuous and let $\{\varphi_n\}$ be a sequence in $\mathcal{F}$. Let us see that $\{\varphi_n\}$ possesses a Cauchy subsequence. Since $C^0(I; \mathbb{R}^N)$ is a Banach space, this will prove that $\mathcal{F}$ is relatively compact.

In order to find this subsequence, we will use what is called a *diagonal extraction process*.

Let $\{r_1, r_2, \dots, r_k, \dots\}$ be a numerotation of the set of rational numbers in I. By assumption, the sequence $\{\varphi_n(r_1)\}$ is bounded in $\mathbb{R}^N$. Thus, there exists a first subsequence, $\{\varphi_{1,n}\}$, such that the $\varphi_{1,n}(r_1)$ converge.

Again, the sequence $\{\varphi_{1,n}(r_2)\}$ is bounded and there exists a second subsequence $\{\varphi_{2,n}\}$ such that the $\varphi_{2,n}(r_2)$ converge. Note that, in fact, the $\varphi_{2,n}(r_1)$ also converge, since $\{\varphi_{2,n}(r_1)\}$ is a subsequence of $\{\varphi_{1,n}(r_1)\}$, that is convergent.

In a third step, we note that $\{\varphi_{2,n}(r_3)\}$ is bounded, whence there exists a new subsequence $\{\varphi_{3,n}\}$ such that the $\varphi_{3,n}(r_i)$ converge for $i = 1, 2, 3$, etc.

In this way, we can extract subsequences $\{\varphi_{1,n}\}$, $\{\varphi_{2,n}\}$, etc., such that, for all $k \geq 1$,

$$\text{the } \varphi_{k,n}(r_i) \quad \text{converge in } \mathbb{R}^N \text{ for } i = 1, \ldots, k.$$

Let us now consider the "diagonal" sequence $\{\varphi_{n,n}\}$. It is a subsequence of the initial sequence $\{\varphi_n\}$ and it is clear that

$$\text{the } \varphi_{n,n}(r_i) \quad \text{converge in } \mathbb{R}^N \text{ for all } i \geq 1,$$

because, for any i, up to a finite amount of terms, $\{\varphi_{n,n}(r_i)\}$ is a subsequence of $\{\varphi_{i,n}(r_i)\}$.

We will check that $\{\varphi_{n,n}\}$ is a Cauchy sequence, which will serve to conclude and will end the proof.

Let $\varepsilon > 0$ be given. By assumption, there exists $\delta > 0$ such that

$$t, s \in I, \ |t - s| \leq \delta \quad \Rightarrow \quad |\varphi_{n,n}(t) - \varphi_{n,n}(s)| \leq \varepsilon/3 \quad \forall n \geq 1.$$

Since I is compact, there exists k_0 such that

$$I \subset \bigcup_{k=1}^{k_0} (r_k - \delta, r_k + \delta).$$

Furthermore, for each $k = 1, \ldots, k_0$, the corresponding $\{\varphi_{n,n}(r_k)\}$ is a Cauchy sequence; whence there exists n_k such that

$$m, n \geq n_k \quad \Rightarrow \quad |\varphi_{m,m}(r_k) - \varphi_{n,n}(r_k)| \leq \varepsilon/3.$$

Let us take

$$n_0 = \max_{1 \leq k \leq k_0} n_k$$

and let us assume that $m, n \geq n_0$. Then, for any $t \in I$, we can find k such that $1 \leq k \leq k_0$ and $t \in (r_k - \delta, r_k + \delta)$ and the following inequalities hold:

$$|\varphi_{m,m}(t) - \varphi_{n,n}(t)| \leq |\varphi_{m,m}(t) - \varphi_{m,m}(r_k)|$$
$$+ |\varphi_{m,m}(r_k) - \varphi_{n,n}(r_k)| + |\varphi_{n,n}(r_k) - \varphi_{n,n}(t)| \leq \varepsilon.$$

Since t is arbitrary in I, this proves that $\|\varphi_{m,m} - \varphi_{n,n}\| \leq \varepsilon$ whenever $m, n \geq n_0$. Since $\varepsilon > 0$ is also arbitrary, we finally have that $\{\varphi_{n,n}\}$ is a Cauchy sequence. $\qquad \square$

Remark 2.5.

(a) The so-called diagonal extraction process is an argument of a general kind. It can be used satisfactorily in many contexts.

On the other hand, in many function spaces, it is frequent (and useful) to find compactness characterization principles similar to Theorem 2.5 that say the following:

A family is relatively compact if and only if it is bounded and, moreover, it satisfies a property directly related to the considered space of functions in a uniform way.

In $C^0(I; \mathbb{R}^N)$, this is just equi-continuity.

(b) In the second part of the previous proof, we did not use that $\mathcal{F}$ is bounded, but only that this set is *pointwise bounded*, that is, that the following property holds:

For every $t \in I$, there exists $M_t > 0$ such that $|\varphi(t)| \leq M_t \quad \forall \varphi \in \mathcal{F}$.

Consequently, the following assertion is true (and the proof is the same):

A family $\mathcal{F} \subset C^0(I; \mathbb{R}^N)$ is relatively compact if and only if it is pointwise bounded and equi-continuous.

This result is known as *Ascoli's Theorem*.

(c) In practice, it is very useful to have in mind conditions that imply equi-continuity.

For example, if $\mathcal{F} \subset C^1(I; \mathbb{R}^N)$ and there exists $M > 0$ such that

$$|\varphi'(t)| \leq M \quad \forall t \in I, \ \forall \varphi \in \mathcal{F},$$

then $\mathcal{F}$ is equi-continuous.

More generally, if there exists a continuous function $h : \mathbb{R}_+ \mapsto \mathbb{R}_+$ with $h(0) = 0$ such that

$$|\varphi(t) - \varphi(s)| \leq h(|t - s|) \quad \forall t, s \in I, \ \forall \varphi \in \mathcal{F},$$

then the family $\mathcal{F}$ is equi-continuous. □

2.5 ε-Approximated Solutions, Peano's Theorem*

This section is devoted to solving (CP) by a *compactness method*.
    The strategy will be the following:

(1) We will construct an interval I and a family of functions $\mathcal{F} = \{\varphi_\varepsilon : \varepsilon > 0\}$ satisfying the integral formulation of (CP) in an approximate way.
(2) Then, we will prove that the φ_ε belong to a compact set in $C^0(I; \mathbb{R}^N)$. To this purpose, we will check that the set $\mathcal{F}$ is bounded and equi-continuous.

(3) Accordingly, we will deduce that $\mathcal{F}$ possesses convergent subsequences and, finally, we will prove that the limit of any of them is a solution to (CP) in I.

Note that the basic topological property (compactness) used in this argument was not used in the proof of Picard's Theorem. Contrarily, the property employed there is the completeness of the space of continuous functions.[3]

We will begin by introducing the concept of approximated solution needed for our purpose:

Definition 2.5. Let $I = [a, b] \subset \mathbb{R}$ be a compact interval containing t_0 and let $\varepsilon > 0$ be given. It will be said that $\varphi_\varepsilon : I \mapsto \mathbb{R}^N$ is an ε-approximated solution to (CP) in I if

(1) $\varphi_\varepsilon \in C^0(I; \mathbb{R}^N)$.
(2) For every $t \in I$, $(t, \varphi_\varepsilon(t)) \in \Omega$ and, furthermore, $\varphi_\varepsilon(t_0) = y_0$.
(3) There exist $a_0, a_1, \ldots, a_m$ such that $a = a_0 < a_1 < \cdots < a_m = b$, φ_ε is continuously differentiable in each $[a_i, a_{i+1}]$ and

$$|\varphi_\varepsilon'(t) - f(t, \varphi_\varepsilon(t))| \leq \varepsilon \quad \forall t \in I \setminus \{a_0, \ldots, a_m\}. \tag{2.31}$$

$\square$

In the next result, we present an integral property satisfied by the ε-approximated solutions:

Lemma 2.6. *Let φ_ε be an ε-approximated solution to* (CP) *in I. Then*

$$\left| \varphi_\varepsilon(t) - y_0 - \int_{t_0}^t f(s, \varphi_\varepsilon(s)) \, ds \right| \leq \varepsilon |t - t_0| \quad \forall t \in I. \tag{2.32}$$

Proof. For example, let us see that (2.32) holds for every $t \in I$ with $t > t_0$.

By assumption, there exist $t_1, \ldots, t_n$ such that $t_0 < t_1 \cdots < t_n = t$, φ_ε is continuously differentiable in each $[t_i, t_{i+1}]$ and

$$|\varphi_\varepsilon'(t) - f(t, \varphi_\varepsilon(t))| \leq \varepsilon \quad \forall t \in I \setminus \{t_0, \ldots, t_n\}. \tag{2.33}$$

---

[3]In fact, an interesting comment must be made here: in practice, the solutions of differential problems of all kinds need arguments ultimately based either on the compactness or the completeness of a well-chosen space of functions.

Thus,

$$\left| \varphi_\varepsilon(t) - \varphi_\varepsilon(t_0) - \int_{t_0}^t f(s, \varphi_\varepsilon(s)) \, ds \right|$$

$$= \left| \sum_{i=0}^{n-1} \left(\varphi_\varepsilon(t_{i+1}) - \varphi_\varepsilon(t_i) - \int_{t_i}^{t_{i+1}} f(s, \varphi_\varepsilon(s)) \, ds \right) \right|$$

$$\leq \sum_{i=0}^{n-1} \left| \varphi_\varepsilon(t_{i+1}) - \varphi_\varepsilon(t_i) - \int_{t_i}^{t_{i+1}} f(s, \varphi_\varepsilon(s)) \, ds \right|$$

$$= \sum_{i=0}^{n-1} \left| \int_{t_i}^{t_{i+1}} (\varphi_\varepsilon'(s) - f(s, \varphi_\varepsilon(s))) \, ds \right|$$

$$\leq \sum_{i=0}^{n-1} \int_{t_i}^{t_{i+1}} |\varphi_\varepsilon'(s) - f(s, \varphi_\varepsilon(s))| \, ds$$

$$\leq \varepsilon \sum_{i=0}^{n-1} |t_{i+1} - t_i| = \varepsilon |t - t_0|.$$

A similar argument leads to the same estimate when $t < t_0$. This proves the lemma. □

We will also need the following result, which furnishes a whole family of ε-approximated solutions:

Lemma 2.7. *Let us assume that $\Omega \subset \mathbb{R}^N$ is a non-empty open set, $f : \Omega \mapsto \mathbb{R}^N$ is continuous and $(t_0, y_0) \in \Omega$. There exist $\delta > 0$, $M > 0$ and a compact set $K \subset \Omega$ with the following property: for every $\varepsilon > 0$, there exists an ε-approximated solution φ_ε to the problem (CP) in I_δ, furthermore satisfying*

$$|\varphi_\varepsilon(t) - \varphi_\varepsilon(s)| \leq M|t - s| \quad \forall t, s \in I_\delta \tag{2.34}$$

and

$$(t, \varphi_\varepsilon(t)) \in K \quad \forall t \in I_\delta. \tag{2.35}$$

Proof. In a first step, we will argue as in the proof of Theorem 2.3.

Thus, the point (t_0, y_0) belongs to the open set Ω and, consequently, there exist a_0 and b_0 such that

$$K := [t_0 - a_0, t_0 + a_0] \times \overline{B}(y_0; b_0) \subset \Omega.$$

Since f is continuous, $|f|$ possesses a maximum M in K that can be assumed positive; then, we take

$$\delta = \min\left(a_0, \frac{b_0}{M}\right). \tag{2.36}$$

Now, for any given $\varepsilon > 0$, we are going to construct an ε-approximated solution in $I_\delta = [t_0 - \delta, t_0 + \delta]$ satisfying (2.34) and (2.35).

Since f is uniformly continuous in K, there exists $\rho > 0$ such that, whenever $(t, y), (t, \tilde{y}) \in K$, $|t - \tilde{t}| \leq \rho$ and $|y - \tilde{y}| \leq \rho$, one has $|f(t, y) - f(\tilde{t}, \tilde{y})| \leq \varepsilon$. Let us set

$$\alpha = \min\left(\rho, \frac{\rho}{M}\right).$$

There exist $t_{-n}, t_{-n+1}, \ldots, t_0, t_1, \ldots, t_{n-1}, t_n$ such that $t_0 - \delta = t_{-n} < t_{-n+1} < \cdots < t_{n-1} < t_n = t_0 + \delta$ and $|t_{i+1} - t_i| < \alpha$ for every i. Accordingly, we set

$$\varphi_\varepsilon(t) := \begin{cases} y_0 & \text{if } t = t_0 \\ \varphi_\varepsilon(t_i) + (t - t_i)f(t_i, \varphi_\varepsilon(t_i)) & \text{if } t \in (t_i, t_{i+1}] \\ \varphi_\varepsilon(t_{-i}) + (t - t_{-i})f(t_{-i}, \varphi_\varepsilon(t_{-i})) & \text{if } t \in [t_{-i-1}, t_{-i}), \end{cases}$$

where $0 \leq i \leq n - 1$. It is customary to say that φ_ε is a *Euler polygonal function*.

It is not difficult to check that φ_ε is well defined and continuous and, furthermore, satisfies (2.34) and (2.35).

For example, let us see by induction that this is true in $[t_0, t_0 + \delta]$:

(a) It is clear that $(t_0, \varphi_\varepsilon(t_0)) \in K$. Thus, φ_ε is well defined in $[t_0, t_1]$,

$$(t, \varphi_\varepsilon(t)) \in K \quad \forall t \in [t_0, t_1]$$

and also

$$|\varphi_\varepsilon(t) - \varphi_\varepsilon(s)| \leq M|t - s| \quad \forall t, s \in [t_0, t_1].$$

(b) Now, let us assume that φ_ε is well defined and continuous in $[t_0, t_i]$, $(t_j, \varphi_\varepsilon(t_j)) \in K$ for $j = 0, 1, \ldots, i$ and $|\varphi_\varepsilon(t) - \varphi_\varepsilon(s)| \leq M|t - s|$ for $t, s \in [t_0, t_i]$. Then $(t_{i+1}, \varphi_\varepsilon(t_{i+1})) \in K$, because $t_{i+1} \in I_\delta \subset [t_0 - a_0, t_0 + a_0]$ and

$$|\varphi_\varepsilon(t_{i+1}) - y_0| \leq |\varphi_\varepsilon(t_{i+1}) - \varphi_\varepsilon(t_i)| + \cdots + |\varphi_\varepsilon(t_1) - \varphi_\varepsilon(t_0)|$$

$$\leq M|t_{i+1} - t_i| + \cdots + |t_1 - t_0| = M|t_{i+1} - t_0| \leq M\delta \leq b_0.$$

In fact, this proves that φ_ε is well defined and continuous in $[t_0, t_{i+1}]$,

$$(t, \varphi_\varepsilon(t)) \in K \quad \forall t \in [t_0, t_{i+1}]$$

and

$$|\varphi_\varepsilon(t) - \varphi_\varepsilon(s)| \le M|t - s| \quad \forall t, s \in [t_0, t_{i+1}].$$

Finally, note that φ_ε is an ε-approximated solution in I_δ.

Indeed, by construction φ_ε is piecewise continuously differentiable and $\varphi_\varepsilon(t_0) = y_0$. If t is different of all the t_i, it is clear that there exists t_j with $\varphi'_\varepsilon(t) = f(t_j, \varphi_\varepsilon(t_j))$ and then

$$|\varphi'_\varepsilon(t) - f(t, \varphi_\varepsilon(t))| = |f(t_j, \varphi_\varepsilon(t_j)) - f(t, \varphi_\varepsilon(t))| \le \varepsilon,$$

since (also by construction)

$$|t_j - t| \le \alpha \le \rho \ \text{ and } \ |\varphi_\varepsilon(t_j) - \varphi_\varepsilon(t)| \le M|t_j - t| \le M\alpha \le \rho. \qquad \square$$

At this point, we can state and prove the main result in this section:

Theorem 2.6 (Peano). *Let us consider the problem* (CP), *where* $f : \Omega \subset \mathbb{R}^{N+1} \mapsto \mathbb{R}^N$ *is continuous and* $(t_0, y_0) \in \Omega$. *There exist* $\delta > 0$ *and a function* $\varphi : I_\delta \mapsto \mathbb{R}^N$ *that solves* (CP) *in* I_δ.

Proof. Let $\delta > 0$, $M > 0$, and $K \subset \Omega$ be as in Lemma 2.7. We know that there exists a family $\{\varphi_\varepsilon : \varepsilon > 0\}$ of ε-approximated solutions in I_δ with the properties (2.34) and (2.35).

This family is relatively compact in $C^0(I_\delta; \mathbb{R}^N)$.

Indeed, from (2.35) we immediately deduce that it is uniformly bounded. On the other hand, in view of (2.34), the family is equi-continuous.

Consequently, there exists a sequence $\{\varepsilon_n\}$ that converges to 0 and a function $\varphi \in C^0(I_\delta; \mathbb{R}^N)$ such that

$$\varphi_{\varepsilon_n} \text{ converges uniformly in } I_\delta \text{ to } \varphi.$$

Let us see that φ solves (CP) in I_δ.

Obviously,

$$(t, \varphi(t)) \in K \quad \forall t \in I_\delta$$

(since this is true for all the φ_{ε_n}, K is compact and φ_{ε_n} converges to φ in $C^0(I_\delta; \mathbb{R}^N)$).

On the other hand, let $t \in I_\delta$ be given. We know that

$$\left| \varphi_{\varepsilon_n}(t) - y_0 - \int_{t_0}^t f(s, \varphi_{\varepsilon_n}(s)) \, ds \right| \le \varepsilon_n |t - t_0| \qquad (2.37)$$

for every $n \geq 1$. If we could affirm that the left-hand side converges to

$$\left| \varphi(t) - y_0 - \int_{t_0}^{t} f(s, \varphi(s)) \, \mathrm{d}s \right|$$

as $n \to +\infty$, it would be immediate to deduce that

$$\varphi(t) = y_0 + \int_{t_0}^{t} f(s, \varphi(s)) \, \mathrm{d}s$$

and, since t is arbitrary in I_δ, we would have that φ is a solution to (CP).

But this is true, since $\varphi_{\varepsilon_n}(t)$ converges to $\varphi(t)$ and $f(\cdot, \varphi_{\varepsilon_n}(\cdot))$ converges uniformly to $f(\cdot, \varphi(\cdot))$ in the interval of end points t_0 and t.

Of course, the latter is a consequence of the continuity of f and the uniform convergence of the φ_{ε_n}.

This ends the proof. □

Let us now come back to ODEs of the higher order.

By introducing again the change of variables that reduces the corresponding Cauchy problem

$$(\text{CP})_n \qquad \begin{cases} y^{(n)} = f(t, y, y', \dots, y^{(n-1)}) \\ y(t_0) = y_0^{(0)}, \ y'(t_0) = y_0^{(1)}, \dots, y^{(n-1)}(t_0) = y_0^{(n-1)} \end{cases}$$

to an "equivalent" problem for a first-order ODS, we find the following:

Theorem 2.7. *Let us consider the Cauchy problem* $(\text{CP})_n$, *where* $\Omega \subset \mathbb{R}^{n+1}$ *is a non-empty open set,* $f : \Omega \mapsto \mathbb{R}$ *is continuous and* $(t_0, y_0^{(0)}, \dots, y_0^{(n-1)}) \in \Omega$. *There exists* $\delta > 0$ *and a function* $\varphi : I_\delta \mapsto \mathbb{R}$ *that solves* $(\text{CP})_n$ *in* I_δ.

Remark 2.6.

(a) A direct comparison of the hypotheses and assertions in Picard's and Peano's Theorems (Theorems 2.3 and 2.6) indicates that

$$\text{Continuity} \Rightarrow \text{Existence}$$

(as usual, weaker assumptions imply weaker conclusions), while

$$\left. \begin{array}{l} \text{Continuity and} \\ \text{Local Lipschitz-continuity with respect to } y \end{array} \right\} \Rightarrow \text{Existence and uniqueness.}$$

(b) Let us emphasize that the nature and structure of the proofs in these results are also different.

In both cases, they rely on properties of the Banach space $C^0(I; \mathbb{R}^N)$.

Thus, as already mentioned, the proof of Picard's Theorem exploits the fact that $C^0(I; \mathbb{R}^N)$ is, for the norm of the uniform convergence, complete (recall that the proof uses Banach's Fixed-Point Theorem, where completeness is absolutely essential).

Contrarily, the proof of Peano's Theorem relies on the characterization of compactness in $C^0(I; \mathbb{R}^N)$ furnished by Ascoli–Arzelà's Theorem.

(c) Euler's polygonal functions, introduced in the proof of Lemma 2.7, provide numerical approximations of the solution to a Cauchy problem like (CP).

Indeed, they are *computable*; each of them is completely described by a finite set of real values and, in order to be determined, the unique nontrivial task is the computation of the values of f at some points.

Here is a simplified description of Euler's polygonal function:

$$y^{i+1} = y^i + (t_{i+1} - t_i) f(t_i, y^i), \quad -n \leq i \leq n - 1, \qquad (2.38)$$

where $t_0 - \delta = t_{-n} < t_{-n+1} < \cdots < t_{n-1} < t_n = t_0 + \delta$.

(d) From the numerical viewpoint, this process can be improved a lot.

Thus, it is possible to obtain much better approximations with a similar computational cost.

For example, this happens when we construct the so-called "improved" Euler polygonal functions, that are given as follows:

$$y^{i+1} = y^i + \frac{t_{i+1} - t_i}{2} \left[f(t_i, y^i) + f(t_{i+1}, y^i + (t_{i+1} - t_i) f(t_i, y^i)) \right]. \quad (2.39)$$

Many other "schemes" can be given. For an introduction to numerical approximation methods for Cauchy problems, see, for instance, Henrici (1962) and Lambert (1991). $\qquad \Box$

2.6 Implicit Differential Equations*

In this section, we will consider Cauchy problems for first-order ODEs that do not have a normal structure.

More precisely, let $\mathcal{O} \subset \mathbb{R}^3$ be a non-empty open set, let $F : \mathcal{O} \mapsto \mathbb{R}$ be a continuous function, and let us consider the ODE

$$F(t, y, y') = 0. \qquad (2.40)$$

Definition 2.6. Let $I \subset \mathbb{R}$ be a non-degenerate interval and let $\varphi : I \mapsto \mathbb{R}$ be a given function. It will be said that φ is a solution to (2.40) in I if the following holds:

(1) φ is continuously differentiable in I.
(2) $(t, \varphi(t), \varphi'(t)) \in \mathcal{O} \quad \forall t \in I$.
(3) $F(t, \varphi(t), \varphi'(t)) = 0 \quad \forall t \in I$. $\square$

By definition, a CP for (2.40) is a system of the form

$$\begin{cases} F(t, y, y') = 0 \\ y(t_0) = y_0, \end{cases} \tag{2.41}$$

where (t_0, y_0) is such that, for some p_0, one has $(t_0, y_0, p_0) \in \mathcal{O}$.

As expected, if I is an interval containing t_0, it will be said that $\varphi : I \mapsto \mathbb{R}$ solves (2.41) in I if it is a solution to (2.40) in that interval and, moreover, $\varphi(t_0) = y_0$.

First of all, we note that the analysis of (2.41) seems in general much more complex than in the case of (CP).

To understand this, let us consider, for example, the Cauchy problem

$$\begin{cases} (y')^2 - t^2 = 0 \\ y(1) = 1. \end{cases} \tag{2.42}$$

The corresponding function F is polynomial and, consequently, posessess maximal regularity; however, (2.42) has two solutions in $\mathbb{R}$, the functions φ and ψ, given by

$$\varphi(t) = \frac{1}{2}(1 + t^2), \quad \psi(t) = \frac{1}{2}(3 - t^2) \quad \forall t \in \mathbb{R}.$$

Observe that the solution can be determined by fixing the value of $y'(1)$ (the *slope* at $t = 1$). But the value of this slope cannot be anything since we have from the ODE written at $t = 1$ that, necessarily, $|y'(1)| = 1$.

However, it must be noted that, in general, fixing the value of the slope does not guarantee uniqueness.

Indeed, the CP

$$\begin{cases} (y')^2 - t^2 = 0 \\ y(0) = 0, \end{cases} \tag{2.43}$$

where the ODE is the same, possesses two solutions in $\mathbb{R}$: the functions φ_1 and φ_2, given by

$$\varphi_1(t) = \frac{t^2}{2} \quad \text{and} \quad \varphi_2(t) = -\frac{t^2}{2} \quad \forall t \in \mathbb{R}.$$

And, in this case, the slopes at $t = 0$ are the same: $\varphi_1'(0) = \varphi_2'(0) = 0$.

In the sequel, we will analyze those cases where, once the point (t_0, y_0) and a "possible slope" p_0 are fixed, (2.41) can be equivalently rewritten as a CP for an ODE with normal structure near (t_0, y_0, p_0) (that is, for solutions defined near $t = t_0$ that satisfy the initial condition $\varphi(t_0) = y_0$ and the additional equality $\varphi'(t_0) = p_0$).

Definition 2.7. Let us consider the ODE (2.40). An integral element of (2.40) is a triplet $(t_0, y_0, p_0) \in \mathcal{O}$ such that $F(t_0, y_0, p_0) = 0$. If (t_0, y_0, p_0) is an integral element, it will be said that it is regular if there exist $\alpha, \beta > 0$ and a function

$$f : \overline{B}((t_0, y_0); \alpha) \mapsto [p_0 - \beta, p_0 + \beta] \tag{2.44}$$

such that $\overline{B}((t_0, y_0); \alpha) \times [p_0 - \beta, p_0 + \beta] \subset \mathcal{O}$, $f \in C^0(\overline{B}((t_0, y_0); \alpha))$,

$$F(t, y, f(t, y)) = 0 \quad \forall (t, y) \in \overline{B}((t_0, y_0); \alpha) \tag{2.45}$$

and

$$\left.\begin{array}{r} (t, y, p) \in \overline{B}((t_0, y_0); \alpha) \times [p_0 - \beta, p_0 + \beta] \\ F(t, y, p) = 0 \end{array}\right\} \Rightarrow p = f(t, y). \tag{2.46}$$

If the integral element (t_0, y_0, p_0) is not regular, it is called singular. $\square$

In any integral element (t_0, y_0, p_0), we usually say that (t_0, y_0) is the point and p_0 is the slope.

It will be seen later that, if (t_0, y_0, p_0) is a regular integral element, then (2.41) possesses at least one local solution $\varphi : I \mapsto \mathbb{R}$ whose slope at t_0 is equal to p_0, that is, such that $\varphi'(t_0) = p_0$.

Let us establish a sufficent condition for the regularity of an integral element:

Theorem 2.8. *Assume that $F : \mathcal{O} \mapsto \mathbb{R}$ is continuous and continuously differentiable with respect to the variable p in $\mathcal{O}$. Let (t_0, y_0, p_0) be an integral element of (2.40) and assume that*

$$\frac{\partial F}{\partial p}(t_0, y_0, p_0) \neq 0.$$

Then (t_0, y_0, p_0) is a regular integral element.

Proof. This result is an almost immediate consequence of the well-known *Implicit Function Theorem*; for example, see Smith (1998) and Spivak (1970).

Indeed, under the previous assumptions, there exist $\alpha, \beta > 0$ and a function $f \in C^0(\overline{B}((t_0, y_0); \alpha))$ such that:

(1) $\overline{B}((t_0, y_0); \alpha) \times [p_0 - \beta, p_0 + \beta] \subset \mathcal{O}$ and

$$f(t, y) \in [p_0 - \beta, p_0 + \beta] \quad \forall (t, y) \in \overline{B}((t_0, y_0); \alpha).$$

(2) $F(t, y, f(t, y)) = 0$ for all $(t, y) \in \overline{B}((t_0, y_0); \alpha)$.
(3) f is "unique" in the following sense: if $(t, y, p) \in \overline{B}((t_0, y_0); \alpha) \times [p_0 - \beta, p_0 + \beta]$ and $F(t, y, p) = 0$, then $p = f(t, y)$.

Obviously, this proves that (t_0, y_0, p_0) is a regular integral element. □

Remark 2.7.

(a) If, under the previous assumptions, F is continuously differentiable in a neighborhood of (t_0, y_0, p_0), it can be ensured that the function f arising in the proof of Theorem 2.8 is also continuously differentiable in $\overline{B}((t_0, y_0); \alpha)$.
 Furthermore, the Implicit Function Theorem affirms that, in that case,

$$\frac{\partial f}{\partial t}(t, y) = -\frac{\frac{\partial F}{\partial t}(t, y, f(t, y))}{\frac{\partial F}{\partial p}(t, y, f(t, y))} \quad \text{and} \quad \frac{\partial f}{\partial y}(t, y) = -\frac{\frac{\partial F}{\partial y}(t, y, f(t, y))}{\frac{\partial F}{\partial p}(t, y, f(t, y))}$$

in a neighborhood of (t_0, y_0); for more details, see Note 11.2.2.

(b) Let (t_0, y_0, p_0) be a singular integral element and let us assume that F is continuously differentiable in a neighborhood of (t_0, y_0, p_0). Then,

$$F(t_0, y_0, p_0) = 0, \quad \frac{\partial F}{\partial p}(t_0, y_0, p_0) = 0.$$

In other words, if F is regular enough, the set S of singular integral elements is contained in the set

$$\Delta_F := \left\{ (t, y, p) \in \mathcal{O} : F(t, y, p) = 0, \quad \frac{\partial F}{\partial p}(t, y, p) = 0 \right\}. \tag{2.47}$$

The set Δ_F is usually called the *discriminant manifold* associated to (2.40).

(c) Theorem 2.8 furnishes a sufficient condition for the regularity of an integral element (t_0, y_0, p_0). It is not necessary.
 This becomes clear by considering the following example.
 Let $\Omega \subset \mathbb{R}^2$ be a non-empty open set, let $g : \Omega \mapsto \mathbb{R}$ be a continuously differentiable function, and set

$$F(t, y, p) = (p - g(t, y)) \sin(p - g(t, y)) \quad \forall (t, y, p) \in \mathcal{O} = \Omega \times \mathbb{R}.$$

Then, any triplet $(t_0, y_0, f(t_0, y_0))$ with $(t_0, y_0) \in \Omega$ is a regular integral element and, however, belongs to the corresponding set Δ_F. □

The main result in this section is the following:

Theorem 2.9. *Let* (t_0, y_0, p_0) *be a regular integral element. Then:*

(1) *There exist* $\delta > 0$ *and a function* $\varphi : I_\delta \mapsto \mathbb{R}$ *that solves* (2.41) *in* I_δ *and furthermore satisfies* $\varphi'(t_0) = p_0$.

(2) *Moreover, if*

$$F \text{ is continuously differentiable in } \mathcal{O} \text{ and } \frac{\partial F}{\partial p}(t_0, y_0, p_0) \neq 0, \quad (2.48)$$

then φ *is unique in a neighborhood of* t_0 *for the slope* p_0. *In other words, if* ψ *is another solution to* (2.41) *in* I_δ *satisfying* $\psi'(t_0) = p_0$, *then* φ *and* ψ *coincide in a neighborhood of* t_0.

Proof. By assumption, there exist $\alpha, \beta > 0$ and $f : \overline{B}((t_0, y_0); \alpha) \mapsto [p_0 - \beta, p_0 + \beta]$ satisfying (2.44)–(2.46). Consequently, it will suffice to prove that the CP

$$\begin{cases} y' = f(t, y) \\ y(t_0) = y_0 \end{cases} \quad (2.49)$$

possesses at least one local solution $\varphi : I_\delta \mapsto \mathbb{R}$ with $\varphi'(t_0) = p_0$. Indeed, if this is proved, by replacing δ by an eventually smaller positive number, we will have

$$(t, \varphi(t), \varphi'(t)) \in \overline{B}((t_0, y_0); \alpha) \times [p_0 - \beta, p_0 + \beta] \subset \mathcal{O} \quad \forall t \in I_\delta$$

and

$$F(t, \varphi(t), \varphi'(t)) = 0 \quad \forall t \in I_\delta.$$

But the existence of a local solution to (2.49) is a consequence of Peano's Theorem (Theorem 2.6). This proves the first part.

Now, let us assume that (2.48) holds.

Then, it can be assumed that the function f in (2.49) is continuously differentiable in $B((t_0, y_0); \alpha)$. Thus, thanks to Picard's Theorem (Theorem 2.3), this Cauchy problem possesses a unique local solution $\varphi : I_\delta \mapsto \mathbb{R}$.

If $\psi : I_\delta \mapsto \mathbb{R}$ is another solution to (2.41) and $\psi'(t_0) = p_0$, then ψ is a new solution to (2.49) in an eventually smaller interval I_ρ (in view of the "uniqueness" of f).

Therefore, φ and ψ must coincide in $I_\delta \cap I_\rho$.

This proves the second part and ends the proof. $\square$

Example 2.2.

(a) Let us consider the ODE

$$4t(y')^2 - (3t - 1)^2 = 0.$$

The set of integral elements is

$$\mathcal{O}_0 = \left\{ \left(t, y, \pm \frac{3t - 1}{2\sqrt{t}} \right) \in \mathbb{R}^3 : t > 0 \right\}$$

and the discriminant manifold is in this case

$$\Delta = \{ (1/3, y, 0) : y \in \mathbb{R} \}.$$

As a consequence, the "possible" singular integral elements are the triplets $(1/3, y, 0)$ with $y \in \mathbb{R}$.

On the other hand, all these integral elements are singular since, for any t sufficiently close to $1/3$ and any $y \in \mathbb{R}$, the point (t, y) "supports" the following two slopes:

$$\frac{3t - 1}{2\sqrt{t}} \quad \text{and} \quad -\frac{3t - 1}{2\sqrt{t}}.$$

(b) Now, let us consider the ODE

$$(y')^2 + ty' - y = 0.$$

In this case,

$$\mathcal{O}_0 = \left\{ \left(t, y, \frac{-t \pm \sqrt{t^2 + 4y}}{2} \right) \in \mathbb{R}^3 : t \in \mathbb{R}, \ y \geq -\frac{t^2}{4} \right\}$$

and the discriminant manifold is $\Delta = \{ (t, -t^2/4, -t/2) : t \in \mathbb{R} \}$. The possible singular integral elements are the triplets $(t, -t^2/4, -t/2)$ with $t \in \mathbb{R}$. All these integral elements are singular, since the points (t, y) with $y < -t^2/4$ do not support any slope. □

Exercises

E.2.1 Prove the following version of Lemma 2.1 for second-order ODEs. Consider the Cauchy problem

$$(*) \qquad \begin{cases} y'' = f(t, y, y') \\ y(t_0) = y_0^{(0)}, \ y'(t_0) = y_0^{(1)}, \end{cases}$$

where $\Omega \subset \mathbb{R}^3$ is a non-empty open set, $f : \Omega \mapsto \mathbb{R}$ is continuous, and $(t_0, y_0^{(0)}, y_0^{(1)}) \in \Omega$. Let I be an interval containing t_0 and let $\varphi : I \mapsto \mathbb{R}$ be a given function. The following assertions are equivalent:

(1) φ is a solution to $(*)$ in I.
(2) $\varphi \in C^1(I)$, $(t, \varphi(t), \varphi'(t)) \in \Omega$ for all $t \in I$ and

$$\varphi(t) = y_0^{(0)} + y_0^{(1)}(t - t_0) + \int_{t_0}^t (t - s)f(s, \varphi(s), \varphi'(s))\,ds \quad \forall t \in I.$$

E.2.2 Prove that the sequence $\{\varphi_n\}$, with $\varphi_n(t) = nte^{-nt^2}$, converges pointwise to zero but does not converge in $C^0([0,1])$.
Hint: Compute or estimate $\|\varphi_n\|_\infty$.

E.2.3 Let $\alpha \in C^0([0,1])$ be a function satisfying $\alpha(t) > 0$ for all $t \in (0,1]$ and set $\|\phi\|_\alpha := \max_{t \in [0,1]} |\alpha(t)\varphi(t)|$ for all $\phi \in C^0([0,1])$.

(1) Prove that $\|\cdot\|_\alpha$ is a norm in the vector space $C^0([0,1])$.
(2) Prove that $\|\cdot\|_\alpha$ is equivalent to the usual norm in $C^0([0,1])$ if and only if $\alpha(0) > 0$.
(3) When is the space $C^0([0,1])$ *complete* for the norm $\|\cdot\|_\alpha$?

E.2.4 Using the sequence

$$\varphi_n(t) = \begin{cases} 1, & \text{if } t \in [0, \frac{1}{2} - \frac{1}{n}] \\ -nt + \dfrac{n}{2}, & \text{if } t \in [\frac{1}{2} - \frac{1}{n}, \frac{1}{2}] \\ 0, & \text{if } t \in [\frac{1}{2}, 1] \end{cases}$$

prove that the normed space $(C^0([0,1]), \|\cdot\|_1)$, where

$$\|\varphi\|_1 = \int_0^1 |\varphi(t)|\,dt \quad \forall \varphi \in C^0([0,1])$$

is not a Banach space.
Hint: Prove that, in $(C^0([0,1]), \|\cdot\|_1)$, φ_n is a Cauchy sequence that does not converge.

E.2.5 Using the sequence

$$\varphi_n(t) = \begin{cases} 0, & \text{if } t \in [-1, 0] \\ nt, & \text{if } t \in [0, \frac{1}{n}] \\ 1, & \text{if } t \in [\frac{1}{n}, 1] \end{cases}$$

prove that the normed space $(C^0([0,1]), \|\cdot\|_2)$, where

$$\|\varphi\|_2 = \left(\int_{-1}^{1} |\varphi(t)|^2 \, dt \right)^{1/2} \quad \forall \varphi \in C^0([-1,1])$$

is not a Banach space.

E.2.6 Let us set

$$A := \{f \in C^0([1,4]) : 5 - 3|t-2| \leq f(t) \leq 5 + \frac{1}{2}|t-2| \ \forall t \in [1,4]\}.$$

Find the sets $\overline{A}$, int A, and ∂A in the Banach space $C^0([1,4])$.

E.2.7 Find the closures, interiors, and boundaries of the sets $M, N, M \cap N$, and $M \cup N$ in the Banach space $C^0([-1,1])$, where

$$\begin{cases} M = \{f \in C^0([-1,1]) : 1 - t^2 \leq f(t) \leq 1 + t^2 \ \forall t \in [-1,1]\} \quad \text{and} \\ N = \{f \in C[-1,1] : t^2 + f(t)^2 \leq 1 \ \forall t \in [-1,1]\}. \end{cases}$$

E.2.8 Let $C_b(\mathbb{R}_+)$ be the vector space formed by the bounded continuous functions $\varphi : [0, +\infty) \mapsto \mathbb{R}$. Let us consider in $C_b(\mathbb{R}_+)$ the mapping $\|\cdot\|_\infty$, given by

$$\|\varphi\|_\infty = \sup_{t \in \mathbb{R}_+} |\varphi(t)| \quad \forall \varphi \in C_b(\mathbb{R}_+).$$

Prove that $(C_b(\mathbb{R}_+), \|\cdot\|_\infty)$ is a Banach space.

E.2.9 Prove that

$$K = \{\varphi \in C^0([0,1]) : \|\varphi\|_\infty = 1, \ 0 \leq \varphi(t) \leq 1 \quad \forall t \in [0,1]\}$$

is a bounded, closed, and non-compact set of $C^0([0,1])$.

E.2.10 Let $I \subset \mathbb{R}$ be a non-empty open interval, assume that $t_0 \in I$ and let $\varphi \in C^0(I)$ be such that

$$\varphi(t) = t^3 + \int_{t_0}^{t} \varphi(s)^2 \log(1 + s^2) \, ds \quad \forall t \in I.$$

Find a CP satisfied by φ.

E.2.11 Let m and M be two real numbers satisfying $0 < m \leq M$. Using the mapping $\mathcal{T} : C^0([0,1]) \mapsto C^0([0,1])$, given by

$$(\mathcal{T}\varphi)(t) = \varphi(t) - \frac{2}{m+M} f(t, \varphi(t)) \quad \forall t \in [0,1], \ \forall \varphi \in C^0([0,1]),$$

prove that, if $f \in C^0([0,1] \times \mathbb{R})$ is continuously differentiable with respect to the variable y in $[0,1] \times \mathbb{R}$ and moreover

$$0 < m \le \frac{\partial f}{\partial y}(t,y) \le M < +\infty \quad \forall (t,y) \in [0,1] \times \mathbb{R},$$

there exists exactly one $\varphi \in C^0([0,1])$ satisfying

$$f(t, \varphi(t)) = 0 \quad \forall t \in [0,1].$$

E.2.12 Let $T_\lambda : C^0([0,1]) \mapsto C^0([0,1])$ be the mapping given by

$$(T_\lambda \varphi)(t) = \lambda \int_0^1 \frac{t^2 + s^2}{1 + |\varphi(s)|} \, ds \quad \forall t \in [0,1], \quad \forall \varphi \in C^0([0,1]).$$

Find the values of λ for which T_λ is a contraction.

E.2.13 Let $T_\lambda : C^0([0,1]) \mapsto C^0([0,1])$ be given by

$$(T_\lambda \varphi)(t) = e^t + \lambda \int_0^1 2e^{t+s} \varphi(s) \, ds \quad \forall t \in [0,1], \quad \forall \varphi \in C^0([0,1]).$$

(1) For which λ is T_λ a contraction? Y T_λ^2?
(2) Determine the fixed-point of T_λ when it exists.

E.2.14 Prove that, in the conditions of (2.1), the estimates (2.11) and (2.12) are fulfilled.

E.2.15 Let $T : C^0([0,1]) \mapsto C^0([0,1])$ be given, with

$$(T\varphi)(t) = 1 + \int_0^t 2s\varphi(s) \, ds \quad \forall t \in [0,1], \quad \forall \varphi \in C^0([0,1]).$$

(1) Is T a contraction? What about T^2?
(2) Determine the fixed points of T.

E.2.16 Let $h \in C^0([0,1])$ be such that $h \ge 0$ in $[0,1]$. Prove that there exists a unique function $\varphi \in C^0([0,1])$ satisfying $\varphi \ge 1$ in $[0,1]$ and

$$\varphi(t) = 1 + \int_0^t h(s) \, |\sin \varphi(s)|^2 \, ds \quad \forall t \in [0,1].$$

E.2.17 Let $h \in C^0([0,1])$ be given. Prove that there exists a unique function $\varphi \in C^0([0,1])$ satisfying

$$\varphi(t) = h(t) + \int_0^t |\sin \varphi(s)|^4 \, ds, \quad \forall t \in [0,1].$$

E.2.18 Let us set $(t_0, y_0) = (0,0)$ and let f be given, with

$$f(t,y) = \begin{cases} -2t & \text{if } y \geq t^2 \\ 2t & \text{if } y \leq 0 \\ -4y/t + 2t & \text{if } 0 < y < t^2. \end{cases}$$

(1) Prove that the corresponding CP possessess exactly one solution in $\mathbb{R}$.

(2) Let $\{\varphi_n\}$ be a sequence of functions defined recursively by

$$\varphi_0(t) \equiv 0$$
$$\varphi_{n+1}(t) = \int_0^t f(s, \varphi_n(s))\, ds \quad \forall t \in \mathbb{R} \quad \forall n \geq 0.$$

Prove that this sequence does not converge (not even pointwise) in any interval of the form $[-\tau, \tau]$, with $\tau > 0$.

(Muller's example, 1927: A Cauchy problem possessing exactly one solution that cannot be obtained as the limit of Picard's sequence.)

E.2.19 Consider the CP

$$\begin{cases} y'' - y = 2e^x \\ y(0) = 1, \ y'(0) = 2. \end{cases}$$

Rewrite this system as a Cauchy problem for a first-order ODS and prove that it possesses exactly one solution in $C^2([-\frac{1}{2}, \frac{1}{2}])$.

Hint: Adapt appropriately the argument of the proof of Picard's Theorem.

E.2.20 Assume that $K \in C^0([a,b] \times [a,b])$ and $f \in C^0([a,b])$ and consider the following equation in $C^0([a,b])$:

$$\varphi(t) = f(t) + \lambda \int_a^t K(t,s)\varphi(s)\, ds \quad \forall t \in [a,b], \quad \forall \varphi \in C^0([a,b]).$$

It is usually said that this is a *Volterra integral equation of the second kind*. Prove that this equation possesses exactly one solution for every λ.

Hint: Set

$$(T_\lambda \varphi)(t) = f(t) + \lambda \int_a^t K(t,s)\varphi(s)\, ds \quad \forall t \in [a,b], \quad \forall \varphi \in C^0([a,b]).$$

Prove that, for every λ, there exists an integer $m \geq 1$ such that T_λ^m is a contraction.

E.2.21 Let $T : C^0([a, b]) \mapsto C^0([a, b])$ be given by

$$(T\varphi)(t) = \int_a^b \alpha(t)\alpha(s)\varphi(s)\,ds \quad \forall t \in [a, b], \quad \forall \varphi \in C^0([a, b]),$$

where $\alpha \in C^0([a, b])$ is a given function.

(1) Find the space $R(T) := \{T\varphi : \varphi \in C^0([a, b])\}$ (the *rank* of T).
(2) Is T a continuous mapping?
(3) Under which conditions is T a contraction?
(4) For $[a, b] = [0, 1]$ and $\alpha(t) \equiv t/2$, solve the integral equation

$$\varphi(t) = t + \frac{1}{4}\int_0^1 ts\varphi(s)\,ds, \quad \forall t \in [0, 1], \quad \forall \varphi \in C^0([0, 1]).$$

E.2.22 Let $a > 0$ be given. Consider the CP

$$\begin{cases} y' = |y|^{a-1}y \\ y(0) = 0. \end{cases}$$

For each a, compute the solution (or solutions) to this problem.

E.2.23 In the following implicit equations, find the supporting set, the singular points, and, if they exist, the singular solutions (a solution $\psi : I \mapsto \mathbb{R}$ is called singular if all the integral triplets $(t, \psi(t), \psi'(t))$ are singular):

(1) $y(y')^2 - 2(t-1)y' + y = 0$. In this case, find also the infinitely many solutions corresponding to the Cauchy data $(2, 1)$.
(2) $t(y')^2 - 2yy' + t = 0$.
(3) $t = y' + \sin y'$. Find also an infinite family of solutions.
(4) $yy' - (y')^2 = e^t$. Find also the solutions corresponding to the Cauchy data $(2, 1)$.
(5) $t^2 + y^2 + (y')^2 = 1$.
(6) $y'((y' - 1)^2 - y^2) = 0$. In this case, which particular property is satisfied by the singular solutions?
(7) $t(y')^2 + t^3 + y^3 = 0$.

Chapter 3

Uniqueness

In this chapter, we will consider again the Cauchy problem for ODSs with a normal structure:

$$\text{(CP)} \qquad \begin{cases} y' = f(t, y) \\ y(t_0) = y_0, \end{cases}$$

where $\Omega \subset \mathbb{R}^{N+1}$ is a non-empty open set, $f : \Omega \mapsto \mathbb{R}^N$ is continuous and locally Lipschitz-continuous with respect to the variable y, and $(t_0, y_0) \in \Omega$.

We will analyze the uniqueness of the solution.

More precisely, we will try to give an answer to the following question:

> Assume that two intervals containing t_0 and two associated solutions to (CP) are given. Do these solutions necessarily coincide in the intersection?

To this purpose, we will use a technical result (important by itself), known as *Gronwall's Lemma*.

3.1 Gronwall's Lemma

The result is the following:

Lemma 3.1 (Gronwall). *Assume that*

$$u(t) \leq M + \int_{t_0}^{t} a(s) u(s) \, ds \quad \forall t \in [t_0, t_1], \qquad (3.1)$$

where $M \in \mathbb{R}$, $a \in C^0([t_0, t_1])$ satisfies $a \geq 0$, and $u \in C^0([t_0, t_1])$. Then

$$u(t) \leq M \exp\left(\int_{t_0}^{t} a(s) \, ds \right) \quad \forall t \in [t_0, t_1]. \qquad (3.2)$$

Proof. Let M, a, and u be in these conditions and set

$$v(t) := \int_{t_0}^t a(s)u(s)\,\mathrm{d}s \quad \forall t \in [t_0, t_1].$$

Then $v \in C^1([t_0, t_1])$ and, by assumption, $u \leq M + v$, whence we also have $v' = au \leq aM + av$, that is,

$$v'(t) - a(t)v(t) \leq Ma(t) \quad \forall t \in [t_0, t_1].$$

After multiplication by $\exp(-\int_{t_0}^t a(s)\,\mathrm{d}s)$, the following is found:

$$\frac{\mathrm{d}}{\mathrm{d}t}\left(v(t)\exp\left(-\int_{t_0}^t a(s)\,\mathrm{d}s\right)\right) = (v'(t) - a(t)v(t))\exp\left(-\int_{t_0}^t a(s)\,\mathrm{d}s\right)$$

$$\leq M\,a(t)\exp\left(-\int_{t_0}^t a(s)\,\mathrm{d}s\right).$$

Integrating in t and taking into account that $v(t_0) = 0$, we see that

$$v(t) \leq M \int_{t_0}^t \left(a(s)\exp\left(\int_s^t a(\sigma)\,\mathrm{d}\sigma\right)\right)\mathrm{d}s = M\exp\left(\int_{t_0}^t a(\sigma)\,\mathrm{d}\sigma\right) - M$$

and, finally,

$$u(t) \leq M + v(t) = M\exp\left(\int_{t_0}^t a(\sigma)\,\mathrm{d}\sigma\right) \quad \forall t \in [t_0, t_1].$$

This proves the lemma. $\qquad\square$

Remark 3.1.

(a) Assume that, instead of (3.1), one has

$$u(t) \leq M + \int_t^{t_1} a(s)u(s)\,\mathrm{d}s \quad \forall t \in [t_0, t_1],$$

where M, a, and u are as before. Then, arguing as in the previous proof, it is not difficult to deduce that

$$u(t) \leq M\exp\left(\int_t^{t_1} a(s)\,\mathrm{d}s\right) \quad \forall t \in [t_0, t_1].$$

(b) We can also obtain estimates of u when the main assumption is a nonlinear integral inequality.

For instance, assume that $\Phi : \mathbb{R} \mapsto \mathbb{R}$ is a non-decreasing strictly positive continuous function and let H be a primitive of $1/\Phi$. Then, from the inequalities

$$u(t) \leq M + \int_{t_0}^t \Phi(u(s))\,\mathrm{d}s \quad \forall t \in [t_0, t_1]$$

we can deduce that

$$u(t) \leq H^{-1}(H(M) + t - t_0) \quad \forall t \in [t_0, t_1].$$

For more details, see for example Buică (2002). □

3.2 Uniqueness of the Solution to the Cauchy Problem

An almost direct consequence of Gronwall's Lemma is the following uniqueness result:

Theorem 3.1. *Assume that f is continuous and locally Lipschitz-continuous with respect to y in Ω and $(t_0, y_0) \in \Omega$. Let I_i be an interval containing t_0 and let φ^i be a solution to* (CP) *in I_i for $i = 1, 2$. Then $\varphi^1 = \varphi^2$ in $I_1 \cap I_2$.*

Proof. Let $t_1 \in I_1 \cap I_2$ be given and assume (for example) that $t_1 > t_0$.
The interval $[t_0, t_1]$ is contained in $I_1 \cap I_2$.
Let us see that φ^1 and φ^2 coincide in $[t_0, t_1]$ (this will be sufficient to achieve the proof).
Let $u : [t_0, t_1] \mapsto \mathbb{R}$ be given by

$$u(t) = |\varphi^1(t) - \varphi^2(t)| \quad \forall t \in [t_0, t_1].$$

Then $u \in C^0([t_0, t_1])$.
Since φ^1 and φ^2 solve (CP) in $[t_0, t_1]$, one has

$$\begin{aligned} u(t) &= \left| \int_{t_0}^t \left(f(s, \varphi^1(s)) - f(s, \varphi^2(s)) \right) ds \right| \\ &\leq \int_{t_0}^t \left| f(s, \varphi^1(s)) - f(s, \varphi^2(s)) \right| ds \end{aligned} \tag{3.3}$$

for all $t \in [t_0, t_1]$.
Let K be the following set:

$$K = K^1 \cup K^2 \quad \text{with} \quad K^i := \{(s, \varphi^i(s)) : s \in [t_0, t_1]\} \quad \text{for} \quad i = 1, 2.$$

Then K is a compact subset of Ω, since it can be written as the finite union of the images of the compact interval $[t_0, t_1]$ by the continuous functions $s \mapsto (s, \varphi^i(s))$.
Consequently, f is globally Lipschitz-continuous with respect to y in K and there exists $L_K > 0$ such that

$$|f(t, y^1) - f(t, y^2)| \leq L_K |y^1 - y^2| \quad \forall (t, y^1), (t, y^2) \in K. \tag{3.4}$$

Taking into account (3.3) and (3.4), we get

$$u(t) \le L_K \int_{t_0}^t |\varphi^1(s)) - \varphi^2(s)| \, ds = L_K \int_{t_0}^t u(s) \, ds \quad \forall t \in [t_0, t_1].$$

At this point, if we apply Lemma 3.1, we deduce that $u \equiv 0$, that is, $\varphi^1 = \varphi^2$ in $[t_0, t_1]$. $\square$

Remark 3.2.

(a) It is possible to establish other uniqueness results for (CP) under slightly weaker assumptions on f; see Cordunneanu (1971) and Hartman (2002).

   On the other hand, H. Okamura proved in 1946 a uniqueness result of a different kind, relying on the existence of an auxiliary function satisfying appropriate properties; see in this respect Note 11.3.

(b) With arguments similar to those in the proof of Theorem 3.1, it is not difficult to deduce the *local continuous dependence* of the solution with respect to initial data.

   Indeed, assume for instance that $(t_0, y_0^i) \in \Omega$ and φ^i is a solution in I_i of the Cauchy problem

$$\begin{cases} y' = f(t, y) \\ y(t_0) = y_0^i \end{cases}$$

for $i = 1, 2$.

   Again, let $t_1 \in I_1 \cap I_2$ be given, with $t_1 \ge t_0$. Then, arguing as before, one has

$$|\varphi^1(t) - \varphi^2(t)| \le |y_0^1 - y_0^2| + L_K \int_{t_0}^t |\varphi^1(s) - \varphi^2(s)| \, ds \quad \forall t \in [t_0, t_1]$$

for some $L_K > 0$, whence

$$|\varphi^1(t) - \varphi^2(t)| \le |y_0^1 - y_0^2| \exp\left(L_K(t - t_0)\right) \quad \forall t \in [t_0, t_1]$$

and

$$\|\varphi^1 - \varphi^2\|_{C^0([t_0, t_1])} \le C|y_0^1 - y_0^2|$$

for $C = e^{L_K(t_1 - t_0)}$.

   This proves that, if we fix the interval $[t_0, t_1]$, and y_0^1 is close to y_0^2, then φ^1 is close to φ^2 (in the sense of $C^0([t_0, t_1])$).

   We will come back to this question later, in Chapter 7.

(c) In Chapter 5, the following situation will be found: $\Omega = (\alpha, \beta) \times \mathbb{R}^N$, with $\alpha, \beta \in \mathbb{R}$, and $\alpha < \beta$, $f : \overline{\Omega} \mapsto \mathbb{R}^N$ is continuous and locally Lipschitz-continuous with respect to y in Ω, and $(t_0, y_0) \in \overline{\Omega}$.

If we extend f as a continuous function (for example to the whole $\mathbb{R}^{N+1}$), we can give a sense to the corresponding Cauchy problem even if $t_0 = \alpha$ or $t_0 = \beta$.

Then, as a consequence of Lemma 3.1, we see that, if I is an interval contained in $[\alpha, \beta]$, there exists *at most* one solution in I.

We leave the proof of this assertion as an exercise to the reader. $\square$

Now, let us assume that $\Omega \subset \mathbb{R}^{n+1}$ is a non-empty open set and let us consider the Cauchy problem

$$(\text{CP})_n \qquad \begin{cases} y^{(n)} = f(t, y, y', \dots, y^{(n-1)}) \\ y(t_0) = y_0^{(0)}, \ y'(t_0) = y_0^{(1)}, \dots, y^{(n-1)}(t_0) = y_0^{(n-1)}, \end{cases}$$

where $f : \Omega \mapsto \mathbb{R}$ is continuous and locally Lipschitz-continuous with respect to the variable $Y = (y, y', \dots, y^{(n-1)})$ and $(t_0, y_0^{(0)}, \dots, y_0^{(n-1)}) \in \Omega$.

The usual change of variable allows to rewrite $(\text{CP})_n$ as an equivalent Cauchy problem for a first-order ODS.

This way, it is easy to obtain the following uniqueness result for $(\text{CP})_n$:

Theorem 3.2. *Assume that $f : \Omega \subset \mathbb{R}^{n+1} \mapsto \mathbb{R}$ is continuous and locally Lipschitz-continuous with respect to the variable Y in Ω and $(t_0, y_0^{(0)}, \dots, y_0^{(n-1)}) \in \Omega$. Let φ^i be a solution in I^i to the Cauchy problem $(\text{CP})_n$ for $i = 1, 2$. Then $\varphi^1 = \varphi^2$ in $I^1 \cap I^2$.*

Exercises

E.3.1 Assume that $\delta > 0, \Omega = (-\delta, \delta) \times \mathbb{R}^N$ and $f_1, f_2 : \Omega \mapsto \mathbb{R}^N$ are continuous and globally Lipschitz-continuous with respect to y. Also, assume that

$$|f_2(t, y) - f_1(t, y)| \le \varepsilon \quad \forall (t, y) \in \Omega.$$

Let $y_0 \in \mathbb{R}^N$ be given and let $L_2 > 0$ be a Lipschitz constant for f_2 with respect to y in Ω. Let $\varphi_i, i = 1, 2$, be a solution in $(-\delta, \delta)$ of the Cauchy problem

$$\begin{cases} y' = f_i(t, y) \\ y(0) = y_0. \end{cases}$$

Prove that

$$|\varphi_2(t) - \varphi_1(t)| \le \varepsilon \delta e^{L_2 \delta} \quad \forall t \in (-\delta, \delta).$$

E.3.2 Consider the nth order ODE
$$y^{(n)} = f(t, y),$$
where $f : D \subset \mathbb{R}^2 \mapsto \mathbb{R}$ is continuous and locally Lipschitz-continuous with respect to the variable y. Let φ and ψ be two solutions to this ODE in the interval I. Assume that $t_0 \in I$ and
$$\varphi^{(j)}(t_0) = \psi^{(j)}(t_0) \quad \forall j = 0, 1, \ldots, n-1$$
and set $u(t) := |\varphi(t) - \psi(t)|$ for all $t \in I$. Apply directly Gronwall's Lemma to the function u and deduce that $\varphi = \psi$.

E.3.3 Consider the nth order ODE in the previous exercise. Let φ and ψ be two solutions in the interval I and suppose that
$$\varphi^{(j)}(t_0) = \psi^{(j)}(t_0) + \varepsilon \quad \forall j = 0, 1, \ldots, n-1,$$
where $t_0 \in I$ and $\varepsilon > 0$. Using Gronwall's Lemma, find a bound of the function $u := |\varphi - \psi|$ in I in terms of ε.

E.3.4 Prove the estimate in Remark 3.1(b).

E.3.5 Assume that $M \in \mathbb{R}$ and $\Phi : \mathbb{R} \times \mathbb{R}_+ \mapsto \mathbb{R}_+$ is a non decreasing continuous function, continuously differentiable with respect to its second argument. Let $u \in C^0([t_0, t_1])$ be given and assume that
$$u(t) \leq M + \int_{t_0}^{t} \Phi(s, u(s)) \, ds \quad \forall t \in [t_0, t_1]. \tag{3.5}$$
Which estimate can be deduced for u in this case?

E.3.6 Let $h, a, u \in C^0([t_0, t_1])$ be given, with $a \geq 0$. Assume that
$$u(t) \leq h(t) + \int_{t_0}^{t} a(s)u(s) \, ds \quad \forall t \in [t_0, t_1]. \tag{3.6}$$
Which estimate can be deduced for u?

E.3.7 Prove that Gronwall's Lemma holds also when a is nonnegative and (for instance) Riemann-integrable, but not necessarily continuous.

E.3.8 Assume that $M > 0$, $a \in C^0([t_0, t_1])$ with $a \geq 0$, $m \in (0, 1)$, and $u \in C^0([t_0, t_1])$. Also, assume that
$$u(t) \leq M + \int_{t_0}^{t} a(s)u(s)^m \, ds \quad \forall t \in [t_0, t_1]. \tag{3.7}$$
Which estimate can be deduced for u? What happens as $m \to 1^-$?

E.3.9 Assume that $\alpha \in (0, 1)$ and $\ell > 0$. Find a *biparametric* family of solutions in $(-\ell, \ell)$ to the Cauchy problem
$$\begin{cases} y' = |y|^\alpha \operatorname{sign} y \\ y(0) = 0. \end{cases}$$
Do other solutions exist? Is this construction possible with $\alpha = 1$? What can be said when the initial condition is $y(0) = y_0$ with $y_0 \neq 0$? Justify the answers.

E.3.10 Consider the Cauchy problem

$$\begin{cases} y' = y(\log |y|)^\alpha \\ y(0) = y_0, \end{cases}$$

where $\alpha \in \mathbb{R}$ and $y_0 \in \mathbb{R}$. Do the following for every α and y_0:

(1) Analyze the uniqueness of the solution.

(2) The explicit computation of the solution(s) in $(-1, 1)$.

E.3.11 Prove the assertion in Remark 3.2(c).

E.3.12 Formulate and prove a result similar to the assertion in Remark 3.2(c) for a Cauchy problem for an ODE of order n

$$y^{(n)} = f(t, y, y', \ldots, y^{(n-1)}),$$

where f is defined in $\overline{\Omega} = [\alpha, \beta] \times \mathbb{R}^n$.

E.3.13 Prove a "discrete" version of Gronwall's Lemma: Let the u^n, a^n be real numbers for $n = 0, 1, \ldots$ with $a^n \geq 0$ for all n. Assume that

$$u^n \leq M + \sum_{k=0}^{n-1} a^k u^k \quad \forall n \geq 0,$$

where $M \geq 0$. Then, find a bound of u^n in terms of M and the a^k.

Chapter 4

The Cauchy Problem: Global Analysis

This chapter is devoted to analyzing the Cauchy problem

(CP)
$$\begin{cases} y' = f(t,y) \\ y(t_0) = y_0 \end{cases}$$

from the viewpoint of global existence.

More precisely, we will try to find a solution to (CP) in the largest possible interval.

Obviously, this interval will depend on the domain Ω of definition of f and also on the behavior of this function.

It is reasonable to expect that, if the values of f grow too much in a region $\Omega_0 \subset \Omega$, the solutions corresponding to initial data in Ω_0 run the risk of blowing up. But this may happen even if f is moderate.

In fact, this was already seen in see Example 1.8(a).

We will prove a result that ensures the existence and uniqueness of a global (or maximal) solution.

Also, we will characterize the maximal solution in terms of the behavior of the associated trajectory "near" the endpoints of the interval where it is defined.

This characterization will allow us to decide, when observing a particular solution, whether it is maximal or not, that is, whether it is possible or not to extend its definition beyond an extreme.

4.1 Existence and Uniqueness of a Maximal Solution

In the sequel, unless otherwise specified, it will be assumed that $f : \Omega \mapsto \mathbb{R}^N$ is continuous and locally Lipschitz-continuous with respect to the variable y in the open set $\Omega \subset \mathbb{R}^{N+1}$.

For each $(t_0, y_0) \in \Omega$, $\mathcal{S}(t_0, y_0)$ will stand for the family of couples (I, φ), where

- $I \subset \mathbb{R}$ is an interval whose interior contains t_0 and
- $\varphi : I \mapsto \mathbb{R}^N$ is a solution to (CP) in I.

In view of Picard's Theorem (Theorem 2.3), it is evident that $\mathcal{S}(t_0, y_0) \neq \emptyset$.

Definition 4.1. Let $(t_0, y_0) \in \Omega$ and $(I, \varphi) \in \mathcal{S}(t_0, y_0)$ be given. It will be said that (I, φ) is right-extendable (resp. left-extendable) if there exists $(J, \psi) \in \mathcal{S}(t_0, y_0)$ such that $\sup I \in \operatorname{int} J$ (resp. $\inf I \in \operatorname{int} J$) and ψ and φ coincides in $I \cap J$.

It will be said that (I, φ) is maximal (or global) if it is neither right-extendable nor left-extendable. $\square$

For a given $(I, \varphi) \in \mathcal{S}(t_0, y_0)$, if there is no possible confusion, it is customary to say that φ is right-extendable, left-extendable or maximal (see Figure 4.1).

The first main result in this chapter is the following:

Theorem 4.1. *Assume that f is continuous and locally Lipschitz-continuous with respect to the variable y in Ω and $(t_0, y_0) \in \Omega$. Then there*

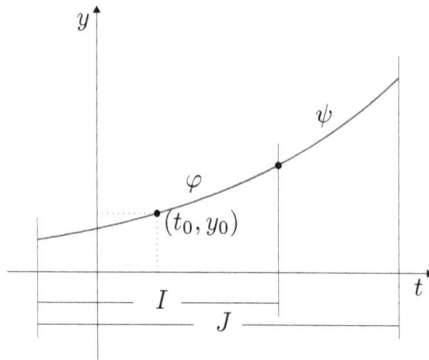

Fig. 4.1 (J, ψ) extends (I, φ) to the right.

exists a unique maximal solution to (CP), $\varphi : I \mapsto \mathbb{R}^N$. *The interval of definition of* φ *is open.*

Proof. Let us first see that, if $\varphi : I \mapsto \mathbb{R}^N$ is a maximal solution, then I is necessarily an open interval.

Let α and β be the endpoints of I and let us assume that, for example, $\beta < +\infty$ and $\beta \in I$. Then $(\beta, \varphi(\beta)) \in \Omega$. Consequently, the auxiliary CP

$$\begin{cases} y' = f(t, y) \\ y(\beta) = \varphi(\beta) \end{cases}$$

possesses a unique solution $\overline{\varphi}$ in a (possibly small) interval $[\beta - \tau, \beta + \tau]$.

Let us set $J = I \cup [\beta, \beta + \tau]$ and let us introduce the function $\psi : J \mapsto \mathbb{R}^N$, with $\psi(t) = \varphi(t)$ if $t \in I$ and $\psi(t) = \overline{\varphi}(t)$ otherwise.

It is not difficult to prove that $\psi \in C^0(J; \mathbb{R}^N)$, $(t, \psi(t)) \in \Omega$ for every $t \in J$ and

$$\psi(t) = y_0 + \int_{t_0}^t f(s, \psi(s)) \, ds \quad \forall t \in J.$$

Therefore, φ is right-extendable, which contradicts the maximality of φ.

This proves that I cannot contain β. In a similar way, it can be proved that I does not contain α.

Hence, I is open.

Let us see that, for every $(t_0, y_0) \in \Omega$, the corresponding CP possesses at least one maximal solution. To this purpose, let us introduce

$$I(t_0, y_0) := \bigcup_{(I, \varphi) \in \mathcal{S}(t_0, y_0)} I.$$

Obviously, $I(t_0, y_0)$ is an interval whose interior contains t_0. Let $\tilde{\varphi} : I(t_0, y_0) \mapsto \mathbb{R}$ be the function defined as follows:

For every $t \in I(t_0, y_0)$, there exists $(I, \varphi) \in \mathcal{S}(t_0, y_0)$ such that $t \in I$. We then set $\tilde{\varphi}(t) = \varphi(t)$.

In view of Theorem 3.1, this definition is correct.

Indeed, if $t \in I \cap J$ with $(I, \varphi), (J, \psi) \in \mathcal{S}(t_0, y_0)$, the functions φ and ψ necessarily coincide at t.

Moreover, $\tilde{\varphi} \in C^0(I(t_0, y_0); \mathbb{R}^N)$.

Indeed, if $t \in \text{int } I(t_0, y_0)$, then $\tilde{\varphi}$ coincides in a neighborhood of t with a continuous function, whence $\tilde{\varphi}$ is continuous at t. If t is an endpoint of $I(t_0, y_0)$, then the interval of ends t_0 and t is a subset of $I(t_0, y_0)$ and, there, $\tilde{\varphi}$ coincides again with a continuous function; consequently, $\tilde{\varphi}$ is also continuous at t in this case.

For similar reasons, we also have that $(t, \tilde{\varphi}(t)) \in \Omega$ for every $t \in I(t_0, y_0)$. Let us see that, for any $t \in I(t_0, y_0)$,

$$\tilde{\varphi}(t) = y_0 + \int_{t_0}^t f(s, \tilde{\varphi}(s)) \, \mathrm{d}s. \tag{4.1}$$

Thus, let us assume that $t \in I(t_0, y_0)$ and let $(I, \varphi) \in \mathcal{S}(t_0, y_0)$ be such that $t \in I$. Let us denote by J_t the closed interval of endpoints t_0 and t. Then $J_t \subset I$ and, therefore, one has $\tilde{\varphi} = \varphi$ in J_t.

In particular, since φ solves (CP) in I, one has

$$\varphi(t) = y_0 + \int_{t_0}^t f(s, \varphi(s)) \, \mathrm{d}s,$$

whence we find that (4.1) holds.

This proves that $\tilde{\varphi}$ is a solution in $I(t_0, y_0)$ to (CP). By construction, $\tilde{\varphi}$ is obviously a maximal solution.

Finally, let us see that, for each $(t_0, y_0) \in \Omega$, there exists at most one maximal solution to (CP).

Let $(I, \varphi), (J, \psi) \in \mathcal{S}(t_0, y_0)$ be given and let us assume that φ and ψ are maximal solutions. In view of Theorem 3.1, it is clear that φ and ψ must coincide in $I \cap J$.

On the other hand, we must have $I = J$.

Indeed, suppose the opposite and consider the open interval $\hat{I} := I \cup J$ and the function $\hat{\varphi}$, with

$$\hat{\varphi}(t) = \begin{cases} \varphi(t) & \text{if } t \in I \\ \psi(t) & \text{if } t \in J. \end{cases}$$

Arguing as before, it is not difficult to prove that this function is well defined and solves (CP) in $\hat{I}$.

Furthermore, since $I \neq J$, $\hat{\varphi}$ must be an extension of φ or ψ. But this is in contradiction with the fact that both φ and ψ are maximal solutions. The consequence is that we necessarily have $I = J$.

Hence, $(I, \varphi) = (J, \psi)$ and the maximal solution is unique. □

Remark 4.1. When $f : \Omega \mapsto \mathbb{R}^N$ is (only) continuous, it can be proved that there exists at least one maximal solution to (CP). For the proof of this result, we need Zorn's Lemma; for more details, see Note 11.4.1. □

4.2   Characterization of the Maximal Solution

For a motivation of the results in this section, let us go back again to Example 1.8(a).

We saw there that the solution to the CP

$$\begin{cases} y' = 1 + y^2 \\ y(0) = 0 \end{cases} \tag{4.2}$$

is

$$\varphi(t) = \tan t \quad \forall t \in (-\pi/2, \pi/2). \tag{4.3}$$

We see that, in this case, f is well-defined, continuous and locally Lipschitz-continuous with respect to y (in fact, f is polynomial) in $\Omega = \mathbb{R}^2$.

From Theorem 4.1, we deduce that φ is the maximal solution.

This shows that, in general, the maximal solution to a (CP) can be defined in an interval *smaller than what could be expected*.

In a general situation, it makes sense to wonder if the values of a maximal solution "necessarily go to infinity", as in (4.3), or contrarily they can stay bounded and even have a limit inside the open set where f is continuous and locally Lipschitz-continuous with respect to y.

The goal of this section is to clarify this and other similar questions.

This will be achieved by establishing a *characterization* of maximality. More precisely, we will find conditions that are necessary and sufficient for a solution $\varphi : I \mapsto \mathbb{R}^N$ to be maximal; these conditions will be related to the behavior of φ near the endpoints of I.

We will need a previous technical result.

Lemma 4.1. *Let (X, d) be a metric space. Assume that $K \subset X$ is compact and $F \subset X$ is closed, with $K, F \neq \emptyset$ and $K \cap F = \emptyset$. Then*

$$d(K, F) := \inf_{k \in K, \ f \in F} d(k, f) > 0.$$

Proof. Let us suppose that $d(K, F) = 0$. Then, for every $n \geq 1$, there exist $k_n \in K$ and $f_n \in F$ such that $d(k_n, f_n) \leq 1/n$.

Since K is compact, the sequence $\{k_n\}$ possesses a subsequence $\{k_{n_\ell}\}$ that converges to some $k \in K$.

Then, k is also the limit of the sequence $\{f_{n_\ell}\}$, since

$$d(k, f_{n_\ell}) \leq d(k, k_{n_\ell}) + d(k_{n_\ell}, f_{n_\ell})$$

and, in this inequality, both terms in the right-hand side go to zero. Since F is closed, we also have $k \in F$.

That is to say, $k \in K \cap F$, which contradicts the assumption $K \cap F = \emptyset$. $\qquad\square$

Observe that, in this lemma, the compactness of K is an essential hypothesis.

Indeed, it is easy to build closed disjoint non-empty sets F_1 and F_2 such that $d(F_1, F_2) = 0$.

For example, in the metric space $\mathbb{R}^2$ (with the usual Euclidean distance), we can take

$$F_1 = \{(x_1, 0) : x_1 \in \mathbb{R}\} \quad \text{and} \quad F_2 = \left\{\left(x_1, \frac{1}{x_1}\right) : x_1 \in \mathbb{R}_+\right\}.$$

In order to achieve our main objectives in this section, we must introduce the concepts of trajectory and half-trajectory.

Definition 4.2. Let $(t_0, y_0) \in \Omega$ and $(I, \varphi) \in \mathcal{S}(t_0, y_0)$ be given. The trajectory of φ is the set

$$\tau(\varphi) = \{(t, \varphi(t)) : t \in I\}.$$

The positive half-trajectory, or half-trajectory to the right, is the set

$$\tau^+(\varphi) = \{(t, \varphi(t)) : t \in I, \ t \geq t_0\}.$$

Similarly, the negative half-trajectory (or half-trajectory to the left) is given by

$$\tau^-(\varphi) = \{(t, \varphi(t)) : t \in I, \ t \leq t_0\}.$$

$\square$

Lemma 4.2. *Assume that $(I, \varphi) \in \mathcal{S}(t_0, y_0)$. The following two conditions are equivalent:*

(1) $\tau^+(\varphi)$ *is bounded and*
$$d(\tau^+(\varphi), \partial\Omega) := \inf_{\substack{(t, y) \in \tau^+(\varphi) \\ (s, z) \in \partial\Omega}} d((t, y), (s, z)) > 0. \tag{4.4}$$

(2) *There exists a compact set K satisfying*
$$\tau^+(\varphi) \subset K \subset \Omega. \tag{4.5}$$

Proof. Let us first assume that $\tau^+(\varphi)$ is bounded and one has (4.4).

Then, the closure of $\tau^+(\varphi)$ is a compact set that contains $\tau^+(\varphi)$. This compact is a subset of Ω.

Indeed, if this were not the case, its intersection with $\partial\Omega$ would be non-empty, which is in contradiction with (4.4).

Conversely, if (4.5) is satisfied for some compact set K, then $\tau^+(\varphi)$ is obviously bounded.

Furthermore, we have in this case
$$d(\tau^+(\varphi), \partial\Omega) \geq d(K, \partial\Omega).$$

But this last quantity is positive, thanks to Lemma 4.1. $\square$

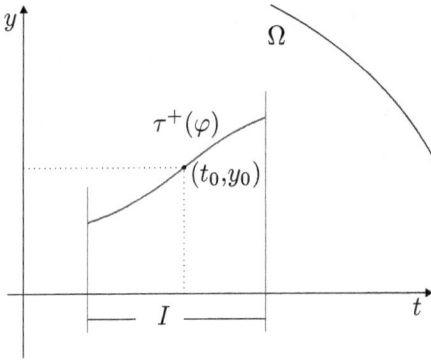

Fig. 4.2   The conditions satisfied in Theorem 4.2.

At this point, we are ready to present a result that characterizes an extendable solution.

Theorem 4.2. *Assume that $f : \Omega \mapsto \mathbb{R}^N$ is continuous and locally Lipschitz-continuous with respect to y, $(t_0, y_0) \in \Omega$ and $(I, \varphi) \in \mathcal{S}(t_0, y_0)$. Then, the following two conditions are equivalent* (Figure 4.2):

(1) φ *is right-extendable.*
(2) $\tau^+(\varphi)$ *is bounded and* (4.4) *is satisfied.*

Proof. Let us first assume that φ is right-extendable.

Let β be the supremum of I. Then $\beta < +\infty$ and there exists an interval of the form $[t_0, \beta + \varepsilon]$ where (CP) possesses exactly one solution $\tilde{\varphi}$, with $\tilde{\varphi}(t) = \varphi(t)$ for every $t \in [t_0, \beta)$.

Obviously,

$$\tau^+(\varphi) \subset K := \{(t, \tilde{\varphi}(t)) : t \in [t_0, \beta + \varepsilon]\},$$

that is a compact subset of Ω.

According to Lemma 4.2, the positive half-trajectory $\tau^+(\varphi)$ is bounded and $d(\tau^+(\varphi), \partial\Omega) \geq d(K, \partial\Omega) > 0$, that is, (4.4) holds.

Now, assume that $\tau^+(\varphi)$ is bounded and (4.4) is fulfilled. Again, the endpoint to the right β must satisfy $\beta < +\infty$.

We know that there exists a compact set K satisfying (4.5). Let us set $M := \max_{(t,y) \in K} |f(t, y)|$.

For any sufficiently large n, we set

$$t_n = \beta - \frac{1}{n}, \quad z_n = \varphi(t_n).$$

Then the (t_n, z_n) belong to $\tau^+(\varphi)$.

The sequence $\{(t_n, z_n)\}$ is bounded. Consequently, there exists a subsequence $\{(t_{n_\ell}, z_{n_\ell})\}$ that converges to a point $(\beta, \overline{z})$ as $\ell \to +\infty$. One has $(\beta, \overline{z}) \in K \subset \Omega$.

Let $\{t'_\ell\}$ be an arbitrary sequence in $[t_0, \beta)$ satisfying $t'_\ell \to \beta$. Then $\varphi(t'_\ell) \to \overline{z}$ since, for any $\ell \geq 1$, one has the following:

$$|\varphi(t_{n_\ell}) - \varphi(t'_\ell)| = \left| \int_{t'_\ell}^{t_{n_\ell}} f(s, \varphi(s)) \, ds \right|$$

$$\leq \left| \int_{t'_\ell}^{t_{n_\ell}} |f(s, \varphi(s))| \, ds \right| \leq M \, |t_{n_\ell} - t'_\ell|.$$

This proves the existence of the lateral limit of φ as t converges from the left to β and, also, that this limit is equal to $\overline{z}$:

$$\exists \lim_{t \to \beta^-} \varphi(t) = \overline{z}. \tag{4.6}$$

In particular, we deduce from (4.6) that

$$\overline{z} = y_0 + \int_{t_0}^{\beta} f(s, \varphi(s)) \, ds. \tag{4.7}$$

Let us consider the auxiliary CP

$$\begin{cases} y' = f(t, y) \\ y(\beta) = \overline{z}. \end{cases} \tag{4.8}$$

In view of Picard's Theorem, there exists $\tau > 0$ such that (4.8) possesses a unique solution ψ in $[\beta - \tau, \beta + \tau]$. Let $\tilde{\varphi} : (\alpha, \beta + \tau] \mapsto \mathbb{R}^N$ be the function defined as follows:

$$\tilde{\varphi}(t) = \begin{cases} \varphi(t) & \text{if } t \in (\alpha, \beta) \\ \psi(t) & \text{if } t \in [\beta, \beta + \tau]. \end{cases}$$

It is then clear that $\tilde{\varphi}$ is well defined and continuous in $(\alpha, \beta + \tau]$ (thanks to (4.6)) and $(t, \tilde{\varphi}(t)) \in \Omega$ for every $t \in (\alpha, \beta + \tau]$.

On the other hand, $\tilde{\varphi}$ is a solution in $(\alpha, \beta + \tau]$ to the CP. This is a consequence of the identities

$$\tilde{\varphi}(t) = y_0 + \int_{t_0}^t f(s, \tilde{\varphi}(s)) \, ds,$$

that hold for all $t \in (\alpha, \beta + \tau]$: if $t \in (\alpha, \beta)$, this is evident; otherwise, if $t \in [\beta, \beta + \tau]$, one has

$$\tilde{\varphi}(t) = \psi(t) = \overline{z} + \int_{\beta}^t f(s, \psi(s)) \, ds$$

$$= y_0 + \int_{t_0}^{\beta} f(s, \varphi(s)) \, ds + \int_{\beta}^t f(s, \psi(s)) \, ds$$

$$= y_0 + \int_{t_0}^t f(s, \tilde{\varphi}(s)) \, ds,$$

where we have used (4.7).

By construction, $\tilde{\varphi}$ is a solution to (CP) that extends φ.

This proves that φ is right-extendable and, consequently, ends the proof. □

Remark 4.2. The fact that a solution φ to (CP) is right-extendable can also be characterized by the properties of the *right terminal set* of φ.

Essentially, this is the set of points $(\beta, \overline{z})$ that can be obtained as limits of sequences $\{(t_n, \varphi(t_n))\}$ where $t_n \to \beta^-$. For the precise definition of the terminal set, see Note 11.4.2. □

Of course, we can also state and prove a result analogous to Theorem 4.2 concerning left-extendable solutions. Using this result and Theorem 4.2, we immediately get the following characterization of a maximal solution.

Theorem 4.3. *Assume that $f : \Omega \subset \mathbb{R}^{N+1} \mapsto \mathbb{R}^N$ is continuous and locally Lipschitz-continuous with respect to y, $(t_0, y_0) \in \Omega$ and $(I, \varphi) \in \mathcal{S}(t_0, y_0)$.*

The following assertions are equivalent:

(1) *φ is the maximal solution to (CP).*
(2) *Either $\tau^+(\varphi)$ is unbounded, or $d(\tau^+(\varphi), \partial\Omega) = 0$; on the other hand, either $\tau^-(\varphi)$ is unbounded, or $d(\tau^-(\varphi), \partial\Omega) = 0$.* □

Thus, roughly speaking, we see that, under the usual existence and uniqueness assumptions, the half-trajectories of a maximal solution can only end either at infinity or on the boundary of Ω.

Remark 4.3. Assume that $\Omega = \mathbb{R}^{N+1}$. Then, if $\varphi : (\alpha, \beta) \mapsto \mathbb{R}^N$ is the maximal solution to (CP), we must have the following:

$$\text{Either } \beta = +\infty, \text{ or } \limsup_{t \to \beta^-} |\varphi(t)| = +\infty;$$

$$\text{Either } \alpha = -\infty, \text{ or } \limsup_{t \to \alpha^+} |\varphi(t)| = +\infty.$$

If $\beta < +\infty$ (resp. $\alpha > -\infty$), it is usually said that φ *blows up to the right* (*resp. to the left*) *in finite time*. ☐

In the particular case of a scalar ODE, with f independent of t, more information can be given on the behavior of the solutions.

More precisely, consider the CP

$$\begin{cases} y' = f(y) \\ y(0) = y_0, \end{cases} \tag{4.9}$$

where $f : \mathbb{R} \mapsto \mathbb{R}$ is locally Lipschitz-continuous with respect to y and $y_0 \in \mathbb{R}$. For simplicity, we have taken $t_0 = 0$.

The following result holds:

Theorem 4.4. *Let $\varphi : I \mapsto \mathbb{R}$ be the maximal solution to (4.9), with $I = (\alpha, \beta)$.*

(1) *If $f(y_0) = 0$, then $I = \mathbb{R}$ and $\varphi(t) = y_0$ for all $t \in \mathbb{R}$. In this case, we say that y_0 is a critical point of f.*

(2) *If $f(y_0) > 0$, then φ is strictly increasing in I. Furthermore:*

(a) *If there exists a critical point greater than y_0, then $\beta = +\infty$ (that is, $I \supset [0, +\infty)$) and*

$$\lim_{t \to +\infty} \varphi(t) = \overline{y},$$

where $\overline{y}$ is the smallest critical point greater than y_0.

(b) *Otherwise, β can be finite or not and one has*

$$\lim_{t \to \beta} \varphi(t) = +\infty.$$

(3) *If $f(y_0) < 0$, similar assertions hold on the left, with the appropriate changes.*

For the proof, see Exercise E.4.15.

Let us now assume that $\Omega \subset \mathbb{R}^{n+1}$ is a non-empty open set and let us consider the CP

$$(\text{CP})_n \qquad \begin{cases} y^{(n)} = f(t, y, y', \ldots, y^{(n-1)}) \\ y(t_0) = y_0^{(0)}, \ y'(t_0) = y_0^{(1)}, \ldots, \ y^{(n-1)}(t_0) = y_0^{(n-1)}, \end{cases}$$

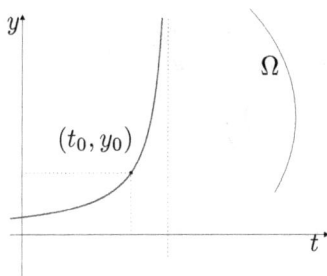

Fig. 4.3   A maximal solution with unbounded $\tau^+(\varphi)$.

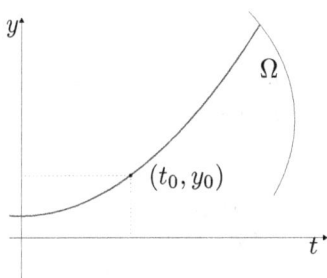

Fig. 4.4   A maximal solution with $d(\tau^+(\varphi), \partial\Omega) = 0$.

where $f : \Omega \mapsto \mathbb{R}$ satisfies the usual existence and uniqueness conditions and $(t_0, y_0^{(0)}, \ldots, y_0^{(n-1)}) \in \Omega$ (Figures 4.3 and 4.4).

Recall that a well-known change of variables allows to reduce $(\text{CP})_n$ to an equivalent CP for a first-order ODS.

This makes it possible to speak of extendable and maximal solutions to $(\text{CP})_n$ and, also, to deduce the following results:

Theorem 4.5. *Let* $f : \Omega \subset \mathbb{R}^{n+1} \mapsto \mathbb{R}$ *be continuous and locally Lipschitz-continuous with respect to the variable* $Y = (y, y', \ldots, y^{(n-1)})$ *and assume that* $(t_0, y_0^{(0)}, \ldots, y_0^{(n-1)}) \in \Omega$. *Then, the Cauchy problem* $(\text{CP})_n$ *possesses exactly one maximal solution* $\varphi : I \mapsto \mathbb{R}$. *The interval of definition* φ *is open.*

For every $(t_0, y_0^{(0)}, \ldots, y_0^{(n-1)}) \in \Omega$, we will denote by

$$\mathcal{S}_n(t_0, y_0^{(0)}, \ldots, y_0^{(n-1)})$$

the family of couples (I, φ) formed by an interval $I \subset \mathbb{R}$ whose interior contains t_0 and a function $\varphi : I \mapsto \mathbb{R}$ that solves $(\text{CP})_n$ in I.

Of course, $\mathcal{S}_n(t_0, y_0^{(0)}, \ldots, y_0^{(n-1)}) \neq \emptyset$.

For any $(I, \varphi) \in \mathcal{S}_n(t_0, y_0^{(0)}, \ldots, y_0^{(n-1)})$, we set

$$\tilde{\tau}^+(\varphi) = \{(t, \varphi(t), \ldots, \varphi^{(n-1)}(t)) : t \in I, \ t \geq t_0\}$$

and

$$\tilde{\tau}^-(\varphi) = \{(t, \varphi(t), \ldots, \varphi^{(n-1)}(t)) : t \in I, \ t \leq t_0\}.$$

It will be said that $\tilde{\tau}^+(\varphi)$ and $\tilde{\tau}^-(\varphi)$ are the *generalized half-trajectories* of φ, respectively, *to the right* and *to the left*, also called positive and negative.

Theorem 4.6. *Assume that $f{:}\Omega \subset \mathbb{R}^{n+1} \mapsto \mathbb{R}$ is continuous and locally Lipschitz-continuous with respect to the variable $Y = (y, y', \ldots, y^{(n-1)})$, $(t_0, y_0^{(0)}, \ldots, y_0^{(n-1)}) \in \Omega$ and $(I, \varphi) \in \mathcal{S}_n(t_0, y_0^{(0)}, \ldots, y_0^{(n-1)})$. The following assertions are equivalent:*

(1) *φ is the maximal solution to $(CP)_n$.*
(2) *Either $\tilde{\tau}^+(\varphi)$ is unbounded, or $d(\tilde{\tau}^+(\varphi), \partial\Omega) = 0$; on the other hand, either $\tilde{\tau}^-(\varphi)$ is unbounded, or $d(\tilde{\tau}^-(\varphi), \partial\Omega) = 0$.*

Example 4.1. Consider the CP

$$\begin{cases} y'' = -2(y')^3 \\ y(1) = 2, \quad y'(1) = \dfrac{1}{2}. \end{cases}$$

The corresponding maximal solution is $\varphi : (-1, +\infty) \mapsto \mathbb{R}$, given by

$$\varphi(t) = \sqrt{2(t+1)} \quad \forall t \in (-1, +\infty).$$

The generalized half-trajectory to the left is the unbounded set

$$\tilde{\tau}^-(\varphi) = \left\{ (t, \sqrt{2(t+1)}, \frac{1}{\sqrt{2(t+1)}}) : t \in (-1, 1] \right\}.$$

We observe however that the half-trajectory to the left $\tau^-(\varphi)$ (in the sense of the Definition 4.2) is bounded as follows:

$$\tau^-(\varphi) = \{(t, \sqrt{2(t+1)}) : t \in (-1, 1]\}.$$

But, obviously, this is not in contradiction with Theorem 4.6.   □

4.3   A Particular Case

In this section, it will be assumed that Ω is a *vertical band*, that is,
$$\Omega = I_0 \times \mathbb{R}^N,$$
with $I_0 = (a, b)$, $-\infty \leq a < b \leq +\infty$.

We have already seen that it can happen that the maximal solution to (CP) is not defined in the whole interval I_0 (recall (4.2) and (4.3)).

However, the following result holds:

Theorem 4.7. *Assume that* $f : \Omega \mapsto \mathbb{R}^N$ *is continuous and globally Lipschitz-continuous with respect to the variable* y *in* $\Omega = I_0 \times \mathbb{R}^N$ *and* $(t_0, y_0) \in \Omega$. *Let* $\varphi : I \mapsto \mathbb{R}^N$ *be the maximal solution to the corresponding CP. Then* $I = I_0$.

Proof. Let us put $I_0 = (a, b)$ and $I = (\alpha, \beta)$, with $-\infty \leq a \leq \alpha < \beta \leq b \leq +\infty$.

We will prove that, under the previous assumptions, $\beta = b$ (in a similar way, it can be proved that $\alpha = a$).

Contrarily, let us assume that $\beta < b$ (in particular, $\beta < +\infty$). We will see that the positive half-trajectory $\tau^+(\varphi)$ is contained in a compact set contained in Ω. This will show that φ cannot be the maximal solution.

Let $L > 0$ be a Lipschitz constant of f with respect to y in Ω. Then
$$|f(t, y^1) - f(t, y^2)| \leq L|y^1 - y^2| \quad \forall (t, y^1), (t, y^2) \in \Omega.$$
Let us introduce $M := \max_{s \in [t_0, \beta]} |f(s, y_0)|$. For every $t \in [t_0, \beta)$, we have
$$|\varphi(t) - y_0| = \left| \int_{t_0}^t f(s, \varphi(s))\, ds \right|$$
$$\leq \int_{t_0}^t |f(s, \varphi(s))|\, ds \leq \int_{t_0}^t |f(s, \varphi(s)) - f(s, y_0)|\, ds$$
$$+ \int_{t_0}^t |f(s, y_0)|\, ds$$
$$\leq L \int_{t_0}^t |\varphi(s) - y_0|\, ds + M(\beta - t_0).$$
Accordingly, we can apply Gronwall's Lemma (Lemma 3.1) and deduce that
$$|\varphi(t) - y_0| \leq M(\beta - t_0)\, e^{L(t - t_0)} \leq M(\beta - t_0)\, e^{L(\beta - t_0)} \quad \forall t \in [t_0, \beta).$$
In other words, if we set $R := M(\beta - t_0)\, e^{L(\beta - t_0)}$, we find that
$$\tau^+(\varphi) \subset K := [t_0, \beta] \times \overline{B}(y_0; R),$$
that is a compact subset of Ω.

This proves the theorem. $\square$

Corollary 4.1. *Assume that $\Omega = (a, b) \times \mathbb{R}^N$ and f is continuous in Ω and globally Lipschitz-continuous with respect to the variable y in any set $(\bar{a}, \bar{b}) \times \mathbb{R}^N$, with $a < \bar{a} < \bar{b} < b$. Let $(t_0, y_0) \in \Omega$ be given and let $\varphi : I \mapsto \mathbb{R}^N$ be the maximal solution to the corresponding CP. Then $I = I_0$.*

Proof. Let $\bar{a}, \bar{b}$ be such that $a < \bar{a} < t_0 < \bar{b} < b$ and let us consider the auxiliary CP

$$\begin{cases} y' = \overline{f}(t, y) \\ y(t_0) = y_0, \end{cases} \tag{4.10}$$

where $\overline{f} : (\bar{a}, \bar{b}) \times \mathbb{R}^N \mapsto \mathbb{R}^N$ is given by

$$\overline{f}(t, y) = f(t, y) \quad \forall (t, y) \in (\bar{a}, \bar{b}) \times \mathbb{R}^N.$$

According to Theorem 4.7, the interval of definition of the maximal solution $\overline{\varphi}$ to (4.10) is $(\bar{a}, \bar{b})$.

On the other hand, $\overline{\varphi}$ also solves (CP) in the same interval. Consequently, the interval of definition of the maximal solution to (CP) contains $(\bar{a}, \bar{b})$.

This proves the desired result, since $\bar{a}$ and $\bar{b}$ are arbitrarily close, respectively, to a and b. $\qquad\square$

Needless to say, the assumption in Corollary 4.1 of global Lipschitz-continuity of f in the sets $(\bar{a}, \bar{b}) \times \mathbb{R}^N$ is essential.

Exercises

In the following exercises, we will frequently find functions f defined in a (not necessarily open) set $G \subset \mathbb{R}^{N+1}$, with values in $\mathbb{R}^N$.

We will call *maximal open set of existence* of f to the largest open set $\Omega_0 \subset G$ where f is continuous (equivalently, this is the union of the open subsets of G where f is continuous).

This makes perfect sense since, if f is continuous in two open sets, it is also continuous in the union of these sets.

By definition, the *maximal domains of existence* of f will be the connected components of Ω_0, that is, the largest open connected subsets of G where f is continuous.

For example, if $G = \{(t, y) \in \mathbb{R}^2 : y \geq 2t^2, \ y \neq 2t^2 + 2\}$ and we have

$$f(t, y) = |2t - y|^{1/3}|2t + y| + |t|^{1/3} + \frac{y^2 - 3t^2 + 1}{y - 2t^2 - 2} + \sqrt{y - 2t^2} \quad \forall (t, y) \in G,$$

we see at once that the maximal open set of existence of f is

$$\Omega_0 = \{(t, y) \in \mathbb{R}^2 : y > 2t^2, \ y \neq 2t^2 + 2\}$$

and the maximal domains of existence are

$$D_1 = \{(t, y) \in \mathbb{R}^2 : 2t^2 < y < 2t^2 + 2\} \quad \text{and} \quad D_1 = \{(t, y) \in \mathbb{R}^2 : y > 2t^2 + 2\}.$$

On the other hand, we will call *maximal open set of existence and uniqueness* of f to the largest open set $\Omega_{00} \subset G$ where f is continuous and locally Lipschitz-continuous with respect to the variable y (in other words, the union of the open subsets of G where f satisfies these properties).

Finally, the *maximal domains of existence and uniqueness* of f will be the connected components of Ω_{00}.

For the previous f, the maximal open set of existence and uniqueness is

$$\Omega_{00} = \{(t, y) \in \mathbb{R}^2 : y > 2t^2, \ y \neq 2t^2 + 2, \ y \neq 2t\},$$

while the maximal domains of existence and uniqueness are

$$D_{01} = \{(t, y) \in \mathbb{R}^2 : 2t^2 < y < 2t\},$$
$$D_{02} = \{(t, y) \in \mathbb{R}^2 : y > 2t^2, 2t < y < 2t^2 + 2\} \quad \text{and} \quad D_{03} = D_2.$$

If $(t_0, y_0) \in \Omega_0$ (resp. $(t_0, y_0) \in \Omega_{00}$), then the corresponding CP possesses at least one maximal solution $\varphi : I \mapsto \mathbb{R}^N$ (resp., exactly one maximal solution).

Obviously, if D is one of the connected components of Ω_0 and $(t_0, y_0) \in D$, then the whole trajectory of φ is contained in D, that is,

$$(t, \varphi(t)) \in D \quad \forall t \in I.$$

The same happens if D is one of the maximal domains of existence and uniqueness.

E.4.1 Consider the ODE

$$y' = 1 + |(t - y)(t - y - 1)| + a\sqrt{3 - y} + bt^2|t|,$$

where $a, b \geq 0$. Do the following:

(1) Find the corresponding maximal domains of existence and uniqueness.
(2) When $a = b = 0$, compute the maximal solutions corresponding to the Cauchy data $(0, 1)$, $(0, -1/2)$, and $(0, -2)$.

E.4.2 Consider the CP, where

$$f(t, y) = \frac{y^2 + 2(y + t + 1) + \dfrac{1}{(t - 1)^2}}{2(1 - t)y - 2t}.$$

Then,

(1) Find the corresponding maximal domains of existence and uniqueness.
(2) For $(t_0, y_0) = (0, 1)$, compute the maximal solution, indicating the corresponding interval of definition.

E.4.3 Consider the CP

$$\begin{cases} y_1' = t \, \max\{y_1, y_2\} + e^{t^2} \\ y_2' = y_1 + y_2 \\ y_1(0) = y_2(0) = 1. \end{cases}$$

Find the interval of definition of the corresponding maximal solution.

E.4.4 Let $f \colon \mathbb{R}^2 \mapsto \mathbb{R}$ be given by $f(t, y) = t \sin y + \cos t$ for all $(t, y) \in \mathbb{R}^2$. Prove that, for every $(t_0, y_0) \in \mathbb{R}^2$, the interval of definition of the maximal solution to the corresponding CP is $\mathbb{R}$.

E.4.5 Find the maximal solution to the CP

$$\begin{cases} y' = \dfrac{-3t^2 - 4ty}{2y + 2t^2} \\ y(0) = 1. \end{cases}$$

E.4.6 Assume that $\Omega = \{(t, y) \in \mathbb{R}^2 : y > 0\}$ and $f : \Omega \mapsto \mathbb{R}$ is continuous and locally Lipschitz-continuous with respect to y in Ω and such that $0 \leq f(t, y) \leq M$ in Ω. For every $(t_0, y_0) \in \Omega$, we consider the corresponding CP.

(1) Indicate the possible negative half-trajectories of the maximal solutions to these problems.
(2) Can we have $\sup I(t_0, y_0) < +\infty$? Why?

E.4.7 Consider the CP

$$\begin{cases} y' = 1 + e^{-ty^2} \\ y(1) = 0. \end{cases}$$

Can we have $\sup I(1, 0) < +\infty$? Why?

E.4.8 Consider the CP

$$\begin{cases} y' = \cos\left(y^2 \log t\right) \\ y(1) = 2. \end{cases}$$

Which is the interval of definition of the corresponding maximal solution?

E.4.9 Assume that $\Omega = \{(t,y) \in \mathbb{R}^2 : t > 0,\ y > 0\}$ and $f(t,y) = \cos^2(\ln y) + \sin^2(ty) + 1$ for every $(t,y) \in \Omega$. Let us consider the CP

$$\begin{cases} y' = f(t,y) \\ y(3) = 2. \end{cases}$$

Let us denote by (α, β) the interval of definition of the maximal solution. Is β equal to $+\infty$? Is α equal to 0?

E.4.10 Consider the CP

$$\begin{cases} y' = \dfrac{1}{t}y + y^3 \\ y(1) = 1. \end{cases}$$

(1) Prove that, in this case, the function f is not globally Lipschitz-continuous with respect to y in the open set $(0, +\infty) \times \mathbb{R}$.

(2) Prove that this problem possesses exactly one maximal solution and compute explicitly this function and its interval of definition.

E.4.11 For every $\alpha \in [0,1]$, we consider the CP

$$\begin{cases} y' = y(y-1) \\ y(0) = \alpha. \end{cases}$$

Without solving explicitly the ODE, prove that there exists a unique maximal solution φ_α to this problem, defined in the whole $\mathbb{R}$, that satisfies $0 \leq \varphi_\alpha \leq 1$.

E.4.12 Assume that $\Omega = (0,2) \times \mathbb{R}$ and

$$f(t,y) = \frac{1}{t-2} \cos(y^2 \log t) \quad \forall (t,y) \in \Omega.$$

For every $(t_0, y_0) \in \Omega$, we consider the corresponding CP. Prove that there always exists a unique maximal solution and determine the corresponding interval $I(t_0, y_0)$.

E.4.13 Consider the ODE

$$y' = \frac{1}{t} \sin^2 \left(1 + \frac{1}{y} \right).$$

(1) Find the associated maximal domains of existence and uniqueness.

(2) Let us denote by $I(1,1)$ the interval of definition of the maximal solution to the CP corresponding to this ODE and the initial condition $y(1) = 1$. Prove that $\sup I(1,1) = +\infty$.

(3) Indicate how the associated negative half-trajectory $\tau^-(\varphi)$ can be.

E.4.14 Let f be given by

$$f(t, y) = \cos^2 \left(\log |y - 2| \right) + \frac{1}{1 + \cos^2 \left(\frac{t}{y-1} \right)}.$$

(1) Find the maximal domains of existence and uniqueness for the ODE $y' = f(t, y)$.

(2) Consider the CP corresponding to f and the initial condition $y(0) = 0$ and denote by (α, β) the interval of definition of the associated maximal solution. Is $\alpha = -\infty$? Is $\beta = +\infty$?

E.4.15 Prove Theorem 4.4.

Hint: First, note that, if $f(y_0) > 0$, there cannot exist $\bar{t} \in I$ such that $f(\varphi(\bar{t})) = 0$; also, deduce that (for the same reason) φ must be strictly increasing on the whole interval I and discuss whether β can be finite or not depending on the existence of critical points greater than y_0.

E.4.16 Consider the ODE

$$y''' = \frac{\sqrt{y - t}}{t^3} + \frac{\log y'}{t^3}.$$

Rewrite this equation as an equivalent ODS of the first order and find the corresponding maximal open sets and domains of existence and uniqueness.

E.4.17 Consider the ODE

$$y'' = \frac{\sqrt{y + y' - t}}{y^3 - t^3} + (y' - 2y)_+^{1/4}.$$

Rewrite this equation as an equivalent ODS of the first order and find the corresponding maximal open sets and domains of existence and uniqueness.

E.4.18 Find the maximal open sets and domains of existence and uniqueness for the following ODEs:

(1) $y' = y^{2/3} + \dfrac{\psi(x)\sqrt{1-t^2}}{t}$, $\quad \psi \in C^5(\mathbb{R})$.

(2) $y' = (6-t)^{1/2}(t^2+y^2-9)^{-1/2} + \log(9-y^2) + \psi(t)$, $\quad \psi \in C^3(\mathbb{R})$.

E.4.19 Find the maximal open sets and domains of existence and uniqueness for the following ODEs and ODSs:

(1) $y_1' = (9 - t^2 - y_1^2 - y_2^2)^{1/2}$, $\quad y_2' = \dfrac{1}{t}$.

(2) $y_1' = |t-1|^3 \ln y_2$, $\quad y_2' = p(t)\dfrac{(2-t-t^2)^{1/2}}{t} + (y_1^2 + y_2^2 - 1)^{2/3}$,
with $p \in C^2(\mathbb{R})$.

(3) $y_1' = (9tx^2)^{1/2}\ln(1+y_2) + \max(\sin t, \cos t)$, $\quad y_2' = (4y_1^2 + 9y_2^2 - 36)^{2/3} + |t|^4 f(y_1, y_2)$, with $f \in C^5(\mathbb{R}_+ \times \mathbb{R}_+)$.

(4) $y'' = \dfrac{\ln(2-t)}{t+1} + e^{|t|}y + \displaystyle\int_1^t f(s)\, ds$, with $f \in C^2(\mathbb{R})$.

(5) $y''' = \dfrac{3f(x, y')}{y'' + 1} + 8y^{1/2} + |t|^5$, with $f \in C^5(\mathbb{R}_+ \times \mathbb{R}_+)$.

E.4.20 Consider the ODE

$$y'' = \frac{1}{1 + 2\log y'}.$$

(1) Find the corresponding maximal open sets and domains of existence and uniqueness.

(2) Let us denote by (α, β) the interval of definition of the maximal solution to the CP corresponding to the initial conditions $y(0) = 0$, $y'(0) = 1$. Find β.

E.4.21 Consider the ODE

$$y' = \frac{2t + y^2}{2 - 2ty}.$$

(1) Find the corresponding maximal open sets and domains of existence and uniqueness.

(2) Find the maximal solution to the CP corresponding to the initial condition $y(0) = 0$, indicating the interval where it is defined.

E.4.22 Consider the ODE

$$y' = \frac{y}{t} + \sin y$$

in $\Omega = (0, +\infty) \times \mathbb{R}$. Prove that, for every $(t_0, y_0) \in \Omega$, the interval of definition of the maximal solution to the CP corresponding to (t_0, y_0) is $(0, +\infty)$.

E.4.23 Consider the ODE

$$y'' = \frac{2y - ty'}{\log(t-1)} + \frac{y^2 + tyy'}{1 + y^4 + (y')^4}$$

in $\Omega = (1, +\infty) \times \mathbb{R} \times \mathbb{R}$. Prove that, for every $(t_0, y_0^{(0)}, y_0^{(1)}) \in \Omega$, the interval of definition of the maximal solution to the corresponding CP is $(1, +\infty)$.

Chapter 5

Cauchy Problems and Linear Systems

This chapter is devoted to the analysis of initial-value problems for linear ODEs and ODSs.

In the sequel, we will denote by $\mathcal{L}(\mathbb{R}^N)$ the linear algebra of real square matrices of order N.

We will consider systems of the form

$$y' = A(t)y + b(t), \tag{5.1}$$

where $A : I \mapsto \mathcal{L}(\mathbb{R}^N)$ and $b : I \mapsto \mathbb{R}^N$ are continuous functions ($I \subset \mathbb{R}$ is an interval).

It will be said that the previous ODS is *linear*. If $b \equiv 0$, we will say that it is a *homogeneous linear ODS*:

$$y' = A(t)y. \tag{5.2}$$

We will also consider in this chapter ODEs like the following:

$$y^{(n)} = a_{n-1}(t)y^{(n-1)} + \cdots + a_1(t)y' + a_0(t)y + b(t), \tag{5.3}$$

where $a_0, a_1, \ldots, a_{n-1}$ and b are continuous real-valued functions in the interval I.

It will be said that (5.3) is a *linear* equation of order n; again, if $b \equiv 0$, this is a *homogeneous linear* ODE of order n.

There exist at least two important reasons to carry out a particularized study of linear ODEs and ODSs:

- First, when an ODE or an ODS is linear, many things can be said about existence, uniqueness, specific properties, etc; many more things than in the general case.

- Also, linear ODEs and ODSs appear "at first approximation" when we try to describe or "model" phenomena of many kinds; in this respect, see Note 11.1.5.

5.1 The Existence and Uniqueness of a Solution

In the following sections, we will use two particular norms in $\mathcal{L}(\mathbb{R}^N)$ that permit to simplify the computations.

The first of them is the *spectral norm*, that is, the matrix norm induced by the Euclidean norm in $\mathbb{R}^N$. It is defined as follows:

$$\|A\|_S := \sup_{z \in \mathbb{R}^N \setminus \{0\}} \frac{|Az|}{|z|} \qquad \forall A \in \mathcal{L}(\mathbb{R}^N). \tag{5.4}$$

The second one is the *maximum norm*, given by

$$\|A\|_\infty := \max_{i,j} |a_{i,j}| \qquad \forall A = \{a_{i,j}\} \in \mathcal{L}(\mathbb{R}^N). \tag{5.5}$$

In the sequel, we will denote by $\mathcal{L}(\mathbb{R}^N)$ indistinctly any of the associated normed spaces.

Obviously, since $\mathcal{L}(\mathbb{R}^N)$ is finite dimensional, the norms $\|\cdot\|_S$ and $\|\cdot\|_\infty$ are equivalent (and also equivalent to any other norm in $\mathcal{L}(\mathbb{R}^N)$).[1]

Let $I \subset \mathbb{R}$ be an interval and assume that $(t_0, y_0) \in I \times \mathbb{R}^N$. We will consider the following Cauchy problem for (5.1):

$$\begin{cases} y' = A(t)y + b(t) \\ y(t_0) = y_0. \end{cases} \tag{5.6}$$

Strictly speaking, we are a little outside of the general framework established in Chapter 4, since we are accepting a situation where t_0 is an endpoint of I. Nevertheless, many previous results still hold for (5.6), as will be shown in what follows.

Our first task will be to establish the existence and uniqueness of a global solution to (5.6). Specifically, the following result holds:

Theorem 5.1. *Let $I \subset \mathbb{R}$ be an interval, let $A : I \mapsto \mathcal{L}(\mathbb{R}^N)$ and $b : I \mapsto \mathbb{R}^N$ be two continuous functions and assume that $(t_0, y_0) \in I \times \mathbb{R}^N$. There exists exactly one solution to (5.6), that is defined in the whole interval I.*

---

[1]Furthermore, it is not difficult to see that $\|\cdot\|_S$ satisfies

$$\|A \cdot B\|_S \le \|A\|_S \|B\|_S \qquad \forall A, B \in \mathcal{L}(\mathbb{R}^N).$$

Proof. If I is open, this result is an immediate consequence of Corollary 4.1.

Indeed, let us set

$$f(t, y) := A(t)y + b(t) \quad \forall I \times \mathbb{R}^N.$$

Obviously, $f : I \times \mathbb{R}^N \mapsto \mathbb{R}^N$ is continuous and locally Lipschitz-continuous with respect to y in $I \times \mathbb{R}^N$.

Let us assume that $I = (\alpha, \beta)$ and let α' and β' be such that $\alpha < \alpha' < t_0 < \beta' < \beta$. Then, for any $(t, y^1), (t, y^2) \in (\alpha', \beta') \times \mathbb{R}^N$, one has

$$\begin{aligned}
|f(t, y^1) - f(t, y^2)| &= |A(t)(y^1 - y^2)| \\
&\leq \|A(t)\|_S \, |y^1 - y^2| \\
&\leq \left(\max_{s \in [\alpha', \beta']} \|A(s)\|_S \right) |y^1 - y^2|.
\end{aligned}$$

Consequently, f is globally Lipschitz-continuous with respect to y in $(\alpha', \beta') \times \mathbb{R}^N$.

Since α' and β' are arbitrary, in view of Corollary 4.1, we see that the maximal solution to (5.6) is defined in the whole I.

This proves the theorem when I is open.

Let us see that the result is also true when I contains an endpoint.

To fix ideas, assume that (for example) I contains its supremum β but not its infimum α. Let $\tilde{A}$ and $\tilde{b}$ be extensions of A and b to the open interval $\tilde{I} := I \cup (\beta, \beta + 1)$, with $\tilde{A} \in C^0(\tilde{I}; \mathcal{L}(\mathbb{R}^N))$ and $\tilde{b} \in C^0(\tilde{I}; \mathbb{R}^N)$.

Consider the auxiliary Cauchy problem

$$\begin{cases} y' = \tilde{A}(t)y + \tilde{b}(t) \\ y(t_0) = y_0. \end{cases} \tag{5.7}$$

Again, thanks to Corollary 4.1, we know that (5.7) possesses exactly one maximal solution $\tilde{\varphi}$ defined in $\tilde{I}$. The restriction of $\tilde{\varphi}$ to I is a solution to (5.6) in this interval.

Consequently, (5.6) possesses at least one solution in I.

On the other hand, with independence of the properties of I, (5.6) possesses at most one solution in I. This is a consequence of Gronwall's Lemma (see Remark 3.2(c) and Exercise E.3.11).

Therefore, the result is also proved in this case. $\qquad \square$

Remark 5.1. Thus, we have seen that the solutions to a linear ODS "cannot blow up in finite time." For example, if $I = [0, +\infty)$ and $t_0 = 0$, the solution to (5.1) is defined for all $t > 0$. As we know, the situation can be completely different for nonlinear ODEs and ODSs. $\qquad \square$

From now on, unless otherwise specified, when we speak of solutions to (5.1), we will be referring to solutions defined in the whole interval I.

A crucial result is the following:

Theorem 5.2. *Let us set* $V := \{\varphi \in C^1(I;\mathbb{R}^N) : \varphi$ *is a solution to* (5.1)$\}$ *and* $V_0 := \{\varphi \in C^1(I;\mathbb{R}^N) : \varphi$ *as a solution to* (5.2)$\}$. *Then the following holds:*

(1) V_0 *is a linear subspace of* $C^1(I;\mathbb{R}^N)$ *of dimension* N.
(2) V *is a linear manifold of* $C^1(I;\mathbb{R}^N)$ *supported by* V_0. *That is, for any solution* φ_p *to* (5.1), *one has*

$$V = \varphi_p + V_0. \tag{5.8}$$

Proof. It is evident that V_0 is a nontrivial linear subspace of $C^1(I;\mathbb{R}^N)$.

It is also easy to check that V is a linear manifold with supporting space V_0; that is, (5.8) holds for any $\varphi_p \in V$.

Let us prove that V_0 is a linear space of dimension N.

Thus, let us fix $t_0 \in I$ and, for each $i = 1, \dots, N$, let us denote by φ^i the solution (in I) of the Cauchy problem

$$\begin{cases} y' = A(t)y \\ y(t_0) = e^i, \end{cases} \tag{5.9}$$

where e^i is the ith vector of the canonical basis of $\mathbb{R}^N$. Then, the following holds:

- If we had

$$\varphi := \alpha_1 \varphi^1 + \cdots + \alpha_N \varphi^N \equiv 0$$

  for some $\alpha_i \in \mathbb{R}$, we would also have in particular that $\varphi(t_0) = 0$, whence all $\alpha_i = 0$. This proves that the φ^i are linearly independent.
- On the other hand, if $\varphi \in V_0$ and $\varphi(t_0) = z_0$, then the components z_{0i} of z_0 satisfy

$$\varphi \equiv z_{01} \varphi^1 + \cdots + z_{0N} \varphi^N,$$

since φ and $z_{01} \varphi^1 + \cdots + z_{0N} \varphi^N$ solve the same Cauchy problem

$$\begin{cases} y' = A(t)y \\ y(t_0) = z_0. \end{cases}$$

This proves that the φ^i span the whole space V_0.

As a consequence, the N functions φ^i form a basis of V_0 and the proof is done. $\qquad\square$

Remark 5.2.

(a) This property (the linear structure of the family of solutions) serves to characterize linear ODSs.

More precisely, if $f : I \times \mathbb{R}^N \mapsto \mathbb{R}^N$ is continuous and locally Lipschitz-continuous with respect to y and the set of solutions in I of the system $y' = f(t, y)$ is a linear manifold, then f is necessarily of the form

$$f(t, y) = A(t)y + b(t) \quad \forall(t, y) \in I \times \mathbb{R}^N,$$

for some $A \in C^0(I; \mathcal{L}(\mathbb{R}^N))$ and $b \in C^0(I; \mathbb{R}^N)$. The proof of this assertion is left as an exercise to the reader.

(b) On the other hand, Theorem 5.2 indicates that linear ODSs are extremely interesting from the practical viewpoint.

The reason is that if we are able to compute N linearly independent solutions to (5.2) and one solution to (5.1), then we will be able to find *all the solutions* to (5.1). □

5.2 Homogeneous Linear Systems

In this section, we will consider the case in which $b \equiv 0$, that is, homogeneous linear ODSs. As already said, the solutions form a subspace of $C^1(I; \mathbb{R}^N)$ of dimension N.

Definition 5.1. A fundamental matrix of (5.2) is any function $F : I \mapsto \mathcal{L}(\mathbb{R}^N)$ satisfying the following:

(1) $F \in C^1(I; \mathcal{L}(R^N))$ and $F'(t) = A(t)F(t)$ for all $t \in I$.
(2) $\det F(t) \neq 0$ for all $t \in I$. □

It is easy to see that a function $F : I \mapsto \mathcal{L}(\mathbb{R}^N)$ is a fundamental matrix of (5.2) if and only if its columns form a basis of V_0, that is, they provide N linearly independent solutions to (5.2).

Consequently, if the matrix function $A : I \mapsto \mathcal{L}(\mathbb{R}^N)$ in (5.2) is continuous, there always exist fundamental matrices.

Moreover, if F and G are fundamental matrices of (5.2) and coincide at some point in I, then $F \equiv G$.

If a fundamental matrix F of (5.2) is known, then the space of solutions is given by

$$V_0 = \{\varphi : \varphi = Fc, \ c \in \mathbb{R}^N\}.$$

This makes easy to compute the unique solution to the Cauchy problem

$$\begin{cases} y' = A(t)y \\ y(t_0) = y_0. \end{cases} \tag{5.10}$$

It is the function φ, with

$$\varphi(t) = F(t) \cdot F(t_0)^{-1} y_0 \quad \forall t \in I. \tag{5.11}$$

Let us enumerate (and prove) some important properties of fundamental matrices:

- Let $F \in C^1(I; \mathcal{L}(R^N))$ be given and let us assume that $F'(t) = A(t)F(t)$ for every $t \in I$ and, also, that there exists $t_0 \in I$ such that $\det F(t_0) \neq 0$. Then F is a fundamental matrix of (5.2).

Indeed, it is clear that the columns of F provide N solutions to (5.2). Let them be denoted by $\varphi^1, \ldots, \varphi^N$. If the linear combination $\alpha_1 \varphi^1 + \cdots + \alpha_N \varphi^N$ vanishes at some $t_1 \in I$, then $\alpha_1 \varphi^1 + \cdots + \alpha_N \varphi^N$ is a solution to the Cauchy problem

$$\begin{cases} y' = A(t)y \\ y(t_1) = 0, \end{cases}$$

whence $\alpha_1 \varphi^1 + \cdots + \alpha_N \varphi^N \equiv 0$ and, in particular, $\alpha_1 \varphi^1(t_0) + \cdots + \alpha_N \varphi^N(t_0) = 0$.

But by assumption this necessarily implies $\alpha_1 = \cdots = \alpha_N = 0$.

Therefore, the columns $\varphi^1(t_1), \ldots, \varphi^N(t_1)$ are linearly independent and, since t_1 is arbitrary in I, we deduce that the φ_i form a basis of V_0 and F is a fundamental matrix.

- Again, assume that $F \in C^1(I; \mathcal{L}(R^N))$ and $F'(t) = A(t)F(t)$ for all $t \in I$. Then we have *Jacobi–Liouville's formula*, as follows:

$$\det F(t) = \det F(t_0) \exp\left(\int_{t_0}^{t} \operatorname{tr} A(s)\,\mathrm{d}s \right) \quad \forall t, t_0 \in I. \tag{5.12}$$

This is an immediate consequence of the identities

$$\frac{\mathrm{d}}{\mathrm{d}t} \det F(t) = (\operatorname{tr} A(t)) \det F(t) \quad \forall t \in I, \tag{5.13}$$

that show that $t \mapsto \det F(t)$ solves a homogeneous ODE.

The proof of (5.13) can be easily obtained by considering the rows of F and writing that the derivative of F is the sum of N determinants, each of them corresponding to the derivatives of one row.

- If F is a fundamental matrix and $P \in \mathcal{L}(\mathbb{R}^N)$ is non-singular, then $G :=$ $F \cdot P$ is a new fundamental matrix.

  Indeed, $G : I \mapsto \mathcal{L}(\mathbb{R}^N)$ is continuously differentiable, $G'(t) = A(t)G(t)$ for every $t \in I$, and $\det G(t) \neq 0$ for all t.
- Conversely, if F and G are fundamental matrices, there exists exactly one non-singular $P \in \mathcal{L}(\mathbb{R}^N)$ such that $G(t) = F(t) \cdot P$ for all t.

The proof is as follows.

Let $t_0 \in I$ be given. We know that $F(t_0)$ and $G(t_0)$ are non-singular matrices; let P be the matrix $P = F(t_0)^{-1} \cdot G(t_0)$. Then G and $F \cdot P$ are two fundamental matrices that coincide at t_0. Therefore, they coincide in the whole I and, certainly, $G = F \cdot P$.

Example 5.1.

(a) Take $I = (0, +\infty)$ and consider the homogeneous linear ODS

$$\begin{cases} y_1' = -y_1 + \dfrac{1}{t^2}\, y_2 \\[2mm] y_2' = \dfrac{2}{t}\, y_2. \end{cases}$$

It is not difficult to compute its solutions: first, we solve the second ODE (that is uncoupled), we then replace in the first equation and we finally solve this one. As a result, we obtain the family of solutions $\varphi(\cdot\,; C_1, C_2)$, with

$$\varphi(t; C_1, C_2) = (C_1 \mathrm{e}^{-t} + C_2, C_2 t^2) \quad \forall t \in (0, +\infty).$$

Consequently, a fundamental matrix is the following function:

$$F(t) = \begin{pmatrix} \mathrm{e}^{-t} & 1 \\ 0 & t^2 \end{pmatrix} \quad \forall t \in (0, +\infty). \tag{5.14}$$

(b) Now, let us take $I = (\pi/4, 3\pi/4)$ and let us consider the system

$$\begin{cases} z_1' = z_2 \\ z_2' = -z_1 \\ z_3' = z_3 \frac{\cos t}{\sin t} + z_1 + z_2. \end{cases}$$

It is again easy in this case to compute the solutions: first, we observe that from the first and second ODE, one has $z_1'' + z_1 = 0$; using this, we compute

z_1 and z_2, we replace in the third ODE and we finally solve this one. The following family is obtained:

$$\psi(t; C_1, C_2) = \begin{pmatrix} C_1 \cos t + C_2 \sin t \\ C_2 \cos t - C_1 \sin t \\ ((C_1 + C_2) \log|\sin t| + (C_2 - C_1)t + C_3) \sin t \end{pmatrix} \quad \forall t \in I.$$

Consequently, a fundamental matrix is given by

$$F(t) = \begin{pmatrix} \cos t & \sin t & 0 \\ -\sin t & \cos t & 0 \\ (\log|\sin t| - t)\sin t & (\log|\sin t| + t)\sin t & \sin t \end{pmatrix} \quad \forall t \in I.$$

We see that $F(t)$ is well defined for all $t \in I$. □

5.3 Non-Homogeneous Linear Systems: Lagrange's Method

Let us consider again the system

$$y' = A(t)y + b(t). \tag{5.1}$$

We will see in this section that the solutions to (5.1) can be computed from the solutions to (5.2), that is, starting from a fundamental matrix.

The technique is known as *method of variations of constants* (or Lagrange's method). It was already presented in Chapter 1 in the particular cases of a first- and second-order linear ODE.

On the other hand, it possesses a lot of generalizations and applications to other contexts; see Note 11.5.1 for more details.

Theorem 5.3. *Let* $A : I \mapsto \mathcal{L}(\mathbb{R}^N)$ *and* $b : I \mapsto \mathbb{R}^N$ *be continuous functions. Let* $F : I \mapsto \mathcal{L}(\mathbb{R}^N)$ *be a fundamental matrix of the linear homogeneous system* (5.2). *Assume that* $t_0 \in I$ *and let us set*

$$\varphi_p(t) = \int_{t_0}^t F(t)F(s)^{-1}b(s)\mathrm{d}s \quad \forall t \in I. \tag{5.15}$$

Then φ_p *is a solution to* (5.1). *Moreover, any other solution is given by* $\varphi_p + Fc$, *where* $c \in \mathbb{R}^N$.

Proof. In order to prove this result, we will try to find a function $h \in C^1(I; \mathbb{R}^N)$ such that the corresponding φ, given by

$$\varphi(t) = F(t) h(t) \quad \forall t \in I, \tag{5.16}$$

solves (5.1).[2]

By imposing (5.1) to the function φ, we easily find that

$$F(t)h'(t) = b(t) \quad \forall t \in I,$$

whence $h' = F^{-1}b$ and h is a primitive of $F^{-1}b$.

In other words, $\varphi = F h$ is a solution to (5.1) if and only if there exists $c \in \mathbb{R}^N$ such that

$$h(t) = \int_{t_0}^t F(s)^{-1}b(s)\,\mathrm{d}s + c \quad \forall t \in I.$$

Consequently, the function φ_p in (5.15) is certainly a solution to (5.1) and any other solution is given by $\varphi_p + Fc$ for some c. $\quad\square$

An immediate consequence of this result is that we can express the solution to the Cauchy problem (5.6) in terms of a fundamental matrix. More precisely, the following holds:

Corollary 5.1. *In the conditions of Theorem 5.3, the unique solution to* (5.6) *is given by*

$$\varphi(t) = F(t)F(t_0)^{-1}y_0 + \int_{t_0}^t F(t)F(s)^{-1}b(s)\,\mathrm{d}s \quad \forall t \in I. \tag{5.17}$$

Example 5.2.

(a) Consider the ODS

$$\begin{cases} y_1' = -y_1 + \dfrac{1}{t^2} y_2 \\ y_2' = \dfrac{2}{t} y_2 + t^2. \end{cases} \tag{5.18}$$

The corresponding homogeneous linear system was analyzed and solved in Example 5.1(a).

---

[2]Remember that the solutions to (5.2) are the functions Fc, with $c \in \mathbb{R}^N$. The fact that we are looking for a solution to (5.1) of the form (5.16) explains that this method must be known as the *method of variations of constants*.

An associated fundamental matrix is given by (5.14). Hence, with the notation of Theorem 5.3 and taking $t_0 = 1$, we have

$$\varphi_p(t) = \begin{pmatrix} t - 2 + e^{-(t-1)} \\ t^2(t-1) \end{pmatrix} \qquad \forall t \in (0, +\infty)$$

and the solutions to (5.18) are the functions

$$\varphi(t; C_1, C_2) = \begin{pmatrix} C_1 e^{-t} + C_2 + e^{-(t-1)} + t - 2 \\ C_2 t^2 + t^2(t-1) \end{pmatrix} \qquad \forall t \in (0, +\infty).$$

(b) Let us set $I = (\pi/4, 3\pi/4)$ and let us consider the ODS

$$\begin{cases} z_1' = z_2 \\ z_2' = -z_1 \\ z_3' = z_3 \dfrac{\cos t}{\sin t} + z_1 + z_2 + \sin t \cos t. \end{cases} \qquad (5.19)$$

The associated homogeneous system was considered in Example 5.1(b). We found there a fundamental matrix.

For $t_0 = \pi/2$, using again Theorem 5.3, the following particular solution is found:

$$\varphi_p(t) = \begin{pmatrix} 0 \\ 0 \\ \sin t \, (\sin t - 1) \end{pmatrix} \qquad \forall t \in I.$$

Consequently, the solutions to (5.19) are the following:

$$\varphi(t; C_1, C_2, C_3)$$
$$= \begin{pmatrix} C_1 \cos t + C_2 \sin t \\ C_2 \cos t - C_1 \sin t \\ ((C_1 + C_2) \log |\sin t| + (C_2 - C_1)t + C_3 + \sin t) \sin t \end{pmatrix} \qquad \forall t \in I.$$

$\square$

5.4 The Exponential of a Matrix

In the particular case in which A is independent of t (that is, the coefficients of the system (5.2) are constant), an explicit fundamental matrix of (5.2) can be found.

To this purpose, it is convenient to speak before of convergent expansions in $\mathcal{L}(\mathbb{R}^N)$ and, in particular, of the exponential series.

Definition 5.2. Assume that the sequence $\{A_n\}$ is given in $\mathcal{L}(\mathbb{R}^N)$. The series of general term A_n is the sequence $\{S_n\}$, where S_n is for each n the so-called nth partial sum

$$S_n = \sum_{k=0}^{n} A_k.$$

□

It is customary to denote the series of general term A_n as follows:

$$\sum_{n \geq 0} A_n. \qquad (5.20)$$

It will be said that the series converges in $\mathcal{L}(\mathbb{R}^N)$ if the corresponding sequence $\{S_n\}$ converges in this normed space.

Obviously, this happens if and only if there exists $S \in \mathcal{L}(\mathbb{R}^N)$ such that, for every $\varepsilon > 0$, there exists n_0 with the following property: if $n \geq n_0$, then

$$\left\| S - \sum_{k=0}^{n} A_k \right\|_{\infty} \leq \varepsilon.$$

Of course, this property is independent of the norm considered in $\mathcal{L}(\mathbb{R}^N)$.

Usually, we say that S is the *sum* of the series and we write that $S = \sum_{n \geq 0} A_n$.[3]

It is immediate that the series $\sum_{n \geq 0} A_n$ converges in $\mathcal{L}(\mathbb{R}^N)$ if and only if each of the $N \times N$ scalar series that can be obtained using the components of the A_n converges in $\mathbb{R}$.

Consequently, if $\sum_{n \geq 0} A_n$ and $\sum_{n \geq 0} B_n$ converge, then the same is true for the series $\sum_{n \geq 0}(A_n + B_n)$ and, in that case, one has

$$\sum_{n \geq 0}(A_n + B_n) = \sum_{n \geq 0} A_n + \sum_{n \geq 0} B_n.$$

Moreover, if $\sum_{n \geq 0} A_n$ converges and $B \in \mathcal{L}(\mathbb{R}^N)$, the series

$$\sum_{n \geq 0} B \cdot A_n \quad \text{and} \quad \sum_{n \geq 0} A_n \cdot B$$

also converge and one has

$$\sum_{n \geq 0} B \cdot A_n = B \cdot \left(\sum_{n \geq 0} A_n \right) \quad \text{and} \quad \left(\sum_{n \geq 0} A_n \right) \cdot B = \sum_{n \geq 0} A_n \cdot B.$$

For the convergence of a series in $\mathcal{L}(\mathbb{R}^N)$, a well-known necessary and sufficient condition, *the Cauchy's criterion*, is many times useful.

---

[3]To some extent, we perceive some ambiguity in the notation. Thus, $\sum_{n \geq 0} A_n$ denotes at the same time the series and, if it exists, its sum.

It reads as follows:

The series $\sum_{n\geq 0} A_n$ converges if and only if for every $\varepsilon > 0$ there exists n_0 such that, if $m > n \geq n_0$, then $\|\sum_{k=n+1}^{m} A_k\|_\infty \leq \varepsilon$.

Definition 5.3. It is said that $\sum_{n\geq 0} A_n$ is absolutely convergent if the series of positive terms $\sum_{n\geq 0} \|A_n\|_\infty$ converges. $\square$

Also, in this definition, we could have used any other norm in $\mathcal{L}(\mathbb{R}^N)$.

It is clear that a series in $\mathcal{L}(\mathbb{R}^N)$ is absolutely convergent if and only if all the $N \times N$ associated scalar series are absolutely convergent (in $\mathbb{R}$).

Therefore, an absolutely convergent series is convergent.

Furthermore, for any absolutely convergent series, one has the *generalized triangular inequality*:

$$\left\| \sum_{n\geq 0} A_n \right\|_\infty \leq \sum_{n\geq 0} \|A_n\|_\infty.$$

Finally, observe that the series $\sum_{n\geq 0} A_n$ is absolutely convergent if and only if the following Cauchy condition is satisfied:

For each $\varepsilon > 0$, there exists n_0 such that, if $m > n \geq n_0$, then $\sum_{k=n+1}^{m} \|A_k\|_\infty \leq \varepsilon$.

Example 5.3. Let $a, b > 0$ be given and assume that

$$A_n := \begin{pmatrix} (-1)^n n^{-a} & 2(-1)^n(n^{-b} - n^{-a}) \\ & (-1)^n n^{-b} \end{pmatrix} \quad \forall n \geq 1.$$

Then, the corresponding series $\sum_{n\geq 0} A_n$ is

(1) Not convergent if $a < 1$ or $b < 1$.
(2) Absolutely convergent if $a > 1$ and $b > 1$.
(3) Convergent (but not absolutely) in the remaining cases.

All this can be easily checked by looking at the scalar series obtained from the components of A_n. $\square$

We will be interested by some particular functions Φ defined in a subset of $\mathcal{L}(\mathbb{R}^N)$ and with values in $\mathcal{L}(\mathbb{R}^N)$, associated to power series. They can be introduced as follows.

Let $\{\alpha_n\}$ be a sequence in $\mathbb{R}$. Let us assume that the corresponding power series $\sum_{n\geq 0} \alpha_n z^n$ possesses a convergence radius $\rho > 0$ and let

$\Phi : (-\rho, \rho) \mapsto \mathbb{R}$ be given by

$$\Phi(z) := \sum_{n \geq 0} \alpha_n z^n \quad \forall z \in (-\rho, \rho)$$

(here, we can eventually have $\rho = +\infty$).

On the other hand, let $A \in \mathcal{L}(\mathbb{R}^N)$ be given, with $\|A\|_S < \rho$. Then, the series $\sum_{n \geq 0} \alpha_n A^n$ converges absolutely in $\mathcal{L}(\mathbb{R}^N)$.

Indeed, it suffices to take into account that, for all $n \geq 0$,

$$\|\alpha_n A^n\|_S \leq |\alpha_n| \, \|A\|_S^n$$

and apply Cauchy's criterion.

Therefore, we can speak of the sum of the series with general term $\alpha_n A^n$. By definition, we set

$$\Phi(A) := \sum_{n \geq 0} \alpha_n A^n.$$

Observe that Φ is well defined in the set

$$\mathcal{P} := \{A \in \mathcal{L}(\mathbb{R}^N) : \|A\|_S < \rho\},$$

that is an open ball in $\mathcal{L}(\mathbb{R}^N)$.

Example 5.4.

(a) Let us take

$$\Phi(z) = e^z \equiv \sum_{n \geq 0} \frac{1}{n!} z^n \quad \forall z \in \mathbb{R}.$$

We are in the previous situation, with $\rho = +\infty$. Accordingly, for any $A \in \mathcal{L}(\mathbb{R}^N)$, we put

$$e^A := \sum_{n \geq 0} \frac{1}{n!} A^n = \mathrm{Id} + A + \frac{1}{2!} A^2 + \cdots .$$

It will be said that e^A is the *exponential of* A.

In this case, $\mathcal{P} = \mathcal{L}(\mathbb{R}^N)$, that is, the function $A \mapsto e^A$ is well defined in the whole space $\mathcal{L}(\mathbb{R}^N)$.

It will be seen very soon that this function plays a fundamental role for the computation of fundamental matrices of linear ODSs when the coefficients are constant.

For the moment, let us just observe the following properties:

$$e^0 = \text{Id} \quad \text{and} \quad e^{\lambda \, \text{Id}} = e^\lambda \, \text{Id} \quad \forall \lambda \in \mathbb{R}.$$

(b) In a similar way, for any $A \in \mathcal{L}(\mathbb{R}^N)$, we put

$$\sin A := \sum_{n \geq 0} \frac{(-1)^n}{(2n+1)!} A^{2n+1}, \quad \cos A := \sum_{n \geq 0} \frac{(-1)^n}{(2n)!} A^{2n}.$$

It is not difficult to check that

$$(\sin A)^2 + (\cos A)^2 = \text{Id}.$$

(c) Now, let us take

$$\Phi(z) = (1 - z)^{-1} \equiv \sum_{n \geq 0} z^n \quad \forall z \in (-1, 1).$$

We are again in the previous situation with $\rho = 1$. Thus, for every $A \in \mathcal{L}(\mathbb{R}^N)$ with $\|A\|_S < 1$, we will write

$$\Phi(A) := \sum_{n \geq 0} A^n.$$

It is easy to check that, if $\|A\|_S < 1$, the matrix $\text{Id} - A$ is non-singular and, moreover, $\Phi(A) = (\text{Id} - A)^{-1}$.

In other words, the two possible definitions of $(\text{Id} - A)^{-1}$ coincide.

(d) Finally, let us set

$$\Phi(z) = \log(1 + z) \equiv \sum_{n \geq 1} \frac{(-1)^{n+1}}{n} z^n \quad \forall z \in (-1, 1).$$

Once more, we are in the good conditions with $\rho = 1$. Accordingly, for every $A \in \mathcal{L}(\mathbb{R}^N)$ with $\|A\|_S < 1$, we will put

$$\log(\text{Id} + A) := \sum_{n \geq 1} \frac{(-1)^{n+1}}{n} A^n = A - \frac{1}{2} A^2 + \frac{1}{3} A^3 - \cdots .$$

$\square$

In the following result, we state and prove an important property of the matrix exponential that will be needed in what follows:

Proposition 5.1. *Let* $A, B \in \mathcal{L}(\mathbb{R}^N)$ *be given, with* $A \cdot B = B \cdot A$. *Then one has*

(1) $A \cdot e^B = e^B \cdot A$.

(2) $e^{A+B} = e^A \cdot e^B = e^B \cdot e^A$.

Proof. Let us first check that, under the assumptions of the theorem, A and e^B commute.

Indeed, one has

$$A \cdot e^B = \sum_{n \geq 0} \frac{1}{n!} A \cdot B^n = \sum_{n \geq 0} \frac{1}{n!} B^n \cdot A = e^B \cdot A.$$

Let us now see that $e^{A+B} = e^A \cdot e^B$. We have

$$\|e^A \cdot e^B - e^{A+B}\|_S = \lim_{k \to +\infty} \|M_k\|_S, \tag{5.21}$$

where

$$M_k = \left(\sum_{n=0}^{k} \frac{1}{n!} A^n \right) \left(\sum_{n=0}^{k} \frac{1}{n!} B^n \right) - \sum_{n=0}^{k} \frac{1}{n!} (A + B)^n$$

for every k.

On the other hand,

$$e^{\|A\|_S} \cdot e^{\|B\|_S} - e^{\|A\|_S + \|B\|_S} = \lim_{k \to +\infty} m_k,$$

where

$$m_k = \left(\sum_{n=0}^{k} \frac{1}{n!} \|A\|_S^n \right) \left(\sum_{n=0}^{k} \frac{1}{n!} \|B\|_S^n \right) - \sum_{n=0}^{k} \frac{1}{n!} (\|A\|_S + \|B\|_S)^n$$

for all k. But m_k can also be written as a finite sum of powers of the form $\|A\|_S^i \|B\|_S^j$ with $0 \leq i, j \leq k$, where all the coefficients are nonnegative:

$$m_k = \sum_{i,j=0}^{k} m_{i,j}^k \|A\|_S^i \|B\|_S^j, \quad \text{with} \quad m_{i,j}^k \geq 0.$$

Since A and B commute, M_k can also be written as a similar sum of powers $A^i B^j$, as follows:

$$M_k = \sum_{i,j=0}^{k} m_{i,j}^k A^i B^j.$$

We deduce that

$$\|M_k\|_S \leq \sum_{i,j=0}^{k} m_{i,j}^k \|A\|_S^i \|B\|_S^j = m_k \quad \forall k \geq 0$$

and, taking into account that $m_k \to 0$ as $k \to +\infty$, we see that the right-hand side of (5.21) is 0.

This ends the proof. $\qquad \square$

An immediate consequence of this result is that e^A is always a non-singular matrix. In fact, for every $A \in \mathcal{L}(\mathbb{R}^N)$, one has

$$\mathrm{Id} = e^0 = e^{A-A} = e^A \cdot e^{-A} = e^{-A} \cdot e^A,$$

whence we also get that $\left(e^A\right)^{-1} = e^{-A}$.

It is also true that, if $e^{A+B} = e^A \cdot e^B$, the matrices A and B necessarily commute, see Rouché and Mawhin (1973).

5.5 Systems with Constant Coefficients

Let us consider the following linear homogeneous ODS where the coefficients are constant:

$$y' = Ay, \tag{5.22}$$

with $A \in \mathcal{L}(\mathbb{R}^N)$.

Theorem 5.4. *Let us set*

$$F(t) = e^{tA} \quad \forall t \in \mathbb{R}.$$

Then F is a fundamental matrix of (5.22).

Proof. We have to prove that $F : \mathbb{R} \mapsto \mathcal{L}(\mathbb{R}^N)$ is continuously differentiable and

$$F'(t) = AF(t) \quad \forall t \in \mathbb{R}. \tag{5.23}$$

Since $F(t)$ is always non-singular, this will suffice.

Let us first prove that F is continuous.

It will be sufficient to check that F is continuous at zero. This is a consequence of the following identity:

$$F(t + h) - F(t) = F(t)\left(F(h) - F(0)\right) = e^{tA}(e^{hA} - \mathrm{Id}), \tag{5.24}$$

that holds for all $t, h \in \mathbb{R}$.

In order to prove the continuity at 0, we observe that

$$\|F(h) - F(0)\| = \|e^{hA} - \mathrm{Id}\| = \lim_{k \to +\infty} \left\| \sum_{n=0}^{k} \frac{h^n}{n!} A^n - \mathrm{Id} \right\|$$

and

$$\left\| \sum_{n=0}^{k} \frac{h^n}{n!} A^n - \mathrm{Id} \right\| \leq \sum_{n=1}^{k} \frac{(|h|\,\|A\|)^n}{n!} = \sum_{n=0}^{k} \frac{(|h|\,\|A\|)^n}{n!} - 1,$$

whence we get

$$\|e^{hA} - \mathrm{Id}\| \le \lim_{k \to +\infty} \left(\sum_{n=0}^{k} \frac{(|h|\,\|A\|)^n}{n!} - 1 \right) = e^{|h|\,\|A\|} - 1.$$

But this last quantity goes to 0 as $h \to 0$. Consequently, $F(h)$ possesses a limit as $h \to 0$ and $\lim_{h\to 0} F(h) = F(0)$.

Let us now see that F is also differentiable at any $t \in \mathbb{R}$ and (5.23) holds, that is,

$$\lim_{h \to 0} \left(\frac{1}{h}(F(t+h) - F(t)) - AF(t) \right) = 0. \tag{5.25}$$

Since

$$\frac{1}{h}(F(t+h) - F(t)) - AF(t) = F(t)\left(\frac{1}{h}(F(h) - F(0)) - AF(0) \right)$$

$$= e^{tA}\left(\frac{1}{h}\left(e^{hA} - \mathrm{Id} \right) - A \right)$$

it will be again sufficient to prove (5.25) at $t = 0$. But

$$\left\| \frac{1}{h}(F(h) - F(0)) - AF(0) \right\| = \lim_{k \to +\infty} \left\| \frac{1}{h}\left(\sum_{n=0}^{k} \frac{h^n}{n!} A^n - \mathrm{Id} \right) - A \right\|$$

$$= \lim_{k \to +\infty} \left\| \sum_{n=2}^{k} \frac{h^{n-1}}{n!} A^n \right\| \le \lim_{k \to +\infty} \sum_{n=2}^{k} \frac{|h|^{n-1}\|A\|^n}{n!}.$$

On the other hand,

$$\frac{1}{|h|}\left(e^{|h|\,\|A\|} - 1 \right) - \|A\|$$

$$= \lim_{k \to +\infty} \left(\frac{1}{|h|}\left[\sum_{n=0}^{k} \frac{(|h|\,\|A\|)^n}{n!} - 1 \right] - \|A\| \right)$$

$$= \lim_{k \to +\infty} \sum_{n=2}^{k} \frac{|h|^{n-1}\|A\|^n}{n!}$$

and this last quantity goes to 0 as $h \to 0$, since the function $t \mapsto e^{t\|A\|}$ is differentiable at 0, with a derivative equal to $\|A\|$.

Therefore,

$$\lim_{h \to 0} \left\| \frac{1}{h}(F(h) - F(0)) - AF(0) \right\| \le \lim_{h \to 0} \frac{1}{|h|}\left(e^{|h|\,\|A\|} - 1 \right) - \|A\| = 0.$$

This yields (5.25) and ends the proof. $\qquad\square$

In view of this result, it seems interesting to indicate a method to compute in practice the exponential of a matrix.

This will be the objective of the next paragraphs.

In the sequel, for any $\lambda \in \mathbb{R}$ and any integer $d \geq 1$, we will denote by $J(\lambda; d)$ the d-dimensional matrix

$$
J(\lambda; d) := \begin{pmatrix}
\lambda & 1 & & & \\
& \lambda & 1 & & \\
& & \ddots & \ddots & \\
& & & \ddots & 1 \\
& & & & \lambda
\end{pmatrix}
$$

(this is the *Jordan block* associated to λ).

On the other hand, for any $\alpha, \beta \in \mathbb{R}$ and any even integer $q \geq 2$, we will denote by $\tilde{J}(\alpha, \beta; q)$ the q-dimensional matrix

$$
\tilde{J}(\alpha, \beta; q) := \begin{pmatrix}
\Lambda & \mathrm{Id} & & & \\
& \Lambda & \mathrm{Id} & & \\
& & \ddots & \ddots & \\
& & & \ddots & \mathrm{Id} \\
& & & & \Lambda
\end{pmatrix}, \quad \text{where} \quad \Lambda := \begin{pmatrix} \alpha & \beta \\ -\beta & \alpha \end{pmatrix}
$$

(this is called the *real Jordan block* associated to α and β).

The following (real) version of *Jordan's Decomposition Theorem* will be used:

Theorem 5.5. *Let $A \in \mathcal{L}(\mathbb{R}^N)$ be given. There exist a non-singular matrix $P \in \mathcal{L}(\mathbb{R}^N)$ and real numbers*

$$
\lambda_1, \ldots, \lambda_m; \quad \alpha_1, \ldots, \alpha_p; \quad \beta_1, \ldots, \beta_p \quad (m + 2p \leq N)
$$

(unique up to the order) such that:

(1) *The eigenvalues of A are the λ_i and $\alpha_k \pm i\beta_k$.*
(2) *$A = P^{-1} \cdot \tilde{J} \cdot P$, where $\tilde{J}$ is the block-diagonal matrix with blocks*

$$
J_1, \ldots, J_m, \tilde{J}_1, \ldots, \tilde{J}_p
$$

and

$$
J_i := J(\lambda_i; d_i), \quad \tilde{J}_k := \tilde{J}(\alpha_k, \beta_k; 2r_k)
$$

for some integers d_i and r_k satisfying $d_1 + \cdots + d_m + 2r_1 + \cdots + 2r_p = N$.

For the proof of Theorem 5.5, see for example Finkbeiner (1966).

It is usually said that $\tilde{J}$ is the *real Jordan canonical form* of A.

Obviously, this is a matrix similar to A that, in some sense, possesses the most simple structure.

It is also usual to say that P is the corresponding *transition* (or change of basis) matrix.

Theorem 5.6. *Conserving the notation introduced in Theorem 5.5, for every $t \in \mathbb{R}$, one has:*

(1) $e^{tA} = P^{-1} \cdot e^{t\tilde{J}} \cdot P$.

(2) $e^{t\tilde{J}}$ *is the block-diagonal matrix of blocks* $e^{tJ_1}, \ldots, e^{tJ_m}, e^{t\tilde{J}_1}, \ldots, e^{t\tilde{J}_p}$.

(3) $e^{tJ_i} = e^{\lambda_i t} K_i$, *where K_i is the d_i-dimensional matrix*

$$
K_i := \begin{pmatrix}
1 & t/1! & t^2/2! & \cdots & t^{(d_i-1)}/(d_i-1)! \\
 & 1 & t/1! & \cdots & t^{(d_i-2)}/(d_i-2)! \\
 & & \ddots & \ddots & \vdots \\
 & & & \ddots & t/1! \\
 & & & & 1
\end{pmatrix}
$$

(4) *Finally,* $e^{t\tilde{J}_k} = e^{\alpha_k t} H_k$, *where H_k is the $2r_k$-dimensional matrix*

$$
H_k := \begin{pmatrix}
\Gamma_k & t/1!\,\Gamma_k & t^2/2!\,\Gamma_k & \cdots & t^{(r_k-1)}/(r_k-1)!\,\Gamma_k \\
 & \Gamma_k & t/1!\,\Gamma_k & \cdots & t^{(r_k-2)}/(r_k-2)!\,\Gamma_k \\
 & & \ddots & \ddots & \vdots \\
 & & & \ddots & t/1!\,\Gamma_k \\
 & & & & \Gamma_k
\end{pmatrix}
$$

and

$$
\Gamma_k := \begin{pmatrix}
\cos(\beta_k t) & \sin(\beta_k t) \\
-\sin(\beta_k t) & \cos(\beta_k t)
\end{pmatrix}.
$$

Proof. The two first assertions are immediate consequences of the definition of the exponential of a matrix.

Indeed, it suffices to take into account that, for each $n \geq 0$, $A^n = P^{-1} \cdot \tilde{J}^n \cdot P$ and $\tilde{J}^n$ is the block-diagonal matrix of blocks $J_1^n, \ldots, J_m^n, \tilde{J}_1^n, \ldots, \tilde{J}_p^n$.

Let us see that, for every λ and every t, $\mathrm{e}^{tJ(\lambda,d)} = \mathrm{e}^{\lambda t}K$, where K is the $d \times d$ matrix

$$K = \begin{pmatrix} 1 & t/1! & t^2/2! & \cdots & t^{(d-1)}/(d-1)! \\ & 1 & t/1! & \cdots & t^{(d-2)}/(d-2)! \\ & & \ddots & \ddots & \vdots \\ & & & \ddots & t/1! \\ & & & & 1 \end{pmatrix}. \qquad (5.26)$$

To this end, let us introduce $U := J(\lambda,d) - \lambda\mathrm{Id}$. Since U and $\lambda\mathrm{Id}$ commute, one has

$$\mathrm{e}^{tJ(\lambda,d)} = \mathrm{e}^{t\lambda\,\mathrm{Id}+tU} = \mathrm{e}^{t\lambda\,\mathrm{Id}} \cdot \mathrm{e}^{tU} = \mathrm{e}^{\lambda t}\mathrm{e}^{tU}.$$

But a simple computation shows that

$$\mathrm{e}^{tU} = \sum_{n=0}^{d-1} \frac{t^n}{n!}\,U^n = K,$$

where K is given by (5.26). Consequently, the third assertion golds good, too.

The proof of the fourth assertion is similar and is left as an exercise. $\square$

A consequence of Theorems 5.4 and 5.6 is that the solutions to (5.22) are functions whose components are sums of products of exponentials by polynomial and trigonometric functions in the variable t.

In other words, these components necessarily have the form

$$\sum_{i=1}^{I} \mathrm{e}^{\alpha_i t}\, p_i(t)\, (c_{1,i}\cos(\beta_i t) + c_{2,i}\sin(\beta_i t)),$$

for some polynomial functions p_i.

Example 5.5. Let A be given by

$$A = \begin{pmatrix} 1 & 2 & -2 \\ 0 & -5 & 2 \\ 0 & -10 & 3 \end{pmatrix}.$$

The eigenvalues are $\lambda = 1$ and $\alpha \pm i\beta = -1 \pm 2i$. An appropriate computation allows to write $A = P^{-1} \cdot \tilde{J} \cdot P$, where

$$P = \begin{pmatrix} 1 & 2 & -1 \\ 0 & 1 & 0 \\ 0 & -2 & 1 \end{pmatrix}, \quad \tilde{J} = \begin{pmatrix} 1 & & \\ & -1 & 2 \\ & -2 & -1 \end{pmatrix}.$$

Consequently,

$$
e^{tA} = P^{-1} \begin{pmatrix} e^t & & \\ & e^{-t}\cos 2t & e^{-t}\sin 2t \\ & -e^{-t}\sin 2t & e^{-t}\cos 2t \end{pmatrix} P
$$

$$
= \begin{pmatrix} e^t & e^t - e^{-t}(2\cos 2t + \sin 2t) & -e^t + e^{-t}\cos 2t \\ 0 & e^{-t}(\cos 2t - 2\sin 2t) & e^{-t}\sin 2t \\ 0 & -5e^{-t}\sin 2t & e^{-t}(\cos 2t + 2\sin 2t) \end{pmatrix}.
$$

Note that a simpler fundamental matrix can be obtained from e^{tA} after some elementary manipulation.

It suffices to first solve the system determined by the last two rows of A and then replace in the ODE determined by the first one. This way, we easily find the following:

$$
F(t) = \begin{pmatrix} e^t & -e^{-t}(2\cos 2t + \sin 2t) & e^{-t}\cos 2t \\ 0 & e^{-t}(\cos 2t - 2\sin 2t) & e^{-t}\sin 2t \\ 0 & -5e^{-t}\sin 2t & e^{-t}(\cos 2t + 2\sin 2t) \end{pmatrix}. \qquad \square
$$

Remark 5.3.

(a) In view of Theorem 5.6, it is reasonable to ask whether it is possible to generalize this construction process to the framework of linear homogeneous ODSs of variable (non-constant) coefficients.

In other words, assume that $A \in C^0(I; \mathcal{L}(\mathbb{R}^N))$ and $t_0 \in I$ and set

$$
F(t) := \exp\left(\int_{t_0}^t A(s)\,\mathrm{d}s \right) \quad \forall t \in I.
$$

Then, the question is the following:

Is F a fundamental matrix of (5.2)*?*

The answer is that in general no. For example, it is relatively easy to check that this does not happen for the ODS considered in Example 5.1, item (a).

(b) The previous results can be viewed as the preamble of the *theory of semigroups of operators*.

For a brief description of the context and objectives and some important applications, see Note 11.5.2. $\qquad \square$

5.6   Linear Equations of Order n

In this section, we consider ODEs of the form

$$y^{(n)} = a_{n-1}(t)y^{(n-1)} + \cdots + a_1(t)y' + a_0(t)y + b(t), \qquad (5.3)$$

where the coefficients a_i and the independent term b are functions in $C^0(I)$ (again, $I \subset \mathbb{R}$ is a non-degenerate interval).

Recall that it is said that (5.3) is a *linear ODE of order n* (of course, with a normal structure). Also, recall that, if $b \equiv 0$, we say that the ODE is linear and homogeneous.

These equations have already been considered in Chapter 1 when $n = 1$ and $n = 2$. In general, for any n, the usual change of variables allows to rewrite a linear ODE as a linear ODS of the first order.

Therefore, the results in the previous sections can be restated for (5.3) and the following assertions are fulfilled:

• First of all, for any $(t_0, y_0^{(0)}, y_0^{(1)}, \ldots, y_0^{(n-1)}) \in I \times \mathbb{R}^n$, the following Cauchy problem makes perfect sense:

$$\begin{cases} y^{(n)} = a_{n-1}(t)y^{(n-1)} + \cdots + a_1(t)y' + a_0(t)y + b(t) \\ y(t_0) = y_0, \ldots, \; y^{(n-1)}(t_0) = y_0^{(n-1)}. \end{cases} \qquad (5.27)$$

There exists a unique solution to this problem in I (if I is open, we are of course speaking of the maximal solution). In the sequel, when we refer to a solution to (5.3), we implicitly mean "a solution in I."

• Let E be the family of solutions to (5.3) and let E_0 be the family of solutions to the associated homogeneous linear ODE

$$y^{(n)} = a_{n-1}(t)y^{(n-1)} + \cdots + a_1(t)y' + a_0(t)y. \qquad (5.28)$$

Then, E_0 is a vector space of finite dimension, equal to n (a subspace of $C^n(I)$) and E is a linear manifold supported by E_0. In other words,

$$E = \varphi_p + E_0,$$

where φ_p is a solution to (5.3).

• By definition, a *fundamental system* of (5.28) is any base of E_0, that is, any collection of n linearly independent solutions to (5.28).

If $\{\varphi_1, \ldots, \varphi_n\}$ is a fundamental system, then for every $t \in I$ the matrix

$$F(t) = \begin{pmatrix} \varphi_1(t) & \varphi_2(t) & \cdots & \varphi_n(t) \\ \varphi_1'(t) & \varphi_2'(t) & \cdots & \varphi_n'(t) \\ \vdots & \vdots & & \vdots \\ \varphi_1^{(n-1)}(t) & \varphi_2^{(n-1)}(t) & \cdots & \varphi_n^{(n-1)}(t) \end{pmatrix}$$

is non-singular.

Of course, F is a fundamental matrix of the first-order ODS equivalent to (5.28); it is usual to say that F is the *Wronsky matrix* (or Wronskian) of the fundamental matrix $\{\varphi_1, \ldots, \varphi_n\}$.

Obviously, in order to determine all the solutions to (5.3), it suffices to compute the n functions of a fundamental system.

Coming back to the notation used in Chapter 1, we see (once again) that the general solution to (5.3) is the sum of a particular solution φ_p and the general solution to the associated homogeneous linear ODE (5.28), as follows:

$$\varphi(\cdot; C_1, \ldots, C_n) = \varphi_p + \varphi_h(\cdot; C_1, \ldots, C_n) = \varphi_p + C_1\varphi_1 + \cdots + C_n\varphi_n.$$

• If a fundamental system $\{\varphi_1, \ldots, \varphi_n\}$ is known, it is possible to determine a solution to the non-homogeneous ODE (5.3) via the *constants variation method*, also called *Lagrange's method*.

This can be carried out as follows. Let us put

$$\varphi = \sum_{i=1}^{n} c_i \, \varphi_i \tag{5.29}$$

for some unknowns $c_i \in C^1(I)$.

Let $t \in I$ be given. Then, we impose the following equalities:

$$c_1'(t)\varphi_1(t) + \cdots + c_n'(t)\varphi_n(t) \qquad = 0$$

$$c_1'(t)\varphi_1'(t) + \cdots + c_n'(t)\varphi_n'(t) \qquad = 0$$

$$\cdots \quad \cdots \quad \cdots \qquad\qquad\qquad \cdots$$

$$c_1'(t)\varphi_1^{(n-1)}(t) + \cdots + c_n'(t)\varphi_n^{(n-1)}(t) = b(t).$$

This is a system of n algebraic linear equations for the n unknowns $c_1'(t), \ldots, c_1'(t)$ that possesses a unique solution. Since this can be done for any $t \in I$, this argument provides the functions c_i' and therefore, up to additive constants, also the functions c_i for $i = 1, \ldots, n$.

It is not difficult to check that the corresponding function φ, given by (5.29), solves (5.3). Indeed, we have

$$\varphi' = \sum_{i=1}^{n} c_i \varphi_i', \quad \varphi'' = \sum_{i=1}^{n} c_i \varphi_i'', \ldots, \quad \varphi^{(n-1)} = \sum_{i=1}^{n} c_i \varphi_i^{(n-1)}$$

and, consequently,

$$
\begin{aligned}
\varphi^{(n)} &= \sum_{i=1}^{n} c_i \varphi_i^{(n)} + \sum_{i=1}^{n} c_i' \varphi_i^{(n-1)} \\
&= \sum_{i=1}^{n} c_i \left(a_{n-1} \varphi_i^{(n-1)} + \cdots + a_1 \varphi_i' + a_0 \varphi_i \right) + b \\
&= a_{n-1} \varphi^{(n-1)} + \cdots + a_1 \varphi' + a_0 \varphi + b.
\end{aligned}
$$

Example 5.6. Consider the linear ODE of third order

$$y''' = 3y'' - 2y' + 2e^{2t} \tag{5.30}$$

and the associated homogeneous ODE

$$y''' = 3y'' - 2y'. \tag{5.31}$$

It is not difficult to compute the solutions to (5.31).

Thus, the change of variable $z = y'$ allows to reduce this equation to a second-order ODE where the variable t does not appear explicitly; then, a new change of variable $p(z) = z'(t)$ leads to a homogeneous first-order ODE that can be easily solved).[4]

Thus, we are led to the fundamental system $\{\varphi_1, \varphi_2, \varphi_3\}$, with

$$\varphi_1(t) \equiv 1, \quad \varphi_2(t) \equiv e^t, \quad \varphi_3(t) \equiv e^{2t}.$$

As a consequence, Lagrange's method leads in this case to the system

$$
\begin{aligned}
c_1'(t) + c_2'(t)e^t + c_3'(t)e^{2t} &= 0 \\
c_2'(t)e^t + 2c_3'(t)e^{2t} &= 0 \\
c_2'(t)e^t + 4c_3'(t)e^{2t} &= 2e^{2t}
\end{aligned}
$$

which provides the functions c_i (up to additive constants). We deduce that the general solution to (5.30) is given by

$$\varphi(t; C_1, C_2, C_3) = C_1 + C_2 e^t + C_3 e^{2t} + \left(t - \frac{3}{2} \right) e^{2t} \quad \forall t \in \mathbb{R},$$

where the C_i are arbitrary constants. $\qquad\square$

---

[4]There is another simpler way to find the solutions to (5.31), since it is an ODE of constant coefficients. This will be seen soon.

We will end this chapter by considering homogeneous linear ODEs with constant coefficients

$$y^{(n)} = a_{n-1} y^{(n-1)} + \cdots + a_1 y' + a_0 y. \tag{5.32}$$

We will indicate what can be done to find a fundamental system for an equation like this one.

Of course, the usual change of variables makes it possible to rewrite (5.32) as an equivalent homogeneous linear ODS with constant coefficients. Then, we can compute an associated fundamental matrix and finally come back to the original variable, which provides the required system of linear independent solutions.

There is however a more simple and fast way to proceed.

The key point is to introduce the so-called *characteristic equation*, given by

$$\lambda^n - a_{n-1}\lambda^{n-1} - \cdots - a_1\lambda - a_0 = 0. \tag{5.33}$$

Let $\lambda_1, \ldots, \lambda_p$ be the real roots of (5.33), whose multiplicities will be denoted by $m_1, \ldots, m_p$. Let $\alpha_1 \pm i\beta_1, \ldots, \alpha_q \pm i\beta_q$ be the pairs of conjugate complex (not real) roots, with multiplicities $r_1, \ldots, r_q$.

Of course, we must have $m_1 + \cdots + m_p + 2r_1 + \cdots + 2r_q = n$.

Then, it is easy to check that a fundamental system of (5.32) is the following:

- $m_1 + \cdots m_p$ solutions associated to the real roots λ_j:

$$e^{\lambda_j t}, te^{\lambda_j t}, \ldots, t^{m_j - 1}e^{\lambda_j t}$$

for $j = 1, \ldots, p$.
- $2r_1 + \cdots 2r_q$ solutions associated to the $\alpha_k \pm i\beta_k$:

$$e^{\alpha_k t}\cos(\beta_k t),\ e^{\alpha_k t}\sin(\beta_k t),\ te^{\alpha_k t}\cos(\beta_k t),\ te^{\alpha_k t}\sin(\beta_k t),$$
$$\cdots \quad \cdots \quad \cdots$$
$$t^{r_k - 1}e^{\alpha_k t}\cos(\beta_k t),\ t^{r_k - 1}e^{\alpha_k t}\sin(\beta_k t)$$

for $k = 1, \ldots, q$.

Example 5.7. Let us consider the ODE $y'' + 5y' + 6y = 5\sin t$. First, we consider the characteristic equation

$$\lambda^2 + 5\lambda + 6 = 0,$$

whose solutions are $\lambda_1 = -3$ and $\lambda_2 = -2$ (both with multiplicity equal to 1). Consequently, a fundamental system is given by $\{\varphi_1, \varphi_2\}$, where

$$\varphi_1(t) \equiv e^{-3t}, \quad \varphi_2(t) \equiv e^{-2t}.$$

The general solution to the original non-homogeneous ODE can now be easily computed via Lagrange's method.

Indeed, it suffices to solve for every t the system

$$c_1'(t)e^{-3t} + c_2'(t)e^{-2t} = 0$$
$$-3c_1'(t)e^{-3t} - 2c_2'(t)e^{-2t} = 5\sin t.$$

After some computations, we get

$$\varphi(t; C_1, C_2) = C_1 e^{-3t} + C_2 e^{-2t} + \frac{1}{2}(\sin t - \cos t) \quad \forall t \in \mathbb{R}. \qquad \square$$

Remark 5.4. Let us set $I = (0, +\infty)$. Any linear ODE

$$y^{(n)} = a_{n-1}(t)y^{(n-1)} + \cdots + a_1(t)y' + a_0(t)y,$$

where the functions a_i are of the form

$$a_i(t) = A_i t^{i-n} \quad \forall t \in (0, +\infty)$$

for some $A_i \in \mathbb{R}$ is called an *Euler equation* (we have already considered equations of this kind for $n = 2$ in Chapter 1).

They can also be written as follows:

$$t^n y^{(n)} = A_{n-1}t^{n-1}y^{(n-1)} + \cdots + A_1 t y' + A_0 y. \qquad (5.34)$$

The change of variable $t = e^\tau$ allows to rewrite (5.34) equivalently as a linear ODE of constant coefficients.

Consequently, for an Euler ODE it is also possible (and easy) to compute a fundamental system. $\qquad \square$

To end this section, we will present a method that can be used to compute a particular solution to a non-homogeneous linear ODE in some cases, as an alternative (and a simplification) to Lagrange's method.

Thus, consider the ODE of constant coefficients

$$y^{(n)} = a_{n-1}y^{(n-1)} + \cdots + a_1 y' + a_0 y + b(t), \qquad (5.35)$$

where the $a_i \in \mathbb{R}$ and b is a sum of products of exponential, polynomial, and trigonometric functions. We will use the so-called *superposition principle* and, then, we will select some particular candidates.

For any $\varphi \in C^n(I)$, let us set

$$L(\varphi) := \varphi^{(n)} - a_{n-1}\varphi^{(n-1)} - \cdots - a_1\varphi' - a_0\varphi.$$

Then, $\varphi_p \in C^n(I)$ is a particular solution to (5.35) in I if and only if

$$L(\varphi_p)(t) = b(t) \quad \forall t \in I.$$

From the fact that (5.35) is linear, we obtain the *superposition principle:* if $\varphi^i \in C^n(I)$ solves $L(y) = b^i$ in I for $i = 1, \ldots, \ell$ and $b = m_1 b^1 + \cdots + m_\ell b^\ell$ for some $m_i \in \mathbb{R}$, then

$$\varphi = m_1 \varphi^1 + \cdots + m_\ell \varphi^\ell$$

solves (5.35) in the same interval.

On the other hand, using the characteristic polynomial associated to (5.35)

$$p(\lambda) = \lambda^n - a_{n-1}\lambda^{n-1} - \cdots - a_1\lambda - a_0 ,$$

we also have

$$L(e^{\lambda t}) = p(\lambda)e^{\lambda t}, \quad L(te^{\lambda t}) = p(\lambda)te^{\lambda t} + p'(\lambda)e^{\lambda t}, \quad \text{etc.}$$

In general, for every m, we get

$$L(t^m e^{\lambda t}) = \sum_{k=0}^{m} \binom{m}{k} p^{(k)}(\lambda) t^{m-k} e^{\lambda t}.$$

Consequently, we see that, if b is of the form

$$b(t) \equiv p_n(t)e^{\alpha t},$$

where $\alpha \in \mathbb{R}$ and p_n is polynomial of degree n, it makes sense to look for a solution of the form

$$\varphi_p(t) = t^s P_n(t)e^{\alpha t},$$

where P_n is also polynomial of degree n, $s = 0$ if α is not a root of the characteristic equation, and s is the multiplicity of α otherwise.

Also, noting that $e^{(\alpha+i\beta)t} = e^{\alpha t}(\cos \beta t + i \sin \beta t)$, we see that it makes sense to look for solutions with a particular structure when b is a sum of products of polynomial and trigonometric functions.

Let us give a list of possible cases:

- $b(t) \equiv p_n(t)e^{\alpha t}$ with $\alpha \in \mathbb{R}$:
  - If α is not a root of the characteristic equation, we look for a particular solution $\varphi_p(t) = P_n(t)e^{\alpha t}$.
  - If α is a root of the characteristic equation of multiplicity s, we look for $\varphi_p(t) = t^s P_n(t)e^{\alpha t}$.
- $b(t) \equiv e^{\alpha t}[p_n(t) \cos \beta t + q_m(t) \sin \beta t]$ con $\alpha, \beta \in \mathbb{R}$:
  - Let us introduce $k := \max\{n, m\}$. If the $\alpha \pm i\beta$ are not roots of the characteristic equation, we look for $\varphi_p(t) = e^{\alpha t}[P_k(t) \cos \beta t + Q_k(t) \sin \beta t]$.

    ○ Otherwise, we look for solutions of the form $\varphi_p(t) = t^s e^{\alpha t} [P_k(t) \cos \beta t + Q_k(t) \sin \beta t]$, where s is the multiplicity of $\alpha \pm i\beta$.

Exercises

E.5.1 Prove the assertion in Remark 5.2, (a).

E.5.2 Prove the fourth assertion in Theorem 5.6.

E.5.3 Consider the linear ODS

$$y' = \begin{pmatrix} -1 + \cos t & 0 \\ \cos t & -1 \end{pmatrix} y + \begin{pmatrix} 0 \\ e^{-t} \end{pmatrix}$$

and find a fundamental matrix of the associated homogeneous linear system. Also, compute the solution corresponding to the Cauchy data $(\pi; 1, 0)$.

E.5.4 Same statement for the system

$$y' = \begin{pmatrix} -2t + 1 & 0 \\ 0 & -2t + 1 \end{pmatrix} y + \begin{pmatrix} 0 \\ 2t - 1 \end{pmatrix}$$

and the Cauchy data $(0; 1, 0)$.

E.5.5 Let $I \subset \mathbb{R}$ be a nontrivial interval and let $A \in C^0(I; \mathcal{L}(R^N))$ and $F \in C^1(I; \mathcal{L}(R^N))$ be such that $F'(t) = A(t)F(t)$ for all $t \in I$. Prove that F satisfies the Jacobi–Liouville formula (5.12).

E.5.6 Let $\varphi \in C^1(\mathbb{R}; \mathbb{R}^N)$ be a solution to the ODS $y' = Ay$, where $A \in \mathcal{L}(\mathbb{R}^N)$. Prove that the function $t \mapsto t\varphi(t)$ solves the non-homogeneous linear ODS $y' = Ay + \varphi(t)$ and apply this result to the resolution of the system

$$y' = \begin{pmatrix} 1 & -1 & 0 \\ 1 & 1 & -1 \\ 2 & 0 & -1 \end{pmatrix} y + \begin{pmatrix} e^t \\ 0 \\ e^t \end{pmatrix}.$$

E.5.7 Let $A \in \mathcal{L}(\mathbb{R}^N)$ be given and assume that $\|A\|_S < 1$. Prove that

$$\exp\left(\log(\mathrm{Id} + A)\right) = \mathrm{Id} + A.$$

E.5.8 Let $A \in \mathcal{L}(\mathbb{R}^N)$ and $b \in C^0(\mathbb{R}_+; \mathbb{R}^N)$ be given. Consider the ODS

$$y' = \frac{1}{x} Ay + b(x), \quad x > 0. \tag{5.36}$$

    (1) Find a fundamental matrix of the associated homogeneous linear ODS.

(2) Find the general solution to (5.36) and also the solution associated to the Cauchy data $(1; 1, 1)$ when

$$A = \begin{pmatrix} 2 & -1 \\ 1 & 0 \end{pmatrix}, \quad b(t) \equiv \begin{pmatrix} t^2 \\ t^2 \end{pmatrix}.$$

E.5.9 Find the general solutions to the following homogeneous linear ODSs, as well as those corresponding to the indicated Cauchy data:

(1)
$$y' = \begin{pmatrix} 4 & -1 & 0 \\ 3 & 1 & -1 \\ 1 & 0 & 1 \end{pmatrix} y; \quad (0; 1, 0, 1).$$

(2)
$$y' = \begin{pmatrix} 0 & 0 & -2 \\ 1 & 1 & -4 \\ 1 & 0 & -4 \end{pmatrix} y; \quad (0; -1, 0, 1).$$

(3)
$$y' = \begin{pmatrix} 0 & 0 & -2 \\ 1 & 1 & -4 \\ 1 & 0 & -3 \end{pmatrix} y; \quad (0; 2, 1, 1).$$

(4)
$$y' = \begin{pmatrix} 4 & -1 & -1 \\ 1 & 2 & -1 \\ 1 & -1 & 2 \end{pmatrix} y; \quad (0; 0, 0, 1).$$

(5)
$$y' = \begin{pmatrix} 2 & -1 & 1 \\ 1 & 2 & -1 \\ 1 & -1 & 2 \end{pmatrix} y; \quad (0; 1, 1, 1).$$

(6)
$$y' = \begin{pmatrix} 0 & 0 & 1 & 0 \\ 0 & 0 & 0 & 1 \\ 1 & -1 & 0 & 0 \\ 0 & -1 & 0 & 0 \end{pmatrix} y; \quad (0; 0, 0, 1/2, 1).$$

E.5.10 Find a fundamental matrix of the homogeneous linear ODS

$$\begin{pmatrix} y_1' \\ y_2' \\ y_3' \end{pmatrix} = \begin{pmatrix} 2t & 0 & 0 \\ 0 & 1 & 1 \\ 0 & 1 & 1 \end{pmatrix} \begin{pmatrix} y_1 \\ y_2 \\ y_3 \end{pmatrix}.$$

E.5.11 Find the general solutions to the ODSs $y' = Ay + b(t)$ for the following matrices A and functions b, as well as those corresponding to the indicated Cauchy data:

(1) $\quad A = \begin{pmatrix} 1 & 1 & 0 \\ 0 & 1 & 1 \\ 1 & 0 & 1 \end{pmatrix}, \quad b(t) = \begin{pmatrix} \sin t \\ 0 \\ \cos t \end{pmatrix}; \quad (0; 1, 0, 1).$

(2) $\quad A = \begin{pmatrix} 1 & 2 & 0 \\ 0 & 1 & 3 \\ 0 & 0 & 1 \end{pmatrix}, \quad b(t) = \begin{pmatrix} 0 \\ 3t/2 \\ 1 \end{pmatrix}; \quad (0; 1, 2, 3).$

(3) $\quad A = \begin{pmatrix} 1 & 1 & -1 \\ -3 & -3 & 3 \\ -2 & -2 & 2 \end{pmatrix}, \quad b(t) = \begin{pmatrix} 1 \\ t \\ 0 \end{pmatrix}; \quad (0; 1, 0, 1).$

(4) $\quad A = \begin{pmatrix} 6 & -6 & 5 \\ 13 & -12 & 9 \\ 6 & -5 & 3 \end{pmatrix}, \quad b(t) = \begin{pmatrix} te^{-t} \\ 0 \\ 0 \end{pmatrix}; \quad (0; 1, 1, 0).$

(5) $\quad A = \begin{pmatrix} 13 & 16 & 16 \\ -5 & -7 & -6 \\ -6 & -8 & -7 \end{pmatrix}, \quad b(t) = \begin{pmatrix} e^{3t} \\ 0 \\ 0 \end{pmatrix}; \quad (0; 1, 0, 0).$

E.5.12 Find the general solutions to the following ODEs:

(1) $ay''' + y'' + 9ay' + 9y = \dfrac{37}{2}e^{-x/a} + 6 - 12\sin^2\dfrac{3x}{2}$, where $a \in \mathbb{R}\setminus\{0\}$.
(2) $y'' - 2y' + 5y = 1 + e^x \cos 2x.$
(3) $y''' = x + \cos x.$
(4) $y''' - 3y'' + 4y' - 2y = e^x \sin 2x \cos 2x.$
(5) $x^2 y'' + 3xy' - 3y = 5x^2.$

E.5.13 Solve the following Cauchy problems:

(1) $y'' - 3y' + 2y = e^{3t}$, $y(0) = y'(0) = 0.$
(2) $y'' + 4y' + 6y = 1 + e^{-t}$, $y(0) = y'(0) = 0.$
(3) $y'' + 4y = 4t$, $y(1) = 3$, $y'(1) = 4.$

E.5.14 Find all the functions $\varphi \in C^2(0, +\infty)$ satisfying

$$t^2 \varphi''(t) + 5t\varphi'(t) + 3\varphi(t) = 8t\log t \quad \forall t \in (0, +\infty).$$

E.5.15 Assume that $I = (a_1, a_2)$ with $-\infty \leq a_1 < a_2 \leq +\infty$, $A \in C^0(I; \mathbb{R}^{N \times N})$, and $b \in C^0(I; \mathbb{R}^N)$ and let F be a fundamental matrix for the ODS $y' = A(t)y$. For every $(t_0, y_0) \in I \times \mathbb{R}^N$, we consider the corresponding Cauchy problem

$$\begin{cases} y' = A(t)y + b(t) \\ y(t_0) = y_0. \end{cases}$$

Let $\phi : I \times I \times \mathbb{R}^N \mapsto \mathbb{R}^N$ be the function that assigns to every $(t; t_0, y_0)$ the value at t of the solution to the previous Cauchy problem.

(1) Prove that ϕ is well defined and determine $\phi(t; t_0, y_0)$ in terms of F.

(2) Prove that ϕ is continuous and possesses continuous partial derivatives with respect to all the variables.

E.5.16 For every $\lambda \in \mathbb{R}^k$, we consider the Cauchy problem

$$\begin{cases} y' = A(t)y + g(\lambda) \\ y(t_0) = y_0 \end{cases}$$

where $g \in C^1(\mathbb{R}^k; \mathbb{R}^N)$ and A and y_0 are as in Exercise E.5.15. Let $\psi : I \times I \times \mathbb{R}^N \times \mathbb{R}^k \mapsto \mathbb{R}^N$ be the function that assigns to every $(t; t_0, y_0, \lambda)$ the value at t of the solution to the previous problem.

(1) Prove that ψ is well defined and determine $\psi(t; t_0, y_0, \lambda)$ in terms of a fundamental matrix F.

(2) Prove that ψ is continuous and possesses continuous partial derivatives with respect to all the variables.

E.5.17 For every $\lambda \in \mathbb{R}^k$, we consider the Cauchy problem

$$\begin{cases} y' = A(\lambda)y + b(t) \\ y(t_0) = y_0, \end{cases}$$

where $A \in C^1(\mathbb{R}^k; \mathbb{R}^{N \times N})$ and b and y_0 are as in Exercise E.5.15. Let $\zeta : I \times I \times \mathbb{R}^N \times \mathbb{R}^k \mapsto \mathbb{R}^N$ be the function that assigns to each $(t; t_0, y_0, \lambda)$ the value at t of the solution to the previous problem.

(1) Construct a family $\{F_\lambda\}$ of fundamental matrices, prove that ζ is well defined and determine $\psi(t; t_0, y_0, \lambda)$ in terms of F_λ.

(2) Prove that ζ is continuous and possesses continuous partial derivatives with respect to all the variables.

Chapter 6

Boundary-Value Problems
for Linear Systems

This chapter is devoted to the analysis of linear ODEs and ODSs complemented with conditions whose structure is different from those found in Cauchy problems.

More precisely, we will consider *boundary conditions.*

They will lead to the so-called *boundary-value problems.* In general terms, we mean problems where we try to compute one or several solutions defined in an interval using information on the behavior of the solution at the endpoints.

Obviously, this goal is different from the objective pursued when solving an initial-value problem.

Our interest will be to study the existence, uniqueness, or multiplicity of the solutions and also to indicate how they can be found.

6.1 First Results

In this chapter, $I = [\alpha, \beta]$ will be a compact interval. We will consider linear ODSs

$$y' = A(x)y + b(x), \quad x \in [\alpha, \beta], \tag{6.1}$$

where $A : I \mapsto \mathcal{L}(\mathbb{R}^N)$ and $b : I \mapsto \mathbb{R}^N$ are continuous functions.[1]

---

[1]In the framework of boundary-value problems, it is intuitively adequate to use the symbol x to denote the independent variable. Indeed, in many of the considered problems, the interpretation as a spatial variable is completely natural.

Definition 6.1. We call boundary-value problem for (6.1) any system of the form

(CPo)
$$\begin{cases} y' = A(x)y + b(x) \\ By(\alpha) + Cy(\beta) = h, \end{cases}$$

where the matrices $B, C \in \mathcal{L}(\mathbb{R}^N)$ and the vector $h \in \mathbb{R}^N$ are given. □

The second equality in (BVP) is usually called the *boundary condition*.

By definition, a solution to (BVP) is any solution φ to the ODS (6.1) (in I) that satisfies the boundary condition, that is, such that

$$B\varphi(\alpha) + C\varphi(\beta) = h.$$

The previous problem is *linear*.

In fact, this is the case for the considered ODS and, moreover, the chosen additional condition also possesses a linear structure.

As will be seen in what follows, this is very advantageous. Among other things, the family of solutions (if they exist) will be again a linear manifold of finite dimension and, consequently, it will suffice to compute a finite amount of solutions in order to know all of them.

It is completely meaningful to consider boundary-value problems where the ODS and/or the boundary condition is nonlinear.

However, their study requires techniques of a higher difficulty; see more details in Note 11.6.1.

By definition, the *homogeneous boundary-value problem* associated to (BVP) is the following:

$$\begin{cases} y' = A(x)y, \quad x \in [\alpha, \beta] \\ By(\alpha) + Cy(\beta) = 0 \end{cases} \tag{6.2}$$

(we say that the homogeneous linear ODS has been "complemented" or "completed" with a *homogeneous* boundary condition).

Obviously, (6.2) always possesses the so-called *trivial* solution $\varphi \equiv 0$. But other solutions can also exist.

In fact, we have the following result:

Proposition 6.1. *Let us set*

$$W_0 = \{\varphi \in C^1(I; \mathbb{R}^N) : \varphi \text{ is a solution to (6.2)}\}.$$

Then:

(1) W_0 *is a linear subspace of* $C^1(I; \mathbb{R}^N)$ *of dimension* $\leq N$.

(2) *Moreover,*

$$\dim W_0 = N - \operatorname{rank}(BF(\alpha) + CF(\beta)), \qquad (6.3)$$

where F is any fundamental matrix of the ODS in (6.2).

Proof. It is clear that W_0 is a linear subspace of the space V_0 of the solutions to the considered homogeneous linear system.

Consequently, its dimension is $\leq N$.

On the other hand, if F is a fundamental matrix, the solutions to (6.2) are obviously the functions $\varphi = Fa$ with $a \in \mathbb{R}^N$ such that

$$0 = B \cdot F(\alpha)a + C \cdot F(\beta)a = (BF(\alpha) + CF(\beta))a.$$

Accordingly, if we denote by $N(BF(\alpha) + CF(\beta))$ the *kernel* or *null subspace* of the linear mapping determined by $BF(\alpha) + CF(\beta)$, one has

$$\dim W_0 = \dim N(BF(\alpha) + CF(\beta)) = N - \operatorname{rank}(BF(\alpha) + CF(\beta)).$$

This ends the proof. □

Observe that $\operatorname{rank}(BF(\alpha) + CF(\beta))$ is independent of the chosen fundamental matrix. Indeed, if G is another fundamental matrix, there exists a non-singular matrix $P \in \mathcal{L}(\mathbb{R}^N)$ such that $G = F \cdot P$ and then

$$BG(\alpha) + CG(\beta) = (BF(\alpha) + CF(\beta)) \cdot P.$$

In view of this proposition, we see that the analysis of (6.2) is equivalent to the study of a linear algebraic system of dimension N.

In particular, either (6.2) only possesses the trivial solution or there exist infinitely many solutions.

On the other hand, looking at the previous proof, we find the way to compute the solutions to (6.2): it suffices to first construct a fundamental matrix F and then solve the homogeneous linear system

$$(BF(\alpha) + CF(\beta))a = 0, \quad a \in \mathbb{R}^N.$$

The family of solutions will be given by the corresponding functions $\varphi = Fa$.

6.2 The Alternative Theorem

The main result in this chapter is the following. It is known as the *Alternative Theorem* for the existence (and uniqueness) of a solution to (BVP):

Theorem 6.1. *Let us consider the boundary-value problem* (BVP) *and the associated homogeneous problem* (6.2). *Let F be a fundamental matrix of the corresponding homogeneous linear ODS and let us set $M := BF(\alpha) + CF(\beta)$. Then:*

(1) *The following three assertions are equivalent:* (6.2) *only possesses the trivial solution*, $\operatorname{rank} M = N$ *and, for every $b \in C^0(I; \mathbb{R}^N)$ and every $h \in \mathbb{R}^N$, the corresponding boundary-value problem* (BVP) *is uniquely solvable.*

(2) *Assume that* (6.2) *possesses nontrivial solutions and let $b \in C^0(I; \mathbb{R}^N)$ and $h \in \mathbb{R}^N$ be given. Then, the corresponding* (BVP) *is solvable if and only if*

$$\operatorname{rank} M = \operatorname{rank} \left(M \,\middle|\, h - CF(\beta) \int_\alpha^\beta F(s)^{-1} b(s) \, ds \right). \qquad (6.4)$$

If this is the case, the solution is not unique.

Proof. In view of Proposition 6.1, (6.2) only possesses the trivial solution if and only if $\operatorname{rank} M = N$.

On the other hand, if $b \in C^0(I; \mathbb{R}^N)$ and $h \in \mathbb{R}^N$ are given and we set

$$\varphi_p(x) := F(x) \int_\alpha^x F(s)^{-1} b(s) \, ds \quad \forall x \in I,$$

the solutions to the corresponding problem (BVP) are the functions $\varphi = Fa + \varphi_p$ satisfying

$$h = B\varphi(\alpha) + C\varphi(\beta) = BF(\alpha)a + C \left(F(\beta)a + F(\beta) \int_\alpha^\beta F(s)^{-1} b(s) \, ds \right),$$

that is,

$$Ma = h - CF(\beta) \int_\alpha^\beta F(s)^{-1} b(s) \, ds. \qquad (6.5)$$

Consequently, (BVP) possesses exactly one solution for every $b \in C^0(I; \mathbb{R}^N)$ and every $h \in \mathbb{R}^N$ if and only if $\operatorname{rank} M = N$.

This proves the first part.

Now, assume that (6.2) possesses nontrivial solutions (that is, $\operatorname{rank} M < N$) and let $b \in C^0(I; \mathbb{R}^N)$ and $h \in \mathbb{R}^N$ be given again. Then, (BVP) is solvable if and only if there exist $a \in \mathbb{R}^N$ satisfying (6.5). But this is equivalent to (6.4).

Therefore, the second part is also true. □

Example 6.1. Assume that $I = [0,1]$,

$$A(x) = \begin{pmatrix} p & 0 & 0 \\ 0 & q & 1 \\ 0 & 0 & q \end{pmatrix},$$

$$b(x) = \begin{pmatrix} 0 \\ e^{qx} \\ e^{qx} \end{pmatrix} \quad \forall x \in I,$$

$$B = \begin{pmatrix} -1 & 0 & 0 \\ 0 & -q & 0 \\ 0 & 0 & 0 \end{pmatrix}, \quad C = \begin{pmatrix} e^{-p} & 0 & 0 \\ 0 & e^{-q} & -e^{-q} \\ 0 & 0 & e^{-q} \end{pmatrix}$$

(p and q are real numbers).

In this case, a fundamental matrix of the homogeneous system is

$$F(x) = e^{xA} = \begin{pmatrix} e^{px} & 0 & 0 \\ 0 & e^{qx} & xe^{qx} \\ 0 & 0 & e^{qx} \end{pmatrix} \quad \forall x \in I.$$

Consequently,

$$M = BF(0) + CF(1) = \begin{pmatrix} 0 & 0 & 0 \\ 0 & 1-q & 0 \\ 0 & 0 & 1 \end{pmatrix}, \quad \operatorname{rank} M = \begin{cases} 2 & \text{if } q \neq 1 \\ 1 & \text{if } q = 1. \end{cases}$$

A particular solution to the ODS is φ_p, with

$$\varphi_p(x) = F(x) \int_0^x F(s)^{-1} b(s)\, ds = \begin{pmatrix} 0 \\ x(1 + x/2)e^{qx} \\ xe^{qx} \end{pmatrix} \quad \forall x \in I.$$

Hence, for any given $b \in C^0(I; \mathbb{R}^3)$ and $h \in \mathbb{R}^3$, the considered problem is solvable (and the solution is not unique) if and only if

$$\operatorname{rank} M = \operatorname{rank} \begin{pmatrix} 0 & 0 & 0 & h_1 \\ 0 & 1-q & 0 & h_2 - \frac{1}{2} \\ 0 & 0 & 1 & h_3 \end{pmatrix}.$$

This means the following:

- If $q \neq 1$, the problem is solvable if and only if $h_1 = 0$. In that case the solutions are the functions

$$\varphi = Fa + \varphi_p, \quad \text{with } a_1 \in \mathbb{R},\ a_2 = \frac{h_2 - 1/2}{1-q},\ a_3 = h_3.$$

- If $q = 1$, there exist solutions if and only if $h_1 = 0$ and $h_2 = 1/2$. Now, the solutions are the following:

$$\varphi = Fa + \varphi_p, \quad \text{with } a_1, a_2 \in \mathbb{R}, \ a_3 = h_3.$$

Remark 6.1. Let us assume that (6.2) only possesses the trivial solution. Then, for every $b \in C^0(I; \mathbb{R}^N)$, there exists exactly one solution φ_b to the problem

$$\begin{cases} y' = A(x)y + b(x), & x \in [\alpha, \beta] \\ By(\alpha) + Cy(\beta) = 0. \end{cases} \tag{6.6}$$

Let Z_0 be the set

$$Z_0 := \{\varphi \in C^1(I; \mathbb{R}^N) : B\varphi(\alpha) + C\varphi(\beta) = 0\}.$$

Then Z_0 is an infinite dimensional linear space of $C^1(I; \mathbb{R}^N)$ and $b \mapsto \varphi_b$ is a linear, continuous, and bijective mapping from $C^0(I; \mathbb{R}^N)$ into Z_0.

This isomorphism is called the *Green operator*. Its description and properties are of great relevance for many reasons; see Note 11.6.2.

Remark 6.2. When $N = 1$, it is easy to rewrite and interpret the Alternative Theorem.

Indeed, a fundamental matrix is given by

$$F(x) = \exp\left(\int_\alpha^x A(s)\,\mathrm{d}s\right) \quad \forall x \in I$$

and, consequently,

$$M = BF(\alpha) + CF(\beta) = B + Ce^{\int_\alpha^\beta A(s)\,\mathrm{d}s}.$$

The following is satisfied:

(1) If $M \neq 0$, (6.2) only posseses the trivial solution and, for every $b \in C^0(I)$ and every $h \in \mathbb{R}$, the corresponding problem (BVP) possesses exactly one solution.
(2) If $M = 0$, (6.2) possesses nontrivial solutions and, for any given $b \in C^0(I)$ and $h \in \mathbb{R}$, the corresponding (BVP) is solvable (not unique) if and only if

$$C\int_\alpha^\beta e^{\int_s^\beta A(\sigma)\,\mathrm{d}\sigma} b(s)\,\mathrm{d}s = h.$$

If $M \neq 0$, the Green operator (the mapping that associates to every $b \in C^0(I)$ the unique solution to (BVP) with $h = 0$) is given by $\mathcal{G} : C^0(I) \mapsto Z_0$, with $\mathcal{G}b = \varphi_b$,

$$\varphi_b(x) = e^{\int_\alpha^x A(s)\,ds} a + \int_\alpha^x e^{\int_s^x A(\sigma)\,d\sigma} b(s)\,ds \quad \forall x \in I,$$

where

$$a = -\frac{C}{M} \int_\alpha^\beta e^{\int_s^\beta A(\sigma)\,d\sigma} b(s)\,ds. \qquad \Box$$

6.3   The Case of a Linear Second-Order ODE

In this section, we will consider linear ODEs of the second order

$$y'' = a_1(x)\, y' + a_0(x)\, y + b(x), \tag{6.7}$$

where $a_0, a_1, b \in C^0(I)$.

The usual change of variables allows to rewrite (6.7) equivalently as a linear ODS. Consequently, it makes sense to consider boundary-value problems for (6.7).

For any given $B, C \in \mathcal{L}(\mathbb{R}^2)$, and $h \in \mathbb{R}^2$, we will analyze the problem

$$(\text{CPo})_2 \qquad \begin{cases} y'' = a_1(x)\, y' + a_0(x)\, y + b(x), \quad x \in [\alpha, \beta] \\[2mm] B\begin{pmatrix} y(\alpha) \\ y'(\alpha) \end{pmatrix} + C\begin{pmatrix} y(\beta) \\ y'(\beta) \end{pmatrix} = h, \end{cases}$$

that can be viewed as a particular case of (BVP).

The associated homogeneous boundary-value problem is

$$\begin{cases} y'' = a_1(x)\, y' + a_0(x)\, y, \quad x \in [\alpha, \beta] \\[2mm] B\begin{pmatrix} y(\alpha) \\ y'(\alpha) \end{pmatrix} + C\begin{pmatrix} y(\beta) \\ y'(\beta) \end{pmatrix} = 0. \end{cases} \tag{6.8}$$

Now, the following assertions can be checked:

- The family H_0 of solutions to (6.8) is a linear subspace of $C^2(I)$ of finite dimension ≤ 2.
- Let $\{\varphi_1, \varphi_2\}$ be a fundamental system of the ODE we find in (6.8), that is, a base of the corresponding space of solutions E_0. A fundamental matrix of the equivalent homogeneous linear ODS is

$$F := \begin{pmatrix} \varphi_1 & \varphi_2 \\ \varphi_1' & \varphi_2' \end{pmatrix}.$$

Consequently, the dimension of H_0 is the following:

$$\dim H_0 = 2 - \text{rank}\left[B\begin{pmatrix} \varphi_1(\alpha) & \varphi_2(\alpha) \\ \varphi_1'(\alpha) & \varphi_2'(\alpha) \end{pmatrix} + C\begin{pmatrix} \varphi_1(\beta) & \varphi_2(\beta) \\ \varphi_1'(\beta) & \varphi_2'(\beta) \end{pmatrix} \right].$$

In this context, the Alternative Theorem takes the following form:

Theorem 6.2. *Let us set $M := BF(\alpha) + CF(\beta)$. Then:*

(1) *The following three assertions are a equivalent: (6.8) only possesses the trivial solution, rank $M = 2$, and for every $b \in C^0(I)$ and every $h \in \mathbb{R}^2$, the corresponding problem* $(BVP)_2$ *possesses exactly one solution.*
(2) *Let us assume that (6.8) possesses nontrivial solutions and let $b \in C^0(I)$ and $h \in \mathbb{R}^2$ be given. Then, the corresponding* $(BVP)_2$ *is (multiply) solvable if and only if*

$$\operatorname{rank} M = \operatorname{rank}\left(M \ \middle| \ h - CF(\beta) \int_\alpha^\beta F(s)^{-1} \begin{pmatrix} 0 \\ b(s) \end{pmatrix} ds \right). \tag{6.9}$$

Example 6.2. Assume that $I = [0, \ell]$ and consider the boundary-value problem

$$\begin{cases} y'' + y = b(x) \\ y(0) = 0, \quad y(\ell) = 0. \end{cases} \tag{6.10}$$

The boundary conditions can be written as in $(BVP)_2$, with

$$B = \begin{pmatrix} 1 & 0 \\ 0 & 0 \end{pmatrix}, \quad C = \begin{pmatrix} 0 & 0 \\ 1 & 0 \end{pmatrix}, \quad h = \begin{pmatrix} 0 \\ 0 \end{pmatrix}.$$

On the other hand, a fundamental system of the associated homogeneous ODE is $\{\varphi_1, \varphi_2\}$, with

$$\varphi_1(x) = \cos x, \quad \varphi_2(x) = \sin x \quad \forall x \in I.$$

Therefore, the equivalent homogeneous system only possesses the trivial solution if and only if rank $M = 2$, where

$$M = BF(0) + CF(\ell) = \begin{pmatrix} 1 & 0 \\ \cos \ell & \sin \ell \end{pmatrix}.$$

Let φ_p be the solution to the nonhomogeneous ODE furnished by Lagrange's method:

$$\varphi_p(x) = \left(-\int_0^x b(s) \sin s \, ds \right) \cos x + \left(\int_0^x b(s) \cos s \, ds \right) \sin x \quad \forall x \in I.$$

We then have the following:

- If ℓ is not an integer multiple of π, rank $M = 2$ and, consequently, for every $b \in C^0(I)$, the corresponding problem (6.10) is uniquely solvable. The solution is given by

$$\varphi(x) = a \sin x + \varphi_p(x) \quad \forall x \in I, \text{ where } a = -\frac{\varphi_p(\ell)}{\sin \ell}.$$

- If ℓ is an integer multiple of π, for any given $b \in C^0(I)$, there exist solutions to the problem (6.10) if and only if $\varphi_p(\ell) = 0$, that is, if and only if

$$\int_0^\ell b(s) \sin s \, ds = 0.$$

In this case, the solutions are the following functions:

$$\varphi(x) = C \sin x + \varphi_p(x) \quad \forall x \in I,$$

where $C \in \mathbb{R}$. □

Remark 6.3. Suppose that (6.8) only possesses the trivial solution. For every $b \in C^0(I)$, there exists exactly one solution ψ_b to the problem

$$\begin{cases} y' = a_1(x)\, y' + a_0(x)\, y + b(x) \\ B \begin{pmatrix} y(\alpha) \\ y'(\alpha) \end{pmatrix} + \begin{pmatrix} y(\beta) \\ y'(\beta) \end{pmatrix} = 0. \end{cases} \quad (6.11)$$

Now, let us introduce the set

$$Y_0 := \{ \varphi \in C^2(I) : B \begin{pmatrix} \varphi(\alpha) \\ \varphi'(\alpha) \end{pmatrix} + C \begin{pmatrix} \varphi(\beta) \\ \varphi'(\beta) \end{pmatrix} = 0 \}.$$

Then Y_0 is an infinite dimensional subspace of $C^2(I)$ and the mapping $b \mapsto \psi_b$ is linear, continuous, and bijective from $C^0(I)$ onto Y_0, again called the *Green operator*; for more details, see Note 11.6.2. □

Remark 6.4. In differential models coming from phenomena arising in other sciences, it is usual to find the so-called *Sturm–Liouville problems*. Here is a particular example, where $a_0 \in C^0([\alpha, \beta])$, and $b_0, b_1, c_0, c_1 \in \mathbb{R}$:

Find couples $(\lambda, \varphi) \in \mathbb{R} \times C^2([\alpha, \beta])$ *such that* φ *is a nontrivial solution to the boundary-value problem*

$$\begin{cases} -y'' + a_0(x)\,y = \lambda y \\ b_0\,y(\alpha) + b_1\,y'(\alpha) = c_0\,y(\beta) + c_1\,y'(\beta) = 0. \end{cases} \tag{6.12}$$

If (λ, φ) is as indicated, it is said that λ is an eigenvalue and φ, an associated eigenfunction of (6.12).

The reader can find in Note 11.6.3 more details on the origin, properties, and usefulness of problems of this kind.   □

Exercises

E.6.1 For the ODS

$$y' = \begin{pmatrix} 1 & 2 & a \\ 0 & 1 & 2 \\ 0 & 0 & 1 \end{pmatrix} y + \begin{pmatrix} e^x \\ 0 \\ 0 \end{pmatrix},$$

where $a \in \mathbb{R}$, analyze the existence of solutions corresponding to the following boundary conditions:

(1) $By(0) + Cy(1) = 0.$
(2) $By(0) + Cy(1) = h.$

Here,

$$h = \begin{pmatrix} 1 \\ 0 \\ 0 \end{pmatrix}, \quad B = \begin{pmatrix} -1 & -2 & -5 \\ 1 & -1 & -2 \\ 0 & 1 & -1 \end{pmatrix}, \quad C = \frac{1}{e}\,\mathrm{Id}.$$

Find the real values a for which these problems possess a unique solution. For $a = 3$, find the solution(s) in both cases (if they exist).

E.6.2 Let $b \in C^0(\mathbb{R})$ be given. Find the values of $a \in \mathbb{R}$ for which the following problem is solvable:

$$\begin{cases} y'' - 3y' + 2y = b(x) \\ y'(0) - y(0) = y'(a) - ey(0) = 0. \end{cases}$$

Compute the solution for $b(x) \equiv 4x - 2$ and $a = 1$.

E.6.3 Let us set

$$A = \begin{pmatrix} 0 & 2 \\ 0 & 1 \end{pmatrix}, \quad b(x) = \begin{pmatrix} e^x \\ 0 \end{pmatrix}, \quad B = \begin{pmatrix} 0 & 2 \\ -1 & 2 \end{pmatrix}, \quad C = \begin{pmatrix} 1 & -2 \\ 1 & -2 \end{pmatrix}.$$

Determine whether or not the following boundary-value problems possess a solution:

$$\begin{cases} y' = Ay \\ By(0) + Cy(1) = 0 \end{cases} \qquad \begin{cases} y' = Ay + b(x) \\ By(0) + Cy(1) = 0 \end{cases} \qquad \begin{cases} y' = Ay + b(x) \\ By(0) + Cy(1) = h. \end{cases}$$

Compute the solution(s) when they exist.

E.6.4 Consider the ODS

$$y' = \begin{pmatrix} t+1 & 2(t-1) \\ 0 & t+1 \end{pmatrix} y + \begin{pmatrix} e^{t^2/2} \\ 0 \end{pmatrix} \qquad (6.13)$$

Do the following:

(1) Find a fundamental matrix for the corresponding homogeneous linear system.
(2) Find the general solution to the associated nonhomogeneous linear system.
(3) Find those $\alpha \in \mathbb{R}$ for which the boundary-value problem associated to (6.13) and the boundary condition

$$\begin{pmatrix} 0 & 0 \\ 0 & \alpha \end{pmatrix} y(-2) + \begin{pmatrix} 1 & 0 \\ 1 & 0 \end{pmatrix} y(0) = 0$$

possesses exactly one solution.
(4) Compute the solutions for $\alpha = 0$ and $\alpha = 1$ (if they exist).

E.6.5 Consider the ODS

$$y' = \begin{pmatrix} 1 & -4 & 1 \\ 0 & -1 & 0 \\ 0 & 0 & 0 \end{pmatrix} y + \begin{pmatrix} e^x \\ 0 \\ e^x \end{pmatrix}. \qquad (6.14)$$

Do the following:

(1) Find the general solution and also the solution corresponding to the Cauchy data $(0; 0, 1, 1)$.
(2) Find those values of $\alpha \in \mathbb{R}$ for which the boundary-value problem associated to (6.13) and the boundary condition

$$\begin{pmatrix} \alpha & -2\alpha & 0 \\ 0 & 0 & 0 \\ 0 & 0 & -\alpha \end{pmatrix} y(0) + \begin{pmatrix} 0 & 0 & 0 \\ 0 & e & 0 \\ 0 & 0 & \alpha \end{pmatrix} y(1) = \begin{pmatrix} 0 \\ 1 \\ 0 \end{pmatrix}$$

is solvable. In particular, if it is possible, solve the problem for $\alpha = 0$.

E.6.6 Prove that the boundary-value problem

$$\begin{cases} y''' + y' = b(x) \\ y(0) + y'(\pi/2) = y'(0) + y''(0) = y''(\pi/2) = 0 \end{cases}$$

possesses exactly one solution $\varphi_b \in C^3([0, \pi/2])$ for every $b \in C^0([0, \pi/2])$.

E.6.7 Consider the boundary-value problem

$$\begin{cases} y'' + y = b(x) \\ y(0) = y'(\pi) = 0 \end{cases}$$

and do the following:

(1) Prove that it is uniquely solvable for every $b \in C^0([0, \pi])$.
(2) Find the unique solution in terms of b.

E.6.8 Consider the boundary-value problem

$$\begin{cases} y'' - y = b(x) \\ y'(0) = 0, \ y(1) + y'(1) = 0 \end{cases} \qquad (6.15)$$

and do the following:

(1) Prove that, for every $b \in C^0([0, 1])$, (16.15) is uniquely solvable.
(2) Find the solution to (16.15) in terms of b.

E.6.9 Consider the boundary-value problem

$$\begin{cases} y'' + y = 1 \\ y(0) + y'(0) = 0, \ y(2\pi) + \alpha y'(2\pi) = 0, \end{cases} \qquad (6.16)$$

where $\alpha \in \mathbb{R}$. The following is required:

(1) Find the general solution to the ODE in (6.16).
(2) Find those values of α for which (6.16) possesses exactly one solution.
(3) Solve (6.16) for $\alpha = 1$.

E.6.10 For the following boundary-value problems, (a) prove that for every continuous function b (in the indicated interval) there exists a unique solution and (b) find the corresponding Green operator.

(1) $x^2 y'' + 3xy' + y = b(x)$, $x \in [1, e]$; $y(1) + y'(1) = 0$, $y'(e) = 0$.

(2) $y''' + y' = b(x)$, $x \in [0, \pi/2]$; $y(0) + y'(\pi/2) = y(0) + y''(0) = y''(\pi/2) = 0$.

(3) $y'' + y = b(x)$, $x \in [0, \pi]$; $y(0) = y'(\pi) = 0$.

(4) $y'' - y = b(x)$, $x \in [0, 1]$; $y'(0) = y(1) + y'(1) = 0$.

E.6.11 Find the eigenvalues λ and eigenfunctions φ of the following Sturm–Liouville problems:

(1) $y'' + \lambda y = 0$, $y(0) = y(\pi) = 0$.

(2) $y'' + \lambda y = 0$, $y'(0) = y'(\pi) = 0$.

(3) $y'' + \lambda y = 0$, $y(-1) = y'(1) = 0$.

(4) $y'' + \lambda y = 0$, $y(0) - y(\ell) = y'(0) - y'(\ell) = 0$.

(5) $y'' - 2y' + \lambda y = 0$, $y(0) = y(1) - y'(1) = 0$.

(6) $y'' - y + \lambda y = 0$, $y(0) = y'(\pi) = 0$.

(7) $y'' + 4y' + y + \lambda y = 0$, $y(0) = 2y(\pi) + y'(\pi) = 0$.

Chapter 7

Some Regularity Results

In this chapter, we will consider again Cauchy problems for first-order ODSs

$$\text{(CP)} \qquad \begin{cases} y' = f(t, y) \\ y(t_0) = y_0 \end{cases}$$

and ODEs of order n

$$\text{(CP)}_n \qquad \begin{cases} y^{(n)} = f(t, y, y', \dots, y^{(n-1)}) \\ y(t_0) = y_0^{(0)}, \ y'(t_0) = y_0^{(1)}, \dots, \ y^{(n-1)}(t_0) = y_0^{(n-1)}. \end{cases}$$

In all of them the function f and the initial data will be assumed to satisfy the usual existence and uniqueness assumptions.

That is, in the case of (CP), it will be assumed that $f : \Omega \mapsto \mathbb{R}^N$ is continuous and locally Lipschitz-continuous with respect to the variable y ($\Omega \subset \mathbb{R}^{N+1}$ is a non-empty open set) and $(t_0, y_0) \in \Omega$.

When considering (CP)_n, it will be assumed that $f : \Omega \mapsto \mathbb{R}$ is continuous and locally Lipschitz-continuous with respect to the variable $Y = (y, y', \dots, y^{(n-1)})$ (Ω is now a non-empty open set in $\mathbb{R}^{n+1}$) and $(t_0, y_0^{(0)}, \dots, y_0^{(n-1)}) \in \Omega$.

Our purpose is twofold:

(1) To prove that the more regular f is considered, the more regular the solution is. This obeys a general principle that holds for the solutions to differential problems.
(2) To identify the way the maximal solutions to (CP) and (CP)_n depend on the initial data.

Obviously, it is expected that maximal solutions corresponding to initial data that are close cannot be two different. But this has to be established rigorously.

7.1 First Results

This section is devoted to proving some first regularity results of the solutions, of a local kind.

We will see for the first time that increasing regularity properties of f have the effect of increasing the regularity of the solutions.

As already mentioned, we will consider the following ODSs and ODEs

$$y' = f(t, y) \tag{7.1}$$

and

$$y^{(n)} = f(t, y, y', \dots, y^{(n-1)}). \tag{7.2}$$

Let $G \subset \mathbb{R}^p$ be a non-empty open set and let $F : G \mapsto \mathbb{R}^q$ be given. It will be said that F is of class C^1 in G if, at any point in G, the components of F possess partial derivatives with respect to all their arguments and, furthermore, these partial derivatives are continuous in G.

By induction, we can also speak of functions of class C^m in G for any m.

More precisely, for any $m \geq 2$, it will be said that F is of class C^m in G if it is of class C^{m-1} and, moreover, any partial derivative of any component of F of order $m-1$ is of class C^1 in G.

Our first regularity result is the following:

Theorem 7.1. *Let $I \subset \mathbb{R}$ be an open interval and assume that $\varphi : I \mapsto \mathbb{R}^N$ is a solution in I to the ODS (7.1). Let $t_0 \in I$ be given and assume that f is of class C^m in a neighborhood of $(t_0, \varphi(t_0))$ (m is an integer ≥ 1). Then φ is of class C^{m+1} in a neighborhood of t_0.*

Proof. We know that $(t, \varphi(t)) \in \Omega$ and $\varphi'(t) = f(t, \varphi(t))$ for all $t \in I$.

The proof is not difficult. It suffices to argue by induction on m.

For $m = 1$, f is of class C^1 in a neighborhood of $(t_0, \varphi(t_0))$ and then φ' is continuously differentiable (that is, φ is of class C^2) in a neighborhood of t_0.

Now, let us assume that the result is true for $m = k$ and let f be of class C^{k+1} in a neighborhood of $(t_0, \varphi(t_0))$. Then φ' is of class C^{k+1}

in a neighborhood of t_0 (since φ' coincides with the composition of two functions that possess continuous partial derivatives up to order $k + 1$).

That means that φ is of class C^{k+2} in a neighborhood of t_0.

This ends the proof. □

Observe that, with the help of the *Chain Rule*, we can compute the derivatives of the solutions to (7.1) of any order in terms of partial derivatives of the components of f, provided they exist.

Indeed, let us assume (for instance) that the assumptions of the previous theorem are satisfied with $m = 2$.

Then, for any $i = 1, \dots, N$ and any t in a neighborhood of t_0, we have

$$\varphi_i''(t) = \frac{d}{dt} f_i(t, \varphi(t)) = \left(\frac{\partial f_i}{\partial t} + \sum_{j=1}^{N} \frac{\partial f_i}{\partial y_j} f_j \right) (t, \varphi(t))$$

and

$$\varphi_i'''(t) = \frac{d^2}{dt^2} f_i(t, \varphi(t)) = \frac{d}{dt} \left[\left(\frac{\partial f_i}{\partial t} + \sum_{j=1}^{N} \frac{\partial f_i}{\partial y_j} f_j \right) (t, \varphi(t)) \right]$$

$$= \left(\frac{\partial^2 f_i}{\partial t^2} + 2 \sum_{j=1}^{N} \frac{\partial^2 f_i}{\partial t \partial y_j} f_j + \sum_{j,k=1}^{N} \frac{\partial^2 f_i}{\partial y_j \partial y_k} f_j f_k \right.$$

$$\left. + \sum_{j=1}^{N} \frac{\partial f_i}{\partial y_j} \left(\frac{\partial f_j}{\partial t} + \sum_{k=1}^{N} \frac{\partial f_j}{\partial y_k} f_k \right) \right) (t, \varphi(t)).$$

As a consequence of Theorem 7.1, we can also deduce local regularity results for the solutions to (7.2):

Corollary 7.1. *Let $I \subset \mathbb{R}$ be an open interval and let $\varphi : I \mapsto \mathbb{R}$ be a solution in I to the ODE (7.2). Let $t_0 \in I$ be given and assume that f is of class C^m in a neighborhood of $(t_0, \varphi(t_0), \varphi'(t_0), \dots, \varphi^{(n-1)}(t_0))$, where $m \geq 1$ is an integer. Then φ is of class C^{m+n} in a neighborhood of t_0.*

Thus, we see that the following general principle is satisfied:

The solutions to an ODS of the first order for which the right-hand side is of class C^m are functions of class C^{m+1}. Also, in the case of an ODE of order n with a right-hand side of class C^m, the solutions are of class C^{m+n}.

7.2   The General Maximal Solution: Continuity

This and the following section are inspired by the arguments and results in [Rouché and Mawhin (1973)].

In order to analyze rigorously the way the solution to (CP) depends on t, t_0 and y_0, we will introduce a new concept: *the general maximal solution*. It is the following:

Definition 7.1. For every $(t_0, y_0) \in \Omega$, we will denote by $I(t_0, y_0)$ the interval of definition of the maximal solution to (CP). Accordingly, we set

$$\Theta := \{(t, t_0, y_0) \in \mathbb{R}^{N+2} : (t_0, y_0) \in \Omega, \ t \in I(t_0, y_0)\}.$$

On the other hand, for every $(t_0, y_0) \in \Omega$, we will denote by $\varphi(t; t_0, y_0)$ the value at t of the unique maximal solution to (CP); in other words, $\varphi(\cdot\,; t_0, y_0)$ will stand for the maximal solution to (CP). Then, it will be said that $\varphi : \Theta \mapsto \mathbb{R}^N$ is the general maximal solution to (7.1).   □

For example, if $I \subset \mathbb{R}$ is an open interval, $A : I \mapsto \mathcal{L}(\mathbb{R}^N)$ and $b : I \mapsto \mathbb{R}^N$ are continuous and

$$f(t, y) = A(t)y + b(t) \quad \forall (t_0, y_0) \in I \times \mathbb{R}^N,$$

we obviously have $\Theta = I \times I \times \mathbb{R}^N$. If $F : I \mapsto \mathcal{L}(\mathbb{R}^N)$ is a fundamental matrix of the corresponding homogeneous linear system, then the general maximal solution φ is given by

$$\varphi(t; t_0, y_0) = F(t)F(t_0)^{-1}y_0 + F(t) \int_{t_0}^{t} F(s)^{-1}b(s)\,\mathrm{d}s$$

$$\forall (t, t_0, y_0) \in I \times I \times \mathbb{R}^N.$$

The main result in this section is the following:

Theorem 7.2. *The set Θ is open and the general maximal solution $\varphi : \Theta \mapsto \mathbb{R}^N$ is continuous.*

Remark 7.1. From a practical viewpoint, the continuity of φ is crucial. In fact, this is the property that allows to view ODEs as a useful tool to model and describe phenomena with origin in sciences.

Indeed, the values of t_0 and y_0 are almost always the result of one or several measures and can be *inexact*. In fact, this is what happens in practice.

If φ were not continuous, there would be no way to guarantee that small perturbations of the initial data and/or the considered values of t generate (only) small perturbations of the values of φ. □

Theorem 7.2 is obviously satisfied in the case of a linear ODS with continuous coefficients. The proof in the general case is given in Section 7.3.

Of course, it is also possible to speak of the general maximal solution to $(CP)_n$ as follows:

Definition 7.2. For every $(t_0, y_0^{(0)}, \ldots, y_0^{(n-1)}) \in \Omega$, the interval of definition of the maximal solution to $(CP)_n$ will be denoted by

$$ I^*(t_0, y_0^{(0)}, \ldots, y_0^{(n-1)}). $$

Let us set

$$ \Theta^* = \{(t, t_0, y_0^{(0)}, \ldots, y_0^{(n-1)}) \in \mathbb{R}^{n+2} : (t_0, y_0^{(0)}, \ldots, y_0^{(n-1)}) \in \Omega, $$
$$ t \in I^*(t_0, y_0^{(0)}, \ldots, y_0^{(n-1)})\} $$

and, for every $(t_0, y_0^{(0)}, \ldots, y_0^{(n-1)}) \in \Omega$, let us denote by $\varphi^*(t; t_0, y_0^{(0)}, \ldots, y_0^{(n-1)})$ the value at t of the unique maximal solution to $(CP)_n$. The function $\varphi^* : \Theta^* \mapsto \mathbb{R}$ is by definition the general maximal solution to (7.2). □

An immediate consequence of Theorem 7.2 is the following:

Theorem 7.3. *The set Θ^* is open and the general maximal solution φ^* : $\Theta^* \mapsto \mathbb{R}$ is continuous.*

7.3 Proof of Theorem 7.2

The following result will be needed:

Proposition 7.1. *Let $(t_0^*, y_0^*) \in \Omega$ be given and let $a, b \in \mathbb{R}$ be such that $a < t_0^* < b$, $I(t_0^*, y_0^*) \supset [a, b]$. Then, for every $\varepsilon > 0$, there exists a neighborhood V_ε of (t_0^*, y_0^*) such that:*

(1) $V_\varepsilon \subset \Omega$.

(2) $I(t_0, y_0) \supset [a, b]$ *for all* $(t_0, y_0) \in V_\varepsilon$.

(3) $\max_{t \in [a,b]} |\varphi(t; t_0, y_0) - \varphi(t; t_0^*, y_0^*)| \leq \varepsilon$ *for every* $(t_0, y_0) \in V_\varepsilon$.

Proof. Let us set

$$K := \{(t, \varphi(t; t_0^*, y_0^*)) \in \mathbb{R}^{N+1} : t \in [a, b]\}.$$

This is a non-empty compact set in Ω. Hence, $d(K, \partial\Omega) > 0$.

We will prove the result for all ε satisfying $0 < \varepsilon < d(K, \partial\Omega)$. Obviously, this will be sufficient.

Thus, let such an ε be given and set

$$A := \{(t, y) \in \mathbb{R}^{N+1} : d((t, y), K) < \varepsilon\}.$$

This set is open (because $(t, y) \mapsto d((t, y), K)$ is continuous) and bounded. Furthermore, it is clear that $K \subset A$.

On the other hand, $\overline{A} = \{(t, y) \in \mathbb{R}^{N+1} : d((t, y), K) \leq \varepsilon\}$ and, in view of the definition of ε, $\overline{A}$ is a new compact subset of Ω.

Let $L > 0$ be a Lipschitz constant for f with respect to the variable y in $\overline{A}$ and let us set

$$V_\varepsilon = \{(t, y) \in \mathbb{R}^{N+1} : t \in [a, b], \ |y - \varphi(t; t_0^*, y_0^*)| < \varepsilon e^{-L(b-a)}\}.$$

It is then clear that V_ε is an open neighborhood of (t_0^*, y_0^*) satisfying

$$K \subset V_\varepsilon \subset A \subset \Omega.$$

Let $(t_0, y_0) \in V_\varepsilon$ be given. Let us see that $I(t_0, y_0) \supset [a, b]$.

Thus, let f_A be the restriction of f to the set A and let φ_A be the maximal solution to the Cauchy problem

$$\begin{cases} y' = f_A(t, y) \\ y(t_0) = y_0. \end{cases} \tag{7.3}$$

Obviously, the interval I_A where the function φ_A is defined satisfies $I_A \subset I(t_0, y_0)$.

Suppose that, for example, $\alpha := \inf I_A \geq a$. For simplicity, let us denote by φ^* the function $\varphi(\cdot\,; t_0^*, y_0^*)$.

Then, for any $t \in (\alpha, t_0]$, we have $(t, \varphi_A(t)), (t, \varphi^*(t)) \in \overline{A}$, and

$$|\varphi_A(t) - \varphi^*(t)| = |y_0 - \varphi^*(t_0) + \int_{t_0}^t (f(s, \varphi_A(s)) - f(s, \varphi^*(s)))ds|$$

$$\leq |y_0 - \varphi^*(t_0)| + L \int_{t_0}^t |\varphi(s) - \varphi^*(s)|ds. \tag{7.4}$$

In view of Gronwall's Lemma, the following is obtained:

$$|\varphi_A(t) - \varphi^*(t)| \le |y_0 - \varphi^*(t_0)| e^{L(t_0-a)} < \varepsilon \quad \forall t \in (\alpha, t_0].$$

But this means that the left half-trajectory of φ_A is inside a compact that is inside A, which contradicts the maximality of φ_A as a solution to (7.3).

Consequently, $\alpha < a$ and, even more, $\inf I(t_0, y_0) < a$.

In a similar way, it is proved that $\sup I(t_0, y_0) > b$. Therefore, $I(t_0, y_0) \supset [a, b]$.

Finally, we observe that

$$|\varphi(t; t_0, y_0) - \varphi(t; t_0^*, y_0^*)| \le \varepsilon \quad \forall (t_0, y_0) \in V_\varepsilon.$$

Indeed, it suffices to argue as in (7.4) and apply again Gronwall's Lemma.

This ends the proof. $\qquad\square$

Now, let us prove Theorem 7.2, that is, that Θ is an open set in $\mathbb{R}^{N+2}$ and the general maximal solution $\varphi : \Theta \mapsto \mathbb{R}^N$ is continuous.

Let $(t^*, t_0^*, y_0^*) \in \Theta$ be given. We have $(t_0^*, y_0^*) \in \Omega$ and t^* obviously belongs to the open interval $I(t_0^*, y_0^*)$. Let $\delta > 0$ be such that $[t^* - \delta, t^* + \delta] \subset I(t_0^*, y_0^*)$. Let us set $a = t^* - \delta$ and $b = t^* + \delta$, let $\varepsilon > 0$ be given, and let V_ε be the corresponding neighborhood of (t_0^*, y_0^*) furnished by Proposition 7.1.

Then $[t^* - \delta, t^* + \delta] \times V_\varepsilon \subset \Theta$. Consequently, Θ is open.

Finally, let $(t^*, t_0^*, y_0^*) \in \Theta$ and $\kappa > 0$ be given. We know that $\varphi(\cdot; t_0^*, y_0^*)$ is continuous. Consequently, there exists $\delta > 0$, such that, if $t \in I(t_0^*, y_0^*)$ and $|t - t^*| \le \delta$, then

$$|\varphi(t; t_0^*, y_0^*) - \varphi(t^*; t_0^*, y_0^*)| \le \frac{\kappa}{2}.$$

Let V be the neighborhood of (t_0^*, y_0^*) constructed as in Proposition 7.1, with $a = t^* - \delta$, $b = t^* + \delta$ and $\varepsilon = \kappa/2$. Then, if $(t, t_0, y_0) \in [t^* - \delta, t^* + \delta] \times V$, one has

$$|\varphi(t; t_0, y_0) - \varphi(t^*; t_0^*, y_0^*)| \le |\varphi(t; t_0, y_0) - \varphi(t; t_0^*, y_0^*)|$$

$$+ |\varphi(t; t_0^*, y_0^*) - \varphi(t^*; t_0^*, y_0^*)|$$

$$\le \kappa.$$

This proves that φ is continuous and ends the proof of Theorem 7.2.

7.4 The General Maximal Solution: Differentiability

According to Theorem 7.2, under the usual existence and uniqueness conditions for (CP), the general maximal solution $\varphi : \Theta \mapsto \mathbb{R}^N$ is continuous.

Obviously, since it solves the ODS, the partial derivatives of its components with respect to the variable t exist. In fact,

$$\frac{\partial \varphi_i}{\partial t}(t; t_0, y_0) = f_i(t, \varphi(t; t_0, y_0)) \quad \forall (t, t_0, y_0) \in \Theta$$

and, consequently, the φ_i are continuously differentiable with respect to t in Θ.

In this section, we investigate whether, under appropriate assumptions, the partial derivatives with respect to t_0 and with respect to the components $y_{0\ell}$ of y_0 also exist.

Let us state the main result:

Theorem 7.4. *Assume that $f : \Omega \mapsto \mathbb{R}^N$ is continuous and the components of f possess continuous partial derivative with respect to all the arguments. Then, the following holds:*

(1) *The general maximal solution satisfies $\varphi \in C^1(\Theta; \mathbb{R}^N)$.*
(2) *Let us introduce w^0, with*

$$w^0(t; t_0, y_0) = \frac{\partial \varphi}{\partial t_0}(t; t_0, y_0) \quad \forall (t, t_0, y_0) \in \Theta.$$

For every $(t_0, y_0) \in \Omega$, $w^0(\cdot; t_0, y_0)$ (that is a function defined in $I(t_0, y_0)$) is the unique maximal solution to the Cauchy problem

$$\begin{cases} w' = A(t; t_0, y_0)w \\ w(t_0) = -f(t_0, y_0), \end{cases} \tag{7.5}$$

where $A(t; t_0, y_0)$ is the $N \times N$ matrix of components

$$A_{ij}(t; t_0, y_0) = \frac{\partial f_i}{\partial y_j}(t, \varphi(t; t_0, y_0)).$$

(3) *For every $\ell = 1, \ldots, N$, let us set*

$$w^\ell(t; t_0, y_0) = \frac{\partial \varphi}{\partial y_{0\ell}}(t; t_0, y_0) \quad \forall (t, t_0, y_0) \in \Theta.$$

Then, for every $(t_0, y_0) \in \Omega$, $w^\ell(\cdot; t_0, y_0)$ (that is again defined in $I(t_0, y_0)$) is the unique maximal solution to the Cauchy problem

$$\begin{cases} w' = A(t; t_0, y_0)w \\ w(t_0) = e^\ell, \end{cases} \qquad (7.6)$$

where e^ℓ is the ℓth vector of the canonical basis of $\mathbb{R}^N$ and $A(t; t_0, y_0)$ is as before.

(4) Finally, the second-order partial derivatives

$$\frac{\partial^2 \varphi_i}{\partial t^2}, \quad \frac{\partial^2 \varphi_i}{\partial t \partial t_0} = \frac{\partial^2 \varphi_i}{\partial t_0 \partial t}, \quad \text{and} \quad \frac{\partial^2 \varphi_i}{\partial t \partial y_{0\ell}} = \frac{\partial^2 \varphi_i}{\partial y_{0\ell} \partial t}$$

exist and are continuous in Θ.

The Cauchy problems (7.5) and (7.6) are usually known as variational problems associated to (CP). This is reminiscent of *Calculus of Variations*.

Remark 7.2. Taking into account Theorems 7.2 and 7.4, it is possible to deduce some additional regularity properties of φ when more restrictive assumptions are imposed on f.

For example, if we assume that f is C^2 in Ω, then $\varphi \in C^2(\Theta; \mathbb{R}^N)$ and, moreover, the following third-order partial derivatives exist and are continuous in Θ :

$$\frac{\partial^3 \varphi_i}{\partial t^3}, \quad \frac{\partial^3 \varphi_i}{\partial t \partial t_0^2}, \quad \frac{\partial^3 \varphi_i}{\partial t \partial t_0 \partial y_{0\ell}}, \quad \frac{\partial^3 \varphi_i}{\partial t \partial y_{0m} \partial y_{0\ell}}, \quad i, \ell, m = 1, \dots, N. \quad \square$$

Of course, it is again possible to state and prove a result similar to Theorem 7.4 for the general maximal solution to $(\mathrm{CP})_n$.

This task is left to the reader.

7.5   Proof of Theorem 7.4*

We know that $\varphi : \Theta \mapsto \mathbb{R}^N$ is continuous and possesses continuous partial derivative with respect to the variable t.

Let $(t^*, t_0, y_0) \in \Theta$ be given. Let us see that φ also possesses partial derivatives with respect to t_0 at (t^*, t_0, y_0) and

$$\frac{\partial \varphi}{\partial t_0}(t^*; t_0, y_0) = w^0(t^*; t_0, y_0),$$

where $w^0(\cdot; t_0, y_0)$ is the maximal solution to (7.5).

For instance, let us suppose that $t^* \geq t_0$. Let $h_0 > 0$ be given such that $[t_0 - h_0, \max(t_0 + h_0, t^*)] \subset I(t_0, y_0)$ and set $a = t_0 - h_0$ and $b = \max(t_0 + h_0, t^*)$.

Thanks to Proposition 7.1, there exists $h_1 > 0$ such that, if $0 \leq |h| \leq h_1$, one has $[a, b] \subset I(t_0 + h, y_0)$.

For every i with $1 \leq i \leq N$ and every h with $0 < |h| \leq h_1$, we set

$$\psi_{i,h}(t) := \frac{1}{h} \left(\varphi_i(t; t_0 + h, y_0) - \varphi_i(t; t_0, y_0) \right) \quad \forall t \in [a, b].$$

Then

$$\psi_{i,h}(t) = \frac{1}{h} \left(\int_{t_0+h}^{t} f_i(s, \varphi(s; t_0 + h, y_0)) \, ds - \int_{t_0}^{t} f_i(s, \varphi(s; t_0, y_0)) \, ds \right)$$

$$= \frac{1}{h} \int_{t_0}^{t} \left(f_i(s, \varphi(s; t_0 + h, y_0)) - f_i(s, \varphi(s; t_0, y_0)) \right) ds$$

$$- \frac{1}{h} \int_{t_0}^{t_0+h} f_i(s, \varphi(s; t_0 + h, y_0)) \, ds$$

$$= \int_{t_0}^{t} \sum_{j=1}^{N} \frac{\partial f_i}{\partial y_j}(s, \varphi(s; \hat{t}_h(s), y_0)) \psi_{j,h}(s) \, ds - f_i(\tilde{t}_h, \varphi(\tilde{t}_h; t_0 + h, y_0)),$$

where $\hat{t}_h(s)$ and $\tilde{t}_h$ belong to the interval of endpoints t_0 and $t_0 + h$.

Consequently, for every $t \in [a, b]$, one has

$$\psi_{i,h}(t) - w_i^0(t; t_0, y_0)$$

$$= \int_{t_0}^{t} \sum_{j=1}^{N} \frac{\partial f_i}{\partial y_j}(s, \varphi(s; \hat{t}_h(s), y_0)) \left(\psi_{j,h}(s) - w_j^0(s; t_0, y_0) \right) ds$$

$$+ \int_{t_0}^{t} \sum_{j=1}^{N} \left(\frac{\partial f_i}{\partial y_j}(s, \varphi(s; \hat{t}_h(s), y_0)) - \frac{\partial f_i}{\partial y_j}(s, \varphi(s; t_0, y_0)) \right)$$

$$\times w_j^0(s; t_0, y_0) \, ds$$

$$- \left[f_i(\tilde{t}_h, \varphi(\tilde{t}_h; t_0 + h, y_0)) - f_i(t_0, y_0) \right].$$

Let K be the following compact set in $\mathbb{R}^{N+1}$:

$$K = \{ (s, \varphi(s; t_0 + h, y_0)) : s \in [a, b], \ h \in [-h_1, h_1] \}.$$

Let ψ_h be the function of components $\psi_{i,h}$ (obviously, ψ_h is well defined and continuous in $[a, b]$). It is not difficult to deduce that

$$|\psi_h(t) - w^0(t; t_0, y_0)| \leq A_1 \int_{t_0}^t |\psi_h(s) - w^0(s)| \, ds$$

$$+ A_{2,h} \int_{t_0}^t |w^0(s; t_0, y_0)| \, ds + A_{3,h}$$

for every $t \in [a, b]$, where

$$A_1 = \max_{(s,y) \in K} \left(\sum_{i,j=1}^N \left| \frac{\partial f_i}{\partial y_j}(s, y) \right|^2 \right)^{1/2},$$

$$A_{2,h} = \max_{s \in [a,b], |h'| \leq |h|} \left(\sum_{i,j=1}^N \left| \frac{\partial f_i}{\partial y_j}(s, \varphi(s; t_0, y_0)) \right. \right.$$

$$\left. \left. - \frac{\partial f_i}{\partial y_j}(s, \varphi(s; t_0 + h', y_0)) \right|^2 \right)^{1/2}$$

and

$$A_{3,h} = \max_{s \in [a,b], |h'| \leq |h|} |f(s, \varphi(s; t_0, y_0)) - f(s, \varphi(s; t_0 + h', y_0))|.$$

Therefore, in view of Gronwall's Lemma, we get that

$$|\psi_h(t) - w^0(t)| \leq A_{4,h} e^{A_1 (b-a)} \quad \forall t \in [a, b],$$

where

$$A_{4,h} = A_{2,h} \int_{t_0}^b |w^0(s; t_0, y_0)| \, ds + A_{3,h} .$$

It is clear that $A_{4,h} \to 0$ as $h \to 0$. This is a consequence of the values of $A_{2,h}$ and $A_{3,h}$ and the continuity of f, its first partial derivatives, and φ. Indeed, note that the functions

$$(s, h') \mapsto f_i(s, \varphi(s; t_0 + h', y_0)) \quad \text{and} \quad (s, h') \mapsto \frac{\partial f_i}{\partial y_j}(s, \varphi(s; t_0 + h', y_0))$$

are uniformly continuous in the compact set $[a, b] \times [-h_1, h_1]$.

Thus, $|\psi_h(t) - w^0(t)| \to 0$ and, in particular, we see that $\varphi(t^*; \cdot, y_0)$ is differentiable at t_0 and its derivative is equal to $w^0(t^*; t_0, y_0)$.

In a similar way, it can be proved that, at each $(t^*, t_0, y_0) \in \Theta$, φ possesses partial derivatives with respect to the $y_{0\ell}$ and moreover

$$\frac{\partial \varphi}{\partial y_{0\ell}}(t^*; t_0, y_0) = w^\ell(t^*; t_0, y_0),$$

where $w^\ell(\,\cdot\,; t_0, y_0)$ is the maximal solution to (7.6).

The other assertions in the statement of Theorem 7.4 are immediate consequences of the properties satisfied by the solutions to the Cauchy problems (7.5) and (7.6).

7.6   Continuity and Differentiability with Respect to Parameters*

Theorems 7.2 and 7.4 can be applied to the analysis of the dependence of the solutions to Cauchy problems with respect to parameters in many different situations.

The results are of major interest from both the theoretical viewpoint and the relevance in applications.

Thus, let $\Omega \subset \mathbb{R}^{N+1}$ be a non-empty open set and let $\tilde{f} : \Omega \times \mathbb{R}^k \mapsto \mathbb{R}^N$ be a given function. We will assume (at least) that $\tilde{f}$ is continuous and locally Lipschitz-continuous with respect to (y, λ).

Then, for every $(t_0, y_0) \in \Omega$ and every $\lambda \in \mathbb{R}^k$, the Cauchy problem

$$\begin{cases} y' = \tilde{f}(t, y, \lambda) \\ y(t_0) = y_0 \end{cases} \tag{7.7}$$

possesses a unique maximal solution, defined in the open interval $I(t_0, y_0, \lambda)$.

We will set

$$\tilde{\Theta} := \{(t, t_0, y_0, \lambda) \in \mathbb{R}^{N+k+2} : (t_0, y_0) \in \Omega, \ \lambda \in \mathbb{R}^k, \ t \in I(t_0, y_0, \lambda)\}$$

and, for every $(t, t_0, y_0, \lambda) \in \tilde{\Theta}$, we will denote by $\tilde{\varphi}(t; t_0, y_0, \lambda)$ the value at t of the maximal solution to (7.7).

We will say that $\tilde{\varphi} : \tilde{\Theta} \mapsto \mathbb{R}^N$ is the *general maximal solution* to (7.7).

As a consequence of Theorems 7.2 and 7.4, we get the following:

Theorem 7.5. *Assume that $\tilde{f} : \Omega \times \mathbb{R}^k \mapsto \mathbb{R}^N$ is continuous and locally Lipschitz-continuous with respect to (y, λ). Then, the corresponding set $\tilde{\Theta} \subset \mathbb{R}^{N+k+2}$ is open and the general maximal solution $\tilde{\varphi} : \tilde{\Theta} \mapsto \mathbb{R}^N$ is continuous.*

Theorem 7.6. *Let us assume that* $\tilde{f} : \Omega \times \mathbb{R}^k \mapsto \mathbb{R}^N$ *is continuous and continuously differentiable with respect to all the arguments. Then the following holds:*

(1) $\tilde{\varphi} \in C^1(\tilde{\Theta}; \mathbb{R}^N)$.

(2) *Let us set*

$$\tilde{w}^0(t; t_0, y_0, \lambda) := \frac{\partial \tilde{\varphi}}{\partial t_0}(t; t_0, y_0, \lambda) \quad \forall(t, t_0, y_0, \lambda) \in \tilde{\Theta},$$

Then, for every $(t_0, y_0, \lambda) \in \Omega \times \mathbb{R}^k$, $\tilde{w}^0(\cdot\,; t_0, y_0, \lambda)$ *is the unique solution to the Cauchy problem*

$$\begin{cases} w' = \tilde{A}(t; t_0, y_0, \lambda)w \\ w(t_0) = -\tilde{f}(t_0, y_0, \lambda), \end{cases} \tag{7.8}$$

where $\tilde{A}(t; t_0, y_0, \lambda)$ *is the matrix of components*

$$\tilde{A}_{ij}(t; t_0, y_0, \lambda) = \frac{\partial \tilde{f}_i}{\partial y_j}(t, \varphi(t; t_0, y_0, \lambda)).$$

(3) *For every* $\ell = 1, \ldots, N$, *let us introduce*

$$\tilde{w}^\ell(t; t_0, y_0, \lambda) := \frac{\partial \tilde{\varphi}}{\partial y_{0\ell}}(t; t_0, y_0, \lambda) \quad \forall(t, t_0, y_0, \lambda) \in \tilde{\Theta}.$$

Then, for every $(t_0, y_0, \lambda) \in \Omega \times \mathbb{R}^k$, $\tilde{w}^\ell(\cdot\,; t_0, y_0, \lambda)$ *is the unique solution to the Cauchy problem*

$$\begin{cases} w' = \tilde{A}(t; t_0, y_0, \lambda)w \\ w(t_0) = e^\ell, \end{cases} \tag{7.9}$$

where e^ℓ *is the* ℓ-*th vector of the canonical basis of* $\mathbb{R}^N$.

(4) *For every* $m = 1, \ldots, k$, *let us set*

$$\tilde{z}^m(t; t_0, y_0, \lambda) := \frac{\partial \tilde{\varphi}}{\partial \lambda_m}(t; t_0, y_0, \lambda) \quad \forall(t, t_0, y_0, \lambda) \in \tilde{\Theta}.$$

Then, for every $(t_0, y_0, \lambda) \in \Omega \times \mathbb{R}^k$, $\tilde{z}^m(\cdot\,; t_0, y_0, \lambda)$ *is the unique solution to the Cauchy problem*

$$\begin{cases} z' = \tilde{A}(t; t_0, y_0, \lambda)z + b(t; t_0, y_0, \lambda) \\ z(t_0) = 0, \end{cases} \tag{7.10}$$

where $b(t; t_0, y_0, \lambda)$ *is the vector of components*

$$b_i(t; t_0, y_0, \lambda) = \frac{\partial \tilde{f}_i}{\partial \lambda_m}(t, \varphi(t; t_0, y_0, \lambda)).$$

(5) *The following second-order partial derivatives exist and are continuous in $\tilde{\Theta}$:*

$$\frac{\partial^2 \tilde{\varphi}_i}{\partial t^2}, \quad \frac{\partial^2 \tilde{\varphi}_i}{\partial t \partial t_0}$$

$$= \frac{\partial^2 \tilde{\varphi}_i}{\partial t_0 \partial t}, \quad \frac{\partial^2 \tilde{\varphi}_i}{\partial t \partial y_{0\ell}} = \frac{\partial^2 \tilde{\varphi}_i}{\partial y_{0\ell} \partial t} \quad \text{and} \quad \frac{\partial^2 \tilde{\varphi}_i}{\partial t \partial \lambda_m} = \frac{\partial^2 \tilde{\varphi}_i}{\partial \lambda_m \partial t}.$$

The proofs are relatively simple. It suffices to take into account that the Cauchy problems (7.7) can be equivalently rewritten in the form

$$\begin{cases} y' = \tilde{f}(t, y, \mu), \quad \mu' = 0 \\ y(t_0) = y_0, \quad \mu(t_0) = \lambda \end{cases} \tag{7.11}$$

and then apply the results in Sections 7.2 and 7.4.

For other regularity results, see Note 11.7.1.

Exercises

E.7.1 Let $\mathcal{O} \subset \mathbb{R}^3$ be a non-empty open set, assume that $F \in C^m(\mathcal{O})$ with $m \geq 1$ and let $(t_0, y_0, p_0) \in \mathcal{O}$ be such that

$$F(t_0, y_0, p_0)) = 0, \quad \frac{\partial F}{\partial p}(t_0, y_0, p_0) \neq 0.$$

Let $\varphi : I \mapsto \mathbb{R}$ be a solution in I to the Cauchy problem

$$\begin{cases} F(t, y, y') = 0 \\ y(t_0) = y_0 \end{cases}$$

for the slope p_0. What can be said on the regularity of φ (with respect to the variable t)?

E.7.2 Let us consider the ODE

$$y''' = e^{|x|}y''(1 + y^2) + \sqrt{\frac{x+1}{x-2}} \, y^{4/3} + \sqrt{1 + (y')^2} + \int_0^x h(s)^3 \, ds,$$

where $h \in C^4(\mathbb{R})$. Determine the associated maximal domains of existence and uniqueness, as well as the regularity of the solution corresponding to the Cauchy data $(3; 0, 1, 1)$.

E.7.3 Consider the ODE

$$y''' = \frac{\sqrt{x - x^2 + 2}}{x - 1} y''(1 + y^2) + (y')^{4/3}\sin|x| + \int_1^x \sqrt{1 + k(\xi)^2} \, d\xi,$$

where $k \in C^1(\mathbb{R})$. Determine the associated maximal domains of existence and uniqueness, as well as the regularity of the solution corresponding to the Cauchy data $(3/2; 0, 0, 1)$.

E.7.4 Consider the ODE

$$y'' = e^{t^2/(1-y)} + \sqrt{\frac{1-y}{1+y}} + y^{5/4} + \sqrt{1 + |y'|^2} + \int_0^x (f(s))_+^3 \, ds,$$

where $f \in C^1(\mathbb{R})$ and $f(0) = 0$. Determine the associated maximal domains of existence and uniqueness, as well as the regularity of the solution corresponding to the Cauchy data $(1; 0, 1)$.

E.7.5 Consider the nth order ODE

$$y^{(n)} = f(t, y, y', \dots, y^{(n-1)}),$$

where $f \in C^0(\Omega)$ ($\Omega \subset \mathbb{R}^{n+1}$ is a non-empty open set). It is assumed that f is continuously differentiable in Ω. Let $\varphi^* : \Theta^* \mapsto \mathbb{R}$ be the general maximal solution to this ODE.

(1) Prove that $\varphi^* \in C^1(\Theta^*)$ and identify the partial derivatives of φ^* with respect to all its arguments.
(2) Determine which derivatives of φ of order n exist and why.

E.7.6 Consider the Cauchy problem

$$\begin{cases} y' = \beta(s)^a \cos(y+t) \\ y(0) = \alpha(s), \end{cases}$$

where $a \in \mathbb{R}$, $\alpha, \beta \in C^2(\mathbb{R})$ and $\beta \geq 0$.

(1) Prove that, for every $s \in \mathbb{R}$, there exists a unique maximal solution, defined in the whole $\mathbb{R}$.
(2) Define appropriately the general maximal solution (expressed in terms of s) and indicate its regularity for every a.

E.7.7 Let $\Omega \subset \mathbb{R}^{N+1}$ be a non-empty open set and assume that $f \in C^2(\Omega; \mathbb{R}^N)$. Let $ph : \Theta \mapsto \mathbb{R}^N$ be the general maximal solution to the ODS

$$y' = f(t, y).$$

Prove that $\varphi \in C^2(\Theta; \mathbb{R}^N)$ and determine the Cauchy problems satisfied by the functions

$$\frac{\partial^2 \varphi}{\partial t_0^2}, \ \frac{\partial^2 \varphi}{\partial t_0 \partial y_{0\ell}} \quad \text{and} \quad \frac{\partial^2 \varphi}{\partial y_{0\ell} \partial y_{0k}}.$$

Justify the existence and continuity of the third-order partial derivatives of φ where at least one differentiation with respect to t is performed.

E.7.8 For every $p \in \mathbb{R}$, let us consider the Cauchy problem

$$\begin{cases} y' = f(t, y, p) \\ y(\tau(p)) = p, \end{cases}$$

where $f \in C^1((-a, a) \times \mathbb{R}^2)$ and $\tau : \mathbb{R} \mapsto (-a, a)$ is a well defined C^2 function.

(1) Prove that, for every $p \in \mathbb{R}$, this problem possesses a unique maximal solution.
(2) Define the general maximal solution to this problem $\tilde{\varphi} = \tilde{\varphi}(t; p)$ (expressed in terms of p).
(3) Prove that $\tilde{\varphi}$ is a function of class C^1. Which second-order partial derivatives of $\tilde{\varphi}$ exist?
(4) Determine the partial derivative of $\tilde{\varphi}$ with respect to p.

E.7.9 For every $(\lambda, \mu) \in \mathbb{R}^2$, we consider the Cauchy problem

$$\begin{cases} y' = f_1(t, y)\lambda + f_2(t, y, \mu) \\ y(t_0) = f_3(\lambda - \mu), \end{cases}$$

where $f_1 \in C^2(\mathbb{R}^2)$, $f_2 \in C^2(\mathbb{R}^3)$, and $f_3 \in C^2(\mathbb{R})$.

(1) Prove that, for every $(t_0, \lambda, \mu) \in \mathbb{R}^3$, the previous problem possesses a unique maximal solution.
(2) Define the general maximal solution to this problem expressed in terms of t_0, λ, and μ: $\tilde{\varphi} = \tilde{\varphi}(t; t_0, \lambda, \mu)$.
(3) Prove that $\tilde{\varphi}$ is of class C^2. Which third-order partial derivatives of $\tilde{\varphi}$ exist?
(4) Determine the first- and second-order derivatives of $\tilde{\varphi}$ with respect to λ.

Chapter 8

Stability Results*

This chapter is devoted to the study of the stability of some ODSs.

Roughly speaking, it will be said that a solution $\varphi : [0, +\infty) \mapsto \mathbb{R}^N$ of an ODS is *stable* if any solution that is "close" to φ at time $t = 0$ keeps close to φ at any time $t \geq 0$.

Thus, stability is a concept that goes beyond continuity with respect to initial data, in the sense that it involves the behavior of the solutions not in a neighborhood of a given $\bar{t}$ but (for instance) in a whole interval of the form $[\bar{t}, +\infty)$.

In what follows, we will present rigorous definitions of stability and other similar notions and we will prove several related results (essentially, necessary and/or sufficient conditions of stability will be established).

8.1 Preliminary Concepts

For clarity, we will restrict ourselves to considering *autonomous* ODSs, that is, systems of the form

$$y' = f(y), \tag{8.1}$$

with $f \in C^1(B(0; R); \mathbb{R}^N)$ (recall that $B(0; R)$ is the open ball in $\mathbb{R}^N$ of radius R centered at the origin) and

$$f(0) = 0. \tag{8.2}$$

For every $y_0 \in B(0; R)$, we will denote by $J(y_0)$ the interval of definition of the maximal solution to the Cauchy problem

$$\begin{cases} y' = f(y) \\ y(0) = y_0 \end{cases} \tag{8.3}$$

(that is, with the notation introduced in Chapter 7, $J(y_0) = I(0, y_0)$).

In particular, thanks to (8.2), one has $J(0) = \mathbb{R}$ and the maximal solution to (8.3) corresponding to $y_0 = 0$ is the function φ^0, with $\varphi^0(t) \equiv 0$ (the trivial solution).

We will set

$$\Xi := \{(t, y_0) \in \mathbb{R}^{N+1} : y_0 \in B(0; R),\ t \in J(y_0)\}$$

and we will denote by $\psi : \Xi \mapsto \mathbb{R}^N$ the function that associates to each $(t, y_0) \in \Xi$ the value at t of the maximal solution to (8.3).

In other words, with the notation of Chapter 7, $\Xi = \{(t, y_0) : (t, 0, y_0) \in \Theta : t_0 = 0\}$ and

$$\psi(t; y_0) = \varphi(t; 0, y_0) \quad \forall (t, y_0) \in \Xi,$$

where φ is the general maximal solution to (8.1).

From the results in Chapter 7, it is immediate that Ξ is a non-empty open set, ψ is of class C^1 in Ξ, and, furthermore, the second-order partial derivatives of ψ

$$\frac{\partial^2 \psi}{\partial t^2} \quad \text{and} \quad \frac{\partial^2 \psi}{\partial t \partial y_{0\ell}} = \frac{\partial^2 \psi}{\partial y_{0\ell} \partial t}$$

exist and are continuous.

Definition 8.1. It is said that φ^0 is a *stable* solution to (8.1) if, for every $\varepsilon > 0$, there exists $\delta > 0$ such that, if $y_0 \in \mathbb{R}^N$ and $|y_0| \leq \delta$, then we necessarily have $J(y_0) \supset [0, +\infty)$ and

$$|\psi(t; y_0)| \leq \varepsilon \quad \forall t \geq 0.$$

It is said that φ^0 is an *attractive* solution to (8.1) if there exists $\eta > 0$ such that, if $y_0 \in \mathbb{R}^N$ and $|y_0| \leq \eta$, then $J(y_0) \supset [0, +\infty)$ and

$$\psi(t; y_0) \to 0 \quad \text{as } t \to +\infty.$$

Finally, it will be said that φ^0 is an *asymptotically stable* solution to (8.1) if it is at the same time stable and attractive. $\qquad\square$

This definition indicates clearly what we have to understand by stability for any other solution $\overline{\varphi} : [0, +\infty) \mapsto \mathbb{R}^N$ to (8.1).

For instance, it will be said that $\overline{\varphi}$ is stable if, for every $\varepsilon > 0$, there exists $\delta > 0$ such that, if $y_0 \in \mathbb{R}^N$ and $|y_0 - \overline{\varphi}(0)| \leq \delta$, then necessarily $J(y_0) \supset [0, +\infty)$ and

$$|\psi(t; y_0) - \overline{\varphi}(t)| \leq \varepsilon \quad \forall t \geq 0.$$

It remains clear too what is an attractive or asymptotically stable solution $\overline{\varphi}$ to (8.1).

In this chapter, we will speak almost exclusively of the stability (and attractiveness) of the trivial solution φ^0.

The stability of any other solution defined for all $t \geq 0$ can be analyzed by applying the results that follow after an appropriate change of variable.

Remark 8.1. It is important to understand well the difference between stability and continuity with respect to the initial conditions.

For example, the trivial solution to the ODE

$$y' = y$$

is not stable (for every $y_0 \neq 0$, one has $J(y_0) \supset [0, +\infty)$ and $\lim_{t \to +\infty} |\psi(t; y_0)| = +\infty$).

However, the results in Chapter 7 concerning the continuity and differentiability of the general maximal solution can be applied to this ODE.

More precisely, we do have the following: for every $\varepsilon > 0$ and every $T > 0$, there exists $\delta > 0$ (depending on ε and T) such that, if $y_0 \in \mathbb{R}^N$ and $|y_0| \leq \delta$, then

$$|\psi(t; y_0)| \leq \varepsilon \quad \forall t \in [0, T]. \qquad \square$$

Remark 8.2. It is not difficult to give examples of ODEs and ODSs whose trivial solutions are stable and not attractive or vice versa.

Thus, in the case of the ODE $y' = 0$, the trivial solution is stable, but not attractive. On the other hand, we leave to the reader the task to find a differential system (for example of dimension 2) whose trivial solution is attractive, but not stable. $\qquad \square$

Remark 8.3. The results that characterize the behavior of the maximal solutions near the endpoints of their intervals of definition show that the trivial solution to (8.1) is stable if and only if for any $\varepsilon > 0$ there exists $\delta > 0$ such that, if $y_0 \in \mathbb{R}^N$ and $|y_0| \leq \delta$, then

$$|\psi(t; y_0)| \leq \varepsilon \quad \forall t \in J(y_0) \cap [0, +\infty).$$

In other words, if this property is satisfied, we can automatically ensure that $J(y_0)$ contains $[0, +\infty)$.

This simple observation abbreviates the stability analysis. □

Example 8.1. Consider the ODE

$$y' = y(y^2 - 1) \tag{8.4}$$

and the maximal solutions φ^0, $\overline{\varphi}$, and φ^-, given by

$$\varphi^0(t) = 0, \quad \varphi^+(t) = 1, \quad \varphi^-(t) = -1 \quad \forall t \in \mathbb{R}.$$

After some elementary computations, it is possible to determine explicitly all the solutions to (8.4). In particular, we have the following for every $y_0 \in \mathbb{R}$:

$$J(y_0) = \begin{cases} (-\infty, \log \dfrac{|y_0|}{(|y_0|^2 - 1)^{1/2}}) & \text{if } |y_0| > 1 \\ \mathbb{R} & \text{otherwise.} \end{cases}$$

When $y_0 \neq 0$, we find that

$$\psi(t; y_0) = (\text{sign } y_0)\left(1 - \left(1 - \frac{1}{|y_0|^2}\right)e^{2t}\right)^{-1/2} \quad \forall t \in J(y_0).$$

Consequently, after an analysis of the behavior of the solutions for $t \geq 0$, we see that the trivial solution φ^0 is asymptotically stable.

In a similar way, it can be checked that the solutions φ^+ and φ^- are neither stable nor attractive. □

Example 8.2. Consider now the ODE

$$y'' = -k \sin y, \tag{8.5}$$

where $k > 0$. The solutions to (8.5) describe the motion of an *ideal pendulum* of length $\ell = g/k$ (g is the acceleration of gravity). Here, $y(t)$ is the angle determined by the vertical axis (downwards directed) and the pendulum at time t (we accept that this angle is 0^0 at time $t = 0$). This ODE can be rewritten in the form

$$\begin{cases} y_1' = y_2 \\ y_2' = -k \sin y_1 \end{cases} \tag{8.6}$$

and possesses two constant solutions, φ^0 and $\overline{\varphi}$, respectively, given by

$$\varphi^0(t) = (0,0), \quad \overline{\varphi}(t) = (\pi, 0) \quad \forall t \in \mathbb{R}. \tag{8.7}$$

A much more realistic situation corresponds to the so-called *pendulum with friction*, that is modeled by the ODE

$$y'' = -k \sin y - \alpha y', \tag{8.8}$$

where $\alpha > 0$. In this case, the equivalent first-order ODS is

$$\begin{cases} y_1' = y_2 \\ y_2' = -k \sin y_1 - \alpha y_2. \end{cases} \tag{8.9}$$

Again, (8.9) possesses exactly two constant solutions, given by (8.7).

The results in the following Sections 8.3 and 8.4 show that φ^0 (resp. $\overline{\varphi}$) is a stable (resp. unstable) solution to (8.6).

They also prove that φ^0 (resp. $\overline{\varphi}$) is asymptotically stable (resp. unstable) regarded as a solution to (8.9).

This coincides with what we expect to have in accordance with our daily experience. □

8.2 Stability of Linear Systems

This section is devoted to analyzing the stability properties of the trivial solution to a linear autonomous ODS, that is, a linear system of constant coefficients

$$y' = Ay, \tag{8.10}$$

where $A \in \mathcal{L}(\mathbb{R}^N)$.

Recall that a fundamental matrix of (8.10) is given by

$$F(t) = e^{tA} \quad \forall t \in \mathbb{R}$$

and, consequently, the maximal solutions to (8.10) are the functions

$$\varphi(t) = e^{tA}c \quad \forall t \in \mathbb{R},$$

where c is an arbitrary vector in $\mathbb{R}^N$.

Therefore, we have in this case that $\Xi = \mathbb{R}^{N+1}$ and

$$\psi(t; y_0) = e^{tA}y_0 \quad \forall (t, y_0) \in \mathbb{R}^{N+1}. \tag{8.11}$$

For clarity, we will first consider the linear homogeneous ODE

$$y' = ay, \tag{8.12}$$

where $a \in \mathbb{R}$. Now, $\Xi = \mathbb{R}^2$ and

$$\psi(t; y_0) = y_0 e^{at} \quad \forall (t, y_0) \in \mathbb{R}^2.$$

It is clear that the stability properties of the trivial solution are determined by the sign of a.

More precisely, the following holds:

- If $a < 0$, the trivial solution to (8.12) is *asymptotically stable*.
- If $a = 0$, it is *stable and not attractive*.
- Contrarily, if $a > 0$, the trivial solution to (8.12) is *unstable*.

These considerations serve as an introduction to the following results:

Theorem 8.1. *The following assertions are equivalent:*

(1) *The trivial solution to (8.10) is stable.*

(2) *There exists a constant $C > 0$ such that*

$$|\psi(t; y_0)| \le C|y_0| \quad \forall t \ge 0, \quad \forall y_0 \in \mathbb{R}^N. \tag{8.13}$$

(3) *Let $\| \cdot \|$ be a norm in the space $\mathcal{L}(\mathbb{R}^N)$. There exists a constant $C > 0$ such that*

$$\|e^{tA}\| \le C \quad \forall t \ge 0. \tag{8.14}$$

(4) *The eigenvalues $\lambda_1, \dots, \lambda_N$ of A satisfy*

$$\begin{cases} \operatorname{Re} \lambda_i \le 0 \quad \forall i = 1, \dots, N \\ \text{If } \operatorname{Re} \lambda_i = 0, \text{ then } \lambda_i \text{ is simple (i.e. has multiplicity 1).} \end{cases} \tag{8.15}$$

Proof. Let us first assume that the trivial solution to (8.10) is stable.

We know that there exists $\delta > 0$ such that, if $y_0 \in \mathbb{R}^N$ and $|y_0| \le \delta$, then

$$|e^{tA} y_0| \le 1 \quad \forall t \ge 0.$$

Hence, if $z_0 \in \mathbb{R}^N$ with $z_0 \ne 0$ and $t \ge 0$, we have that

$$\left| e^{tA} \left(\frac{\delta}{|z_0|} z_0 \right) \right| \le 1,$$

whence

$$|e^{tA} z_0| \le \frac{1}{\delta} |z_0|.$$

This proves (8.13).

Secondly, note that, if the inequalities (8.13) hold, since $\psi(t; y_0) \equiv e^{tA} y_0$, one has

$$\|e^{tA}\|_S \le C \quad \forall t \ge 0.$$

In this respect, let us recall that $\| \cdot \|_S$ is the *spectral norm*, that is, the matrix norm induced by the Euclidean norm in $\mathbb{R}^N$; see (5.4).

Since all the norms in $\mathcal{L}(\mathbb{R}^N)$ are equivalent, we deduce that, independently of the choice of the norm $\|\cdot\|$, there exists a (new) constant $C > 0$ for which one has (8.14).

Now, let us assume that the property indicated in part 3 is satisfied. Let $\tilde{J}$ be a real canonical form of A. There exists a non-singular matrix $P \in \mathcal{L}(\mathbb{R}^N)$ such that $A = P^{-1} \cdot \tilde{J} \cdot P$ and

$$e^{tA} = P^{-1} \cdot e^{t\tilde{J}} \cdot P \quad \forall t \in \mathbb{R}.$$

Consequently, there exists a constant $C > 0$ such that

$$\|e^{t\tilde{J}}\|_\infty = \|P \cdot e^{tA} \cdot P^{-1}\|_\infty \leq C \|P\|_\infty \|P^{-1}\|_\infty \quad \forall t \geq 0. \tag{8.16}$$

(recall that $\|\cdot\|_\infty$ is given by (5.5)).

Taking into account the results in Chapter 5, we can ensure that the non-zero components of $t \mapsto e^{t\tilde{J}}$ are of the form

$$t^n e^{\lambda t}, \quad t^n e^{\alpha t} \cos(\beta t), \quad \text{or} \quad t^n e^{\alpha t} \sin(\beta t), \tag{8.17}$$

where λ is a real eigenvalue of A, α and β are, respectively, the real and imaginary parts of a complex eigenvalue, and n is non-negative integer, less or equal to the multiplicity minus one.

In view of (8.16) and the definition of $\|\cdot\|_\infty$, it becomes clear that the real part of any eigenvalue of A must be non-positive.

It is also clear that, if the real part of an eigenvalue is zero, its multiplicity cannot be greater than 1 (otherwise, we could have terms as those above with $\lambda = 0$ or $\alpha = 0$ and $n \geq 1$).

This proves (8.15), that is, the property stated in part 4 holds.

Finally, let us assume that we have (8.15) and let us see that the trivial solution to (8.10) is stable.

For every $y_0 \in \mathbb{R}^N$, we have (8.11).

Let $\tilde{J}$ be a real canonical form of A and let $P \in \mathcal{L}(\mathbb{R}^N)$ be non-singular and such that $A = P^{-1} \cdot \tilde{J} \cdot P$. Then, there exist positive constants C_1 (only depending on A) and C_2 (only depending on N) such that

$$|\psi(t; y_0)| \leq \|e^{tA}\|_S |y_0| = \|P^{-1} \cdot e^{t\tilde{J}} \cdot P\|_S |y_0|$$

$$\leq C_1 \|e^{t\tilde{J}}\|_S |y_0| \leq C_1 C_2 \|e^{t\tilde{J}}\|_\infty |y_0|$$

for all $t \in \mathbb{R}$.

From the properties (8.15) satisfied by the eigenvalues of A, we deduce that, for a new constant C_3, one has

$$\|e^{t\tilde{J}}\|_\infty \leq C_3 \quad \forall t \geq 0.$$

As a consequence, there exists a positive constant $M = C_1 C_2 C_3$ (only depending on N and A) such that

$$|\psi(t; y_0)| \le M|y_0| \quad \forall t \ge 0$$

and this holds for all $y_0 \in \mathbb{R}^N$.

Using these inequalities, it is now very easy to prove that the trivial solution is stable. □

Theorem 8.2. *The following assertions are equivalent:*

(1) *The trivial solution to (8.10) is asymptotically stable, i.e. stable and attractive.*
(2) *The trivial solution to (8.10) is attractive.*
(3) *For every $y_0 \in \mathbb{R}^N$, one has $\psi(t; y_0) \to 0$ as $t \to +\infty$.*
(4) *One has*

$$e^{tA} \to 0 \quad en \ \mathcal{L}(\mathbb{R}^N) \quad as \ t \to +\infty. \tag{8.18}$$

(5) *The eigenvalues $\lambda_1, \dots, \lambda_N$ of A satisfy*

$$\mathrm{Re}\ \lambda_i < 0 \quad \forall i = 1, \dots, N. \tag{8.19}$$

Proof. Evidently, the first assertion implies the second one.

Let us now assume that the trivial solution is attractive. Then, there exists $\eta > 0$ such that, if $y_0 \in \mathbb{R}^N$ and $|y_0| \le \eta$, one has

$$\psi(t; y_0) = e^{tA} y_0 \to 0 \quad as \ t \to +\infty.$$

Consequently, if $z_0 \in \mathbb{R}^N$ and $z_0 \ne 0$, one also has

$$e^{tA}\left(\frac{\eta}{|z_0|} z_0\right) \to 0 \quad as \ t \to +\infty,$$

whence $e^{tA} z_0 \to 0$.

This proves that the second assertion implies the third one.

Let us now assume that $e^{tA} y_0 \to 0$ as $t \to +\infty$ for every $y_0 \in \mathbb{R}^N$.

In particular, if we denote (as usual) by e_j the jth vector of the canonical basis in $\mathbb{R}^N$, we find that $e^{tA} e_j \to 0$ as $t \to +\infty$.

Let $\varepsilon > 0$ be given. For every $j = 1, \dots, N$, there exists t_j such that, if $t \ge t_j$, we necessarily have $|e^{tA} e_j| \le \varepsilon/\sqrt{N}$.

Let us set $t_0 := \max_j t_j$. Then, if $t \ge t_0$ and $v = \sum_j v_j e_j \in \mathbb{R}^N$, we have

$$|e^{tA} v| = \left|\sum_{j=1}^N v_j e^{tA} e_j\right| \le \left(\sum_{j=1}^N |e^{tA} e_j|^2\right)^{1/2} |v| \le \varepsilon|v|.$$

In other words, if $t \geq t_0$, then

$$\|e^{tA}\|_S \leq \varepsilon.$$

Since $\varepsilon > 0$ is arbitrary, we have proved (8.18).

Thus, we see that the third assertion implies the fourth one.

Now, assume that (8.18) holds.

Let $\tilde{J}$ be a real canonical form of A. It is then clear that $e^{t\tilde{J}} \to 0$ in $\mathcal{L}(\mathbb{R}^N)$ as $t \to +\infty$.

The components of $e^{t\tilde{J}}$ are of the form (8.17), where λ is an eigenvalue of A and α and β are, respectively, the real and imaginary parts of a complex eigenvalue.

Consequently, we necessarily have (8.19).

Finally, let us suppose that the eigenvalues of A satisfy (8.19).

Thanks to Theorem 8.1, we know that the trivial solution to (8.10) is stable.

On the other hand, taking into account that the components of the solutions to (8.10) are linear combinations of the functions arising in (8.17), we see that $\psi(t; y_0) \to 0$ as $t \to +\infty$ for all $y_0 \in \mathbb{R}^N$.

Therefore, the trivial solution to (8.10) is also attractive. $\qquad\square$

Example 8.3. Consider the second-order ODE

$$z'' = -kz - \alpha z' - g,$$

where k and g are positive constants and α is a nonnegative constant. Let us analyze the stability of the following particular solution

$$\psi_0(t) = -\frac{g}{k} \quad \forall t \in \mathbb{R}. \tag{8.20}$$

The change of variable $z = y - g/k$ allows to rewrite the ODE in the form

$$y'' = -ky - \alpha y'.$$

The usual change of variables $(y_1 = y, y_2 = y')$ allows to rewrite the ODE as a first-order system:

$$\begin{cases} y_1' = y_2 \\ y_2' = -ky_1 - \alpha y_2. \end{cases} \tag{8.21}$$

The task is thus reduced to identifying the stability properties of the trivial solution to (8.21). Taking into account the previous results, it will suffice to determine the eigenvalues of the corresponding coefficient matrix. We have two different possible scenarios:

(1) If $\alpha = 0$, the eigenvalues are $\pm i\sqrt{k}$ and, consequently, the trivial solution to (8.21) is in this case stable but not attractive.
(2) If $\alpha > 0$, the eigenvalues possess a negative real part equal to $-\alpha$. Hence, the trivial solution to (8.21) is asymptotically stable. □

Remark 8.4. In the context of some important applications, it becomes interesting to speak of exponential stability.

Let us consider again the autonomous system (8.1). It is said that the trivial solution φ_0 is *exponentially stable* if there exist $C, \alpha > 0$, and $\gamma \in (0, \rho)$ such that, if $|y_0| \leq \gamma$, then $I(y_0) \supset [0, +\infty)$ and

$$|\psi(t; y_0)| \leq Ce^{-\alpha t}|y_0| \quad \forall t \geq 0.$$

Theorem 8.2 shows that, if the considered ODS is linear, the trivial solution is asymptotically stable if and only if it is exponentially stable.

Indeed, the first assertion in Theorem 8.2 implies the fifth one.

Conversely, let $\tilde{J}$ be a real canonical form of A. Then, there exist constants C and α such that the absolute values of all components of $e^{t\tilde{J}}$ are $\leq Ce^{-\alpha t}$ for all $t \geq 0$. Obviously, this implies the exponential stability of the trivial solution. □

8.3　Comparison and First Approximation

In this section, we will prove stability results for some (nonlinear in general) ODSs. The fundamental task will be to compare with the solutions to another more simple ODS or ODE.

In a first result, let us consider the ODS (8.1) and the ODE

$$z' = h(z), \tag{8.22}$$

where $h \in C^1(-R, R)$ and $h(0) = 0$.

Theorem 8.3. *Assume that*

$$\begin{cases} h \text{ is non-decreasing and} \\ |f(y)| \leq h(|y|) \quad \forall y \in B(0; R). \end{cases} \tag{8.23}$$

Then, if the trivial solution to (8.22) is stable, so is the trivial solution to (8.1).

On the other hand, if the trivial solution to (8.22) is attractive, the same happens to the trivial solution to (8.1).

Proof. For every $z_0 \in (-R, R)$, we will denote by $J^*(z_0)$ the interval of definition of the maximal solution to (8.22) that satisfies $z(0) = z_0$. We will also put

$$\Xi^* := \{(t, z_0) \in \mathbb{R}^2 : z_0 \in (-R, R),\ t \in J^*(z_0)\}$$

and we will denote by ψ^* the function that, to each $(t, z_0) \in \Xi^*$, associates the value at t of the previous maximal solution.

Let us assume that we already have

$$\begin{cases} (t, y_0) \in \Xi,\ t \geq 0 \ \Rightarrow\ (t, |y_0|) \in \Xi^* \\ |\psi(t; y_0)| \leq \psi^*(t; |y_0|) \quad \forall (t, y_0) \in \Xi \quad \text{with } t \geq 0. \end{cases} \tag{8.24}$$

Then the stability of the trivial solution to (8.22) implies the same property for the trivial solution to (8.1). The same can be said for attractiveness.

In order to establish (8.24), it suffices to check that, if $y_0 \in B(0; R)$,

$$|\psi(t; y_0)| \leq \psi^*(t; |y_0|) \quad \forall t \in J(y_0) \cap J^*(|y_0|) \cap [0, +\infty).$$

Thus, let $\bar{t} \in J(y_0) \cap J^*(|y_0|)$ be given, with $\bar{t} > 0$ and set

$$u(t) = (|\psi(t; y_0)| - \psi^*(t; |y_0|))_+ \quad \forall t \in [0, \bar{t}].$$

Let us see that $u \equiv 0$.

We observe that, for every $t \in J(y_0)$ with $t \geq 0$,

$$|\psi(t; y_0)| \leq |y_0| + \int_0^t |f(\psi(s; y_0))|\,\mathrm{d}s$$

$$\leq |y_0| + \int_0^t h(|\psi(s; y_0)|)\,\mathrm{d}s.$$

On the other hand, since h is non-decreasing and locally Lipschitz-continuous with respect to the variable z, there exists a constant $L > 0$ such that

$$h(|\psi(s; y_0)|) - h(\psi^*(s; |y_0|)) \leq L(|\psi(s; y_0)| - \psi^*(s; |y_0|))_+ \quad \forall s \in [0, \bar{t}].$$

Consequently,

$$u(t) \leq L \int_0^t u(s)\,\mathrm{d}s \quad \forall t \in [0, \bar{t}]. \tag{8.25}$$

Indeed, if $|\psi(t; y_0)| \leq \psi^*(t; |y_0|)$, then $u(t) = 0$ and the previous inequality is trivial. Otherwise, we have

$$u(t) = |\psi(t; y_0)| - \psi^*(t; |y_0|)$$

$$\leq \int_0^t [h(|\psi(s; y_0)|) - h(\psi^*(s; |y_0|))] \, ds$$

$$\leq L \int_0^t (|\psi(s; y_0)| - \psi^*(s; |y_0|))_+ \, ds,$$

whence we have again (8.25).

Now, using Gronwall's Lemma, we deduce that $u \leq 0$.

This ends the proof. $\qquad\square$

We will consider now some ODSs that are in some sense "close" to linear ODSs. More precisely, we will deal with

$$y' = Ay + g(y), \tag{8.26}$$

where $A \in \mathcal{L}(\mathbb{R}^N)$ and $g \in C^1(B(0; R); \mathbb{R}^N)$ is *superlinear* at zero.

The following result allows to compare the stability properties of the trivial solutions to (8.26) and (8.10):

Theorem 8.4. *Assume that*

$$\frac{g(y)}{|y|} \to 0 \quad as \ |y| \to 0. \tag{8.27}$$

Then, if the trivial solution to (8.10) is asymptotically stable, this is also true for the trivial solution to (8.26).

Proof. Let us assume that the trivial solution to (8.10) is exponentially stable. Then, there exist $C, \alpha > 0$ such that

$$\|e^{tA}\|_S \leq Ce^{-\alpha t} \quad \forall t \geq 0.$$

Let $0 < \gamma < \alpha/C$ be given. In view of (8.27), there exists $\mu \in (0, R)$ such that

$$|y| \leq \mu \Rightarrow |g(y)| \leq \gamma|y|.$$

Let us fix $\varepsilon > 0$ and take

$$\delta = \min\left(\frac{\varepsilon}{2}, \frac{\varepsilon}{2C}, \mu\right).$$

Then, if $y_0 \in B(0; R)$ and $|y_0| \leq \delta$, the interval of definition of the maximal solution to (8.26) satisfying the initial condition $y(0) = y_0$ contains $[0, +\infty)$

and satisfies

$$|\psi(t; y_0)| \leq \varepsilon \quad \forall t \geq 0.$$

Indeed, if this were not the case, since $|y_0| \leq \varepsilon/2$, there would exist $\bar{t} > 0$ such that $J(y_0) \supset [0, \bar{t}]$, $|\psi(\bar{t}; y_0)| = \varepsilon$, and $|\psi(s; y_0)| < \varepsilon$ for all $s \in [0, \bar{t})$. Observe that

$$\psi(t; y_0) = e^{tA} y_0 + \int_0^t e^{(t-s)A} g(\psi(s; y_0)) \, ds \quad \forall t \in J(y_0).$$

Therefore, if $t \in [0, \bar{t}]$, one has

$$|\psi(t; y_0)| \leq |e^{tA} y_0| + \int_0^t |e^{(t-s)A} g(\psi(s; y_0))| \, ds$$

$$\leq Ce^{-\alpha t} |y_0| + C \int_0^t e^{-\alpha(t-s)} |g(\psi(s; y_0))| \, ds$$

$$\leq Ce^{-\alpha t} |y_0| + C\gamma \int_0^t e^{-\alpha(t-s)} |\psi(s; y_0)| \, ds.$$

Let us introduce u, with $u(t) = e^{\alpha t} |\psi(t; y_0)|$ for all $t \in [0, \bar{t}]$. Then $u : [0, \bar{t}] \mapsto \mathbb{R}$ is continuous and

$$u(t) \leq C|y_0| + C\gamma \int_0^t u(s) \, ds \quad \forall t \in [0, \bar{t}].$$

Consequently, from Gronwall's Lemma, we deduce that

$$u(t) \leq C|y_0| e^{C\gamma t} \leq \frac{\varepsilon}{2} e^{C\gamma t} \quad \forall t \in [0, \bar{t}]$$

and, in particular,

$$|\psi(\bar{t}; y_0)| = e^{-\alpha \bar{t}} u(\bar{t}) \leq \frac{\varepsilon}{2} e^{-(\alpha - C\gamma)\bar{t}} \leq \frac{\varepsilon}{2}.$$

But this is in contradiction with the equality $|\psi(\bar{t}; y_0)| = \varepsilon$. Thus, we have proved that the trivial solution to (8.26) is stable.

Let us repeat the previous argument with $\varepsilon = 1$ (for example). We deduce that, for $\delta = \min(1/2, 1/(2C), \mu)$ and $y_0 \in B(0; R)$ with $|y_0| \leq \delta$, one has $J(y_0) \supset [0, +\infty)$ and

$$|\psi(t; y_0)| \leq \frac{1}{2} e^{-(\alpha - C\gamma)t} \quad \forall t \geq 0,$$

whence $\psi(t; y_0) \to 0$ as $t \to +\infty$.

As a consequence, the trivial solution to (8.26) is also attractive. This ends the proof. $\qquad \square$

Example 8.4. Consider the ODS

$$\begin{cases} y_1' = y_2 - y_1^3 \\ y_2' = -2y_1 - 2y_2 - y_2^3. \end{cases}$$

We can apply to this system Theorem 8.4 with

$$A = \begin{pmatrix} 0 & 1 \\ -2 & -2 \end{pmatrix} \quad \text{and} \quad g(y) \equiv (-y_1^3, -y_2^3).$$

Since the eigenvalues of A have real part -1 and g satisfies (8.27), we deduce that the trivial solution to this system is asymptotically stable. □

Example 8.5. In general, it can happen that the trivial solution to (8.10) is stable, but the trivial solution to (8.26) is not. For example, this is the situation found for the linear ODE $y' = 0$ and the perturbed equation $y' = y^2$. □

Corollary 8.1. *Consider the ODS* (8.1), *where* $f \in C^1(B(0; R); \mathbb{R}^N)$ *satisfies* (8.2). *Let us set* $A = f'(0)$ *and let us assume that all the eigenvalues of A possess negative real part. Then, the trivial solution to* (8.1) *is asymptotically stable.*

Proof. It suffices to take into account that f can be written in the form

$$f(y) = f(0) + Ay + g(y) = Ay + g(y) \quad \forall y \in B(0; R),$$

where g satisfies (8.27).

A direct application of Theorem 8.4 leads to the desired result. □

It is usual to say that Theorem 8.4 and Corollary 8.1 are stability results *at first approximation.*

Example 8.6. Consider the ODS

$$\begin{cases} y_1' = -y_1 - y_2 (\sin y_1 + \cos y_2) \\ y_2' = y_1 (1 + \log(1 + |y_2|)) - y_2^3. \end{cases} \tag{8.28}$$

This is a system of the form (8.1), where $f(y) \equiv (-y_1 - y_2, y_1 - y_2^3)$ and

$$f'(0) = \begin{pmatrix} -1 & -1 \\ 1 & 0 \end{pmatrix}.$$

With $A = f'(0)$, we see that the eigenvalues of A have real part equal to $-1/2$. Consequently, we deduce from Corollary 8.1 that the trivial solution to (8.28) is also asymptotically stable. □

8.4 Liapunov's Stability Method

In this section, we present the fundamentals of Liapunov's method for the analysis of the stability of an ODS.

To this purpose, we must introduce a new concept: the *Liapunov function*.

Definition 8.2. Let $V : \overline{B}(0; R) \mapsto \mathbb{R}$ be a given function, with $V(0) = 0$.

(1) It will be said that V is a definite positive function in $\overline{B}(0; R)$ if $V(y) > 0$ for every $y \in \overline{B}(0; R)$ with $y \neq 0$. It will be said that V is definite negative if $-V$ is definite positive.

(2) It will be said that V is a Liapunov function for the ODS (8.1) if there exists $R' \in (0, R)$ such that V is continuously differentiable and definite positive in $\overline{B}(0; R')$ and, moreover,

$$\dot{V}(y) := \sum_{j=1}^{N} \frac{\partial V}{\partial y_j}(y) \cdot f_j(y) \leq 0 \quad \forall y \in \overline{B}(0; R'). \tag{8.29}$$

$\square$

The most relevant property of the Liapunov functions is that they decrease along the trajectories determined by the solutions.

Indeed, let us assume that V is a definite positive and continuously differentiable in $\overline{B}(0; R')$ and, also, that $\dot{V} \leq 0$ in $\overline{B}(0; R')$. Let $\varphi : I \mapsto \mathbb{R}^N$ be a solution to (8.1) in the interval I, with $\varphi(t) \in \overline{B}(0; R')$ for every $t \in I$. Then, the function $t \mapsto V(\varphi(t))$ is well defined and is continuously differentiable in I and satisfies

$$\frac{\mathrm{d}}{\mathrm{d}t} V(\varphi(t)) = \sum_{j=1}^{N} \frac{\partial V}{\partial y_j}(\varphi(t)) f_j(\varphi(t)) = \dot{V}(\varphi(t)) \leq 0 \quad \forall t \in I. \tag{8.30}$$

The main result in this section, usually known as *Liapunov's Theorem,* is the following:

Theorem 8.5 (Liapunov). *Consider the ODS (8.1), where $f \in C^1(B(0; R); \mathbb{R}^N)$ satisfies (8.2).*

(1) *If there exists a Liapunov function V for (8.1), then the trivial solution to this system is stable.*

(2) *If, moreover, the corresponding function $\dot{V}$ is definite negative in a ball $\overline{B}(0; R')$, then the trivial solution is asymptotically stable.*

For the proof, we will need the following result, whose proof will be furnished in what follows:

Lemma 8.1. *Assume that* $V \in C^0(\overline{B}(0;R))$ *satisfies* $V(0) = 0$ *and is definite positive in* $\overline{B}(0;R)$. *Then, there exists a function* $a : [0,R] \mapsto \mathbb{R}$ *with the following properties*:

(1) $a \in C^0([0,R])$, a *is strictly increasing and* $a(0) = 0$.
(2) $V(y) \geq a(|y|)$ *for all* $y \in \overline{B}(0;R)$.

Proof of Theorem 8.5. Let us assume that there exists a Liapunov function V for (8.1).

We know that there exists $R' \in (0,R)$ such that V is continuously differentiable and definite positive and satisfies $\dot{V} \leq 0$ in $\overline{B}(0;R')$.

Let $a \in C^0([0,R'])$ be strictly increasing and such that $a(0) = 0$ and $V(y) \geq a(|y|)$ for all y.

Let $\varepsilon > 0$ be given and let $\delta_\varepsilon > 0$ be such that, if $y \in B(0;R')$ and $|y| \leq \delta_\varepsilon$, one has $0 \leq V(y) \leq a(\varepsilon)$.

Then, for $|y_0| \leq \delta_\varepsilon$, we have

$$a(|\psi(t;y_0)|) \leq V(\psi(t;y_0)) \leq a(\varepsilon) \quad \forall t \in J(y_0) \cap [0,+\infty). \qquad (8.31)$$

Indeed, it suffices to take into account that the function $t \mapsto V(\psi(t;y_0))$ is defined and continuously differentiable in $J(y_0) \cap [0,+\infty)$, possesses a nonpositive derivative, and satisfies $V(y_0) \leq a(\varepsilon)$.

Since a is strictly increasing, we deduce from (8.31) that $|\psi(t;y_0)| \leq \varepsilon$ for every $t \in J(y_0) \cap [0,+\infty)$. The consequence is that the trivial solution to (8.1) is stable.

Now, let us assume that there exists a Liapunov function V as the one above, with $\dot{V}$ definite negative in $\overline{B}(0;R)$ and let us see that the trivial solution to (8.1) is attractive.

Let us repeat the previous argument and let us take $\delta = \delta_1$. Let us see that

$$y_0 \in B(0;R), \quad |y_0| \leq \delta \implies V(\psi(t;y_0)) \to 0 \quad \text{as } t \to +\infty. \qquad (8.32)$$

Indeed, for such a y_0 we know that $J(y_0) \supset [0,+\infty)$ and $t \mapsto V(\psi(t;y_0))$ is nonnegative and nonincreasing. Consequently,

$$\exists \ell := \lim_{t \to +\infty} V(\psi(t;y_0)) \geq 0.$$

Let us assume that $\ell > 0$.

Then, there exist $t_0 > 0$ and $\mu > 0$ such that, for every $t \geq t_0$, we have $|\psi(t; y_0)| \geq \mu$. As a consequence, for any $t \geq t_0$ we would have

$$V(\psi(t; y_0)) = V(y_0) + \int_0^t \dot{V}(\psi(s; y_0)) \, \mathrm{d}s$$

$$\leq V(y_0) - \int_{t_0}^t b(|\psi(s; y_0)|) \, \mathrm{d}s$$

$$\leq V(y_0) - b(\mu)(t - t_0).$$

But this contradicts the positivity of V and is thus impossible. Hence, $\ell = 0$.

Taking into account (8.32), we see that, if $|y_0| \leq \delta$, then $a(|\psi(t; y_0)|) \to 0$ as $t \to +\infty$. Finally, since a is strictly increasing, we deduce that, for these y_0, $\psi(t; y_0) \to 0$.

This ends the proof. $\qquad\square$

Proof of Lemma 8.1. As a first step, let us take

$$\tilde{a}(r) := \min_{r \leq |y| \leq R} V(y) \quad \forall r \in [0, R].$$

Then, the function $\tilde{a} : [0, R] \mapsto \mathbb{R}$ is well defined, continuous, and non-decreasing and satisfies $\tilde{a}(0) = 0$ and $V(y) \geq \tilde{a}(|y|) > 0$ for all $y \in \overline{B}(0; R)$ with $y \neq 0$.

Now, let us take

$$a(r) = \frac{1}{R} \int_0^r \tilde{a}(s) \, \mathrm{d}s \quad \forall r \in [0, R].$$

Then, it is not difficult to check that the function $a : [0, R] \mapsto \mathbb{R}$ satisfies all the desired properties (and is moreover continuously differentiable in $[0, R]$). $\qquad\square$

Remark 8.5. The assumption "$\dot{V}$ is definite negative in a ball $\overline{B}(0; R')$" in Theorem 8.5 can be changed by the following:

There exists $R' \in (0, R)$ such that $\{0\}$ is the unique invariant set of $E = \{y \in \overline{B}(0; R') : \dot{V}(y) = 0\}$.

Here, it is said that $A \subset B(0; R)$ is invariant if $\psi(t; y) \in A$ for all $t \in J(y) \cap [0, +\infty)$ and every $y \in A$.

This is a consequence of *La Salle's Theorem,* a result allowing to deduce attractiveness properties of the invariant sets; see Rouché and Mawhin (1973) for more details. $\qquad\square$

Example 8.7. Let us consider again the ODEs (8.5) and (8.8) and the corresponding equivalent systems (8.6) and (8.9). Let us see that φ^0 is a stable solution to (8.6) and an asymptotically stable solution to (8.9).

Indeed, let us take $0 < R < 2\pi$ and set

$$V(y_1, y_2) = \frac{1}{2}|y_2|^2 + k(1 - \cos y_1) \quad \forall (y_1, y_2) \in \overline{B}(0; R).$$

From a physical view point, V must be interpreted as the total energy (that is, the sum of the kinetic and potential energies) associated to a "state" (y_1, y_2) determined by an angle y_1 and an angular velocity y_2.

We have that V is well defined and is continuously differentiable in $\overline{B}(0; R)$ and is positive in that ball. The function $\dot{V}$ associated to (8.6) is given by

$$\dot{V}(y_1, y_2) = 0 \quad \forall (y_1, y_2) \in \overline{B}(0; R).$$

Consequently, V is a Liapunov function for (8.6) and the trivial solution to this system is stable.

On the other hand, the function $\dot{V}$ associated to (8.9) is

$$\dot{V}(y_1, y_2) = -\alpha |y_2|^2 \quad \forall (y_1, y_2) \in \overline{B}(0; R).$$

Evidently, this function is not a definite negative in any $\overline{B}(0; R')$. Nevertheless, it is easy to check that the unique invariant subset of the corresponding set $E = \{y \in \overline{B}(0; R) : \dot{V}(y) = 0\}$ is $\{0\}$. Therefore, thanks to (8.5), the trivial solution to (8.9) is asymptotically stable. $\qquad \square$

Remark 8.6. More generally, it makes sense to consider a system with N freedom degrees, described by the second-order ODS

$$q_i'' + \frac{\partial U}{\partial q_i}(q) + \Phi_i(q') = 0, \quad i = 1, \dots, N, \tag{8.33}$$

where $q = (q_1, \dots, q_N)$, $U \in C^2(B(0; R))$, and the $\Phi_i \in C^1(B(0; R))$. We interpret that the q_i are the *generalized coordinates* of the particles of the system.

On the particles we find the action of "conservative" forces, generated by the *potential* $U = U(q)$ and also "dissipative" forces of components $\Phi_i = \Phi_i(q')$. This ODS can be rewritten in the form

$$\begin{cases} q_i' = p_i \\ p_i' = -\dfrac{\partial U}{\partial q_i}(q) - \Phi_i(p), \end{cases} \tag{8.34}$$

where the p_i must be interpreted as *generalized velocities*.

Let us assume that

$$\frac{\partial U}{\partial q_i}(0) = 0 \quad \text{and} \quad \Phi_i(0) = 0 \quad \text{for } i = 1, \ldots, N \tag{8.35}$$

and, also, that there exists $R' \in (0, R)$ such that

$$U(q) > U(0) \quad \forall q \in \overline{B}(0; R') \setminus \{0\} \tag{8.36}$$

and

$$\sum_{i=1}^{N} \Phi_i(p) \, p_i \geq 0 \quad \forall p \in \overline{B}(0; R'). \tag{8.37}$$

Then, the trivial solution to (8.34) is stable.

Indeed, under these conditions, there exists a Liapunov function for (8.34); it suffices to take

$$V(q, p) = \frac{1}{2}|p|^2 + U(q) - U(0) \quad \forall (q, p) \in \overline{B}((0.0); R'). \tag{8.38}$$

We observe that

$$\dot{V}(q, p) = -\sum_{i=1}^{N} \Phi_i(p) \, p_i \quad \forall (q, p) \in \overline{B}((0.0); R'). \tag{8.39}$$

Now, let us assume that (8.35) and (8.36) are satisfied, $q = 0$ is the unique point in $\overline{B}(0; R')$ where all the partial derivatives of U vanish and

$$\sum_{i=1}^{N} \Phi_i(p) p_i > 0 \quad \forall p \in \overline{B}(0; R') \setminus \{0\}. \tag{8.40}$$

Then, taking into account Remark 8.5, we can easily deduce that the trivial solution to (8.34) is asymptotically stable.

Again, (8.38) can be viewed as the *total energy* of the system.

When the $\Phi_i \equiv 0$, we have thanks to (8.39) that V is constant along the trajectories; it is then said that the physical system under consideration is *conservative*.

Under the assumption (8.40), V decreases along the trajectories and it is said that the system is *dissipative*.

For more details on the motivation and interpretation of the stability properties for systems of this kind, see Rouché and Mawhin (1973). □

Remark 8.7. Theorem 8.5 and Remark 8.5 suggest the following strategy to prove the stability of the trivial solution to an autonomous ODS: find a Liapunov function for the ODS.

This is called (*direct*) *Liapunov's method.* The weakness of this method is that there is no general rule to construct Liapunov functions.   □

At this point, let us present a sufficient condition of instability:

Theorem 8.6. *Consider the ODS* (8.1), *where* $f \in C^1(B(0; R); \mathbb{R}^N)$ *satisfies* (8.2). *Assume that there exist* $R' \in (0, R]$ *and* $V \in C^1(\overline{B}(0; R'))$ *such that*

(1) $V(0) = 0$.
(2) *For every* $\delta \in (0, R')$ *there exists* $y_\delta \in \overline{B}(0; \delta)$ *with* $V(y_\delta) > 0$.
(3) $\dot{V}$ *is definite positive in* $\overline{B}(0; R')$.

Then, the trivial solution to (8.1) *is unstable.*

Proof. Let us assume the opposite. Then, there exists $\delta \in (0, R')$ such that, if $y_0 \in \mathbb{R}^N$ and $|y_0| \leq \delta$, necessarily $J(y_0) \supset [0, +\infty)$ and

$$|\psi(t; y_0)| \leq \frac{R'}{2} \quad \forall t \geq 0.$$

In particular, there must exist $M > 0$ such that

$$|V(\psi(t; y_0))| \leq M \quad \forall y_0 \in \overline{B}(0; \delta), \ \forall t \geq 0. \tag{8.41}$$

By assumption, there exists $y_\delta \in \overline{B}(0; \delta)$ such that $V(y_\delta) > 0$. Let us see that, contrarily to (8.41), the fact that $\dot{V}$ is definite positive implies

$$\lim_{t \to +\infty} V(\psi(t; y_\delta)) = +\infty, \tag{8.42}$$

which is an absurdity.

Indeed, we know that there exists a function $a : [0, R'] \mapsto \mathbb{R}$, continuous and strictly increasing, such that $a(0) = 0$ and $V(y) \geq a(|y|)$ for all $y \in \overline{B}(0; R')$. Then

$$\frac{\mathrm{d}}{\mathrm{d}t} V(\psi(t; y_\delta)) = \dot{V}(\psi(t; y_\delta)) \geq a(|\psi(t; y_\delta)|) \geq 0 \quad \forall t \geq 0, \tag{8.43}$$

whence

$$V(\psi(t; y_\delta)) \geq V(y_\delta) > 0 \quad \forall t \geq 0.$$

We claim that there exists $\alpha > 0$ with

$$|\psi(t; y_\delta)| \geq \alpha \quad \forall t \geq 0. \tag{8.44}$$

Indeed, if this were not the case, for every $n \geq 1$ there would exist $t_n \geq 0$ such that $|\psi(t_n; y_\delta)| \leq 1/n$. We would then have

$$0 < V(y_\delta) \leq V(\psi(t_n; y_\delta)), \quad \text{with } V(\psi(t_n; y_\delta)) \to V(0) = 0,$$

which is impossible.

Taking into account (8.43) and (8.44), we easily find that

$$\frac{\mathrm{d}}{\mathrm{d}t} V(\psi(t; y_\delta)) \geq a(\alpha) > 0$$

and

$$V(\psi(t; y_\delta)) \geq V(y_\delta) + a(\alpha)t \quad \forall t \geq 0,$$

which yields (8.42),

This ends the proof.

Example 8.8. Let us consider again the ODEs (8.5) and (8.8) and the ODSs (8.6) and (8.9). Now, let us see that the function $\overline{\varphi}$, given by

$$\overline{\varphi}(t) = (\pi, 0) \quad \forall t \in \mathbb{R}$$

is an unstable solution to (8.6) and also that when $\alpha > 0$ is sufficiently small, the same can be said for (8.9). To this purpose, let us introduce the change of variables

$$y_1 = z_1 + \pi, \quad y_2 = z_2.$$

The ODSs (8.6) and (8.9) can then be rewritten as follows:

$$\begin{cases} z_1' = z_2 \\ z_2' = -k \sin(z_1 + \pi) \end{cases} \tag{8.45}$$

and

$$\begin{cases} z_1' = z_2 \\ z_2' = -k \sin(z_1 + \pi) - \alpha z_2. \end{cases} \tag{8.46}$$

Again, let us assume that $0 < R < 2\pi$ and let us set

$$V(z_1, z_2) = \frac{1}{2} |z_2|^2 + k(1 - \cos(z_1 + \pi)) + z_1 z_2 \quad \forall (z_1, z_2) \in \overline{B}(0; R).$$

Then, V is well defined and continuously differentiable in $\overline{B}(0; R)$. Furthermore, the function $\dot{V}$ associated to (8.45) is given by

$$\dot{V}(z_1, z_2) = k z_1 \sin z_1 + |z_2|^2 \quad \forall (z_1, z_2) \in \overline{B}(0; R).$$

Consequently, the assumptions of Theorem 8.6 are satisfied and the trivial solution to (8.45) is unstable; in other words, $\overline{\varphi}$ is an unstable solution to (8.6).

On the other hand, the function $\dot{V}$ associated to (8.46) is

$$\dot{V}(z_1, z_2) = kz_1 \sin z_1 + (1 - \alpha)|z_2|^2 - \alpha z_1 z_2 \quad \forall (z_1, z_2) \in \overline{B}(0; R).$$

It is not difficult to check that, if

$$0 < \alpha < 1, \quad \frac{\alpha^2}{1 - \alpha} < k,$$

there exists $R' \in (0, R]$ such that $\dot{V}$ is definite positive in $\overline{B}(0; R')$. Consequently, if α is sufficiently small, $\overline{\varphi}$ is also an unstable solution to (8.9). □

Remark 8.8. Consider the non-autonomous ODS

$$y' = f(t, y), \tag{8.47}$$

where $f : \mathbb{R} \times B(0; R) \subset \mathbb{R}^{N+1} \mapsto \mathbb{R}^N$ is (for example) continuously differentiable in the set $\mathbb{R} \times B(0; R)$ and

$$f(t, 0) \equiv 0.$$

Again, the function $\varphi^0 : \mathbb{R} \mapsto \mathbb{R}^N$ with $\varphi^0(t) \equiv 0$ is a maximal solution.

In this framework, it makes sense to analyze the stability properties of the function φ^0. The definitions are a little more complex, but play more or less the same roles; on the other hand, we can establish results similar to those above; for details, see Note 11.8.1. □

Remark 8.9. The analysis of the stability of an autonomous ODS can also be achieved by adopting other viewpoints.

Thus, it is possible to carry out a global study of the solutions by analyzing the behavior of the *orbits,* deducing the existence of *attractors* and *limit cycles,* etc.; see Note 11.8.2. □

Exercises

E.8.1 Analyze the stability properties of the trivial solution in the case of the following ODSs:

$$(1) \quad y' = \begin{pmatrix} -1 & 0 & 0 \\ 0 & -2 & 1 \\ 0 & 0 & -2 \end{pmatrix} y; \qquad (2) \quad y' = \begin{pmatrix} -2 & 0 & 0 \\ 0 & -1 & 1 \\ 0 & 0 & 1 \end{pmatrix} y;$$

$$(3) \quad y' = \begin{pmatrix} -1 & 2 & 0 \\ 0 & -1 & 2 \\ 0 & 0 & -1 \end{pmatrix} y; \qquad (4) \quad y' = \begin{pmatrix} -1 & 0 & 0 \\ 1 & 0 & 2 \\ 0 & 0 & 0 \end{pmatrix} y;$$

$$(5) \quad y' = \begin{pmatrix} 0 & 0 & 0 \\ 0 & -1 & -1 \\ 0 & 1 & -1 \end{pmatrix} y; \quad (6) \quad y' = \begin{pmatrix} 1 & 0 & 0 \\ 0 & -1 & -1 \\ 0 & 1 & -1 \end{pmatrix} y.$$

In the case of the ODSs (2) and (6), determine a necessary and sufficient condition that must be satisfied by a maximal solution to have

$$\lim_{t \to +\infty} \varphi(t) = 0.$$

E.8.2 Analyze the stability of the trivial solution to the ODE $y' = -y^2$. Also, analyze the stability of the solution $\varphi(t) = e^{-2t}$ to the ODE $y'' + 4y' + 4y = 0$.

E.8.3 Determine if the trivial solution is stable, asymptotically stable, or unstable in the case of the ODE $y'' + 2ky' + a^2 y = 0$, where $ka \in \mathbb{R}$ and $k > 0$.

E.8.4 Determine if the trivial solution is stable, asymptotically stable, or unstable in the case of the ODS

$$y' = \begin{pmatrix} 0 & 0 & 0 & 0 \\ 0 & 0 & 0 & 0 \\ 0 & 0 & -1 & 3 \\ 0 & 0 & 0 & -1 \end{pmatrix} y.$$

E.8.5 Consider the ODS

$$y' = \begin{pmatrix} -1 & e^{2t} \\ 0 & -1 \end{pmatrix} y.$$

Prove that there exists $\varepsilon > 0$ such that, for any $\eta > 0$, there exist initial data y_η and times $t_\eta > 0$ such that

$$|y_\eta| \le \eta, \quad t_\eta \in J(y_\eta), \quad \text{and} \quad |\psi(t_\eta; y_\eta)| > \varepsilon.$$

Explain why this does not contradict the fact that the matrix of coefficients possesses the eigenvalue -1 of multiplicity 2 for all t.

E.8.6 Consider the ODE $y' = y(1 - y)$. Prove that the trivial solution is unstable and that the solution $\varphi(t) \equiv 1$ is asymptotically stable.

E.8.7 Analyze the stability properties of the trivial solution in the case of the following ODSs:

$$(1) \begin{cases} y_1' = -y_2 - y_1^3 \\ y_2' = y_1 - y_2^3; \end{cases} \quad (2) \begin{cases} y_1' = -y_1 y_2^4 \\ y_2' = y_2 y_1^4; \end{cases}$$

$$(3) \begin{cases} y_1' = -y_1 - y_2 \\ y_2' = y_1 - y_2^3. \end{cases}$$

E.8.8 Analyze the stability of the trivial solution in the case of the following ODSs using, when it is indicated, the corresponding function V:

(1) $\begin{cases} y_1' = -y_1 + y_2^2 \\ y_2' = -y_2 + y_1 y_2; \end{cases}$

(2) $\begin{cases} y_1' = -6y_2 - \frac{1}{4}y_1 y_2^2 \\ y_2' = 4y_1 - \frac{1}{6}y_2 \end{cases}$ $V(y_1, y_2) \equiv 2y_1^2 + 3y_2^2;$

(3) $y'' + \alpha y' + \beta y^3 + y = 0$, con $\alpha, \beta > 0$;

(4) $\begin{cases} y_1' = 2y_1 + 8\sin y_2 \\ y_2' = 2 - e^{y_1} - 3y_2 - \cos y_2; \end{cases}$ (5) $\begin{cases} y_1' = -4y_2 - y_1^3 \\ y_2' = 3y_1 - y_2^3. \end{cases}$

E.8.9 Prove that the trivial solution to the ODE $y'' + (1-y^2)y^2 y' + \frac{1}{2}y = 0$ is stable.

E.8.10 Is the trivial solution to $y'' + y^2 y' + y = 0$ stable? Justify the answer.

E.8.11 Let $A \in \mathcal{L}(\mathbb{R}^N)$, let the $\lambda_1, \ldots, \lambda_N$ be the eigenvalues of A, and assume that

$$\alpha > \max\{\operatorname{Re}(\lambda_i),\ 1 \le i \le N\}.$$

Let φ be a solution to the ODS $y' = Ay$ and let $\tau \in \mathbb{R}$ be given. Prove that there exists $C > 0$ such that

$$|\varphi(t)| \le Ce^{\alpha t}, \quad \forall t \ge \tau.$$

E.8.12 Consider the second-order ODE $y'' + \alpha^2 y = f(t)$, with $\alpha \ne 0$ and $f \in C^0(\mathbb{R}_+)$ such that

$$\int_0^\infty |f(t)|\, dt < +\infty.$$

Prove that, if $\varphi : \mathbb{R}_+ \mapsto \mathbb{R}$ solves in $(0, +\infty)$ this equation, then

$$\sup_{t \in \mathbb{R}_+} |\varphi(t)| < +\infty \quad \text{and} \quad \sup_{t \in \mathbb{R}_+} |\varphi'(t)| < +\infty.$$

E.8.13 Analyze the stability of the trivial solution in the case of the following ODSs, using the function V when it is indicated (here, a, α, and β are real constants):

(1) $\begin{cases} y_1' = -y_2 + y_1(a^2 - y_1^2 - y_2^2) \\ y_2' = y_1 + y_2(a^2 - y_1^2 - y_2^2); \end{cases}$ (2) $\begin{cases} y_1' = -4y_2 - y_1^3 \\ y_2' = 3y_1 - y_2^3; \end{cases}$

(3) $\begin{cases} y_1' = -y_1 + y_2^2 \\ y_2' = -y_2 + y_1 y_2; \end{cases}$

(4) $\begin{cases} y_1' = -6y_2 - \frac{1}{4}y_1 y_2^2 \\ y_2' = 4y_1 - \frac{1}{6}y_2 \end{cases}$ $V(y_1, y_2) \equiv 2y_1^2 + 3y_2^2;$

(5) $\begin{cases} y_1{}' = y_2 + y_1(y_1^2 + y_2^2)^{1/2}(y_1^2 + y_2^2 - 1)^2 \\ y_2' = -y_1 + y_2(y_1^2 + y_2^2)^{1/2}(y_1^2 + y_2^2 - 1)^2; \end{cases}$

(6) $\begin{cases} y_1' = y_1 + y_2^3 \\ y_2' = y_2 + y_1^2; \end{cases}$

(7) $y'' + \alpha y' + \beta y^3 + y = 0;$ (8) $\begin{cases} y_1' = 2y_1 + 8\sin y_2 \\ y_2' = 2 - e^{y_1} - 3y_2 - \cos y_2; \end{cases}$

(9) $\begin{cases} y_1' = 3y_1 + y_2^3 \\ y_2' = -4y_2 + y_1^3 \end{cases}$ $V(y_1, y_2) \equiv 4y_1^2 - 3y_2^2.$

E.8.14 Let us consider the ODS

$$\begin{cases} y_1' = -y_1 + 2y_1^3 y_2^2 \\ y_2' = -y_2. \end{cases} \tag{8.48}$$

(1) Prove that the trivial solution is asymptotically stable.
(2) Prove that their set $D = \{(y_1, y_2) \in \mathbb{R}^2 : y_1^2 + y_2^2 \leq \frac{1}{\sqrt{2}}\}$ is positively invariant; in other words, with $\psi(\cdot; (z_1, z_2))$ being the maximal solution to (8.48) satisfying $y_1(0) = z_1$, $y_2(0) = z_2$, prove that

$$(z_1, z_2) \in D \Rightarrow \psi(t; (z_1, z_2)) \in D \quad \forall t \geq 0.$$

E.8.15 Analyze the stability of the *critical points* (i.e. the constant solutions $\overline{\varphi}(t) \equiv z$ with $f(z) = 0$) in the case of the ODS

$$\begin{cases} y_1' = y_1 - y_1 y_2 \\ y_2' = -y_2 + y_1 y_2. \end{cases}$$

To this end, use the function $V(y_1, y_2) \equiv y_1 + y_2 - \log(1 + y_1)(1 + y_2)$.

E.8.16 Consider the ODE $y'' + a(y^2 - 1)y' + y = 0$, where $a < 0$. Rewrite this equation as a first-order ODS and determine if the associated trivial solution is stable and/or attractive.

E.8.17 Let $g \in C^1(-k, k)$ be given with $k > 0$ and assume that $\xi g(\xi) > 0$ whenever $\xi \neq 0$. Analyze the stability of the trivial solution in the case of the ODEs $y'' + g(y) = 0$ and $y'' + y' + g(y) = 0$.

E.8.18 Prove that the trivial solution to the ODS

$$\begin{cases} y_1' = y_2 \\ y_2' = -\sin y_1 - y_2 \cos y_1 \end{cases}$$

is asymptotically stable.

Chapter 9

The Method of Characteristics for First-Order Linear and Quasi-Linear Partial Differential Equations

The goal of this chapter is to present the method of characteristics for partial differential equations of the first order.

A partial differential equation is an identity where we find an unknown function that depends on several real variables together with some of its partial derivatives.

As before, we say that the order of the equation is the maximal order of the derivatives of the unknown (we will use the acronym PDE in the sequel).

Again, these equations provide a lot of motivations in science and engineering and their analysis and exact or approximate resolution is of major interest from many viewpoints.

The study of PDEs is a much more complex subject and there is no way to give a significative set of results in a reduced amount of chapters.

There is, however, a connection of first-order PDEs and ODSs: we can associate to any given PDE a particular ODS (the characteristic system), whose solutions can be used to construct the solutions to the PDE. This key fact allows to formulate and solve, at least locally, the Cauchy problem for a PDE.

It will be seen in this chapter that these ideas are natural when we interpret the PDE as a law governing the behavior of a physical phenomenon in several different situations.

9.1   The Concepts and Some Preliminary Results

In this chapter, we find *partial differential equations* (PDEs) of the first order. We will analyze related initial-value problems, also known as Cauchy problems, from the viewpoint of the solution's existence and uniqueness.

The inclusion of PDEs of the first order in this book is justified by the resolution method, usually known as the *method of characteristics*.

For any given function $u = u(x_1, \ldots, x_N)$, if they exist, the partial derivatives of u will be denoted in several ways:

$$\frac{\partial u}{\partial x_i}, \ \partial_i u, u_{x_i}, \ \text{etc.}$$

Actually, several different notations will be used also for the independent variables: t, x, y, etc.

For simplicity, we will speak only of linear or quasi-linear PDEs.

A PDE of the first order is said to be quasi-linear if it is of the form

$$\sum_{i=1}^{N} f_i(x, u)\partial_i u = g(x, u)$$

for some functions f_i and g.

For PDEs with a more general structure, it is necessary to carry out a more advanced study. Nevertheless, we will give some related information. For other results, see for instance Sneddon, 1957 and John, 1991; see also Note 11.9.

First of all, let us present some fundamental concepts.

Definition 9.1. Let $N \geq 1$ be an integer and let $\mathcal{O} \subset \mathbb{R}^{2N+1}$ be a nonempty open set. A PDE of the first order is an equality of the form

$$F(x_1, \ldots, x_N, u, \partial_1 u, \ldots, \partial_N u) = 0, \tag{9.1}$$

where $F : \mathcal{O} \mapsto \mathbb{R}$ is a continuous function. $\qquad\square$

In (9.1), it is usual to say that the x_i are the *independent variables* and u is the *dependent variable* or unknown. Briefly, we write

$$F(x, u, Du) = 0, \tag{9.2}$$

where all the x_i (resp. the $\partial_i u$) have been put together in x (resp. Du).

If $\mathcal{O}$ is of the form $\mathcal{O} = U \times \mathbb{R}^N$, where U is an open set in $\mathbb{R}^{N+1}$ and

$$F(x, u, p) = \sum_{i=1}^{N} f_i(x, u)p_i - g(x, u) \quad \forall(x, u, p) \in \mathcal{O} \qquad (9.3)$$

for some continuous functions $f_i, g : U \mapsto \mathbb{R}$, it will be said that the PDE is *quasi-linear*.

On the other hand, if $\mathcal{O}$ is of the form $\mathcal{O} = U_0 \times \mathbb{R}^{N+1}$, where U_0 is an open set in $\mathbb{R}^N$ and

$$F(x, u, p) = \sum_{i=1}^{N} a_i(x)p_i + a_0(x)u - b(x) \quad \forall(x, u, p) \in \mathcal{O} \qquad (9.4)$$

for some continuous functions $a_i, a_0, b : U_0 \mapsto \mathbb{R}$, it is said that the PDE is *linear*.

Usually, the linear and quasi-linear PDEs are written in the form

$$\sum_{i=1}^{N} a_i(x)\, \partial_i u + a_0(x)u = b(x) \qquad (9.5)$$

and

$$\sum_{i=1}^{N} f_i(x, u)\partial_i u = g(x, u), \qquad (9.6)$$

respectively.

In general, any PDE that is not quasi-linear is said to be *nonlinear*.

The motivations of first-order PDEs can be found in many problems with origins in a lot of fields in mathematics (the properties of curves and surfaces, the control and optimization of differential systems, etc.) and other sciences (classical mechanics and Hamiltonian systems, gas dynamics, imaging, cell evolution, transport phenomena, etc.).

Some of them will be presented now; see also Note 11.9.1.

Example 9.1. (a) The PDE

$$u_t + uu_x = 0, \qquad (9.7)$$

where the independent variables are x and t, is known as *Burgers's equation*. This is a quasi-linear PDE.

Among other phenomena, this PDE serves to describe the evolution of the particles of an ideal gas, if we assume that the motion is one-directional.

We interpret that $u(t, x)$ is the velocity of the gas particle located at x at time t.

This is in fact the simplest example of *motion equation* of a fluid; more details can be found in Chorin, 1975.

(b) Let the vector $V \in \mathbb{R}^n$ be given and consider the PDE

$$u_t + V \cdot \nabla u = 0$$

in the $n + 1$ independent variables $t, x_1, \ldots, x_n$. Here, ∇u is the *gradient* of u, that is, the n-dimensional vector of components $\partial_1 u, \ldots, \partial_n u$.

We are dealing here with the *transport equation*.

In this PDE, we interpret that $u = u(t, x)$ is the density or concentration of a substance filling a region in $\mathbb{R}^n$ during a time interval that is being *transported* at velocity V.

Obviously, this is a linear PDE.

(c) More generally, for any given open set $G \subset \mathbb{R}^{n+1} \times \mathbb{R} \times \mathbb{R}^n$ and any continuous function $H : G \mapsto \mathbb{R}$, we can consider the PDE

$$u_t + H(x, t, u, \nabla u) = 0.$$

This is the so-called *Hamilton–Jacobi equation* associated to the *Hamiltonian* $H = H(x, t, u, p)$.

This is in general a nonlinear PDE of great relevance in *classical mechanics* and *control theory*, among other areas; see Rund (1966) and Trélat (2005) for more details.

(d) The PDE with two independent variables

$$u_x^2 + u_y^2 = 1$$

is called the *eikonal equation*. This is a relatively simple example of nonlinear PDE with origin in *geometric optics*; see for instance (Speck, 2016). □

In this chapter, we will mainly consider the quasi-linear PDE (9.6).

First of all, we will specify what is a solution to this equation:

Definition 9.2. Let $D \subset \mathbb{R}^N$ be a non-empty open set and let $u : D \mapsto \mathbb{R}$ be a given function. It will be said that u is a (classical) solution to (9.6) in D if

(1) $u \in C^1(D)$.
(2) For every $x \in D$, $(x, u(x)) \in U$.
(3) For every $x \in D$, one has $\sum_{i=1}^{N} f_i(x, u(x)) \partial_i u(x) = g(x, u(x))$. □

In the previous definition, we speak of "classical" solutions. This means that we ask u to be continuously differentiable and satisfy (9.6) in the usual pointwise sense.

However, for some reasons that will be found in what follows, it is convenient sometimes to consider other (less strong) solution concepts. This will lead to the so-called *weak* or *generalized* solutions.

In order to understand which is the structure of the family of solutions to a first-order PDE, let us first see what happens in the case of some elementary linear PDEs.

Example 9.2. (a) Let us consider the PDE

$$\partial_1 u + \partial_2 u = 0. \tag{9.8}$$

Let $D \subset \mathbb{R}^2$ be an open set and let ϕ be an arbitrary function in $C^1(\mathbb{R})$. Then, the function u, given by

$$u(x_1, x_2) = \phi(x_1 - x_2) \quad \forall (x_1, x_2) \in D,$$

solves (9.8) in D.

(b) Consider now the PDE

$$\partial_1 u + \partial_2 u - \partial_3 u = 0. \tag{9.9}$$

Let $D \subset \mathbb{R}^3$ be a non-empty open set and, now, let ϕ be an arbitrary function in $C^1(\mathbb{R} \times \mathbb{R})$. Then, if we put

$$u(x_1, x_2, x_3) = \phi(x_1 + x_3, x_1 + x_3) \quad \forall (x_1, x_2, x_3) \in D,$$

we see that u is a solution in D of (9.9). $\qquad \square$

These examples seem to suggest that the solutions of a linear PDE of first-order with N independent variables (9.5) form a family of functions of the form

$$u = \phi(Z_1(x), \dots, Z_{N-1}(x))$$

that depend on the choice of an arbitrary function $\phi = \phi(z_1, \dots, z_{N-1})$ of $N - 1$ variables.

This family can be called the general solution (or general integral) to (9.5).

Consequently, it is appropriate to consider problems where we require the solution to satisfy a first-order PDE together with some additional conditions that allow to select within the family of solutions one single function.

In this chapter, this is achieved by introducing the so-called *initial-value problem* (or Cauchy problem) for (9.6).

For the formulation, we will previously need to speak of *hyper-surfaces of class C^k in $\mathbb{R}^N$*:

Definition 9.3. Let $k \geq 1$ be an integer. A hyper-surface of class C^k in $\mathbb{R}^N$ is any set S of the form

$$S = \{\xi(s_1, \ldots, s_{N-1}) : (s_1, \ldots, s_{N-1}) \in \mathcal{O}\}, \tag{9.10}$$

where $\mathcal{O}$ is an open neighborhood of 0 in $\mathbb{R}^{N-1}$, $\xi \in C^k(\mathcal{O}; \mathbb{R}^N)$, and

$$\text{rank} \left[\frac{\partial \xi}{\partial s_1}(s) \big| \cdots \big| \frac{\partial \xi}{\partial s_{N-1}}(s) \right] = N - 1 \quad \forall s = (s_1, \ldots, s_{N-1}) \in \mathcal{O},$$

that is, the $N - 1$ "tangential" vectors

$$\frac{\partial \xi}{\partial s_1}(s), \ldots, \frac{\partial \xi}{\partial s_{N-1}}(s)$$

are linearly independent for all s. $\square$

When $N = 2$, we are speaking of a curve arc of class C^k; the nonzero vector

$$\xi'(s) = \frac{d\xi}{ds}(s)$$

is in this case a tangent vector to the curve at the point $\xi(s)$.

When $N = 3$, we are considering a piece of surface of class C^k; now, the vectors

$$\frac{\partial \xi}{\partial s_1}(s_1, s_2) \quad \text{and} \quad \frac{\partial \xi}{\partial s_2}(s_1, s_2)$$

span the tangent plane S at $\xi(s_1, s_2)$ for every (s_1, s_2).

Definition 9.4. Let us consider the quasi-linear PDE (9.6), where the functions f_i and g are (at least) continuous in U ($U \subset \mathbb{R}^{N+1}$ is a non-empty open set). An initial-value problem (or Cauchy problem) for (9.6) is a system of the form

$$\text{(IVP)} \qquad \begin{cases} \displaystyle\sum_{i=1}^{N} f_i(x, u)\, \partial_i u = g(x, u) \\ u|_S = u_0, \end{cases}$$

where S is a hyper-surface of class C^1 in $\mathbb{R}^N$, given by (9.10) and $u_0 \in C^1(\mathcal{O})$. $\square$

The second equality in (IVP) is known as the *initial condition* or *Cauchy condition*. Obviously, it indicates that the values of the unknown u are known on S.

It is clear that (IVP) generalizes, at least formally, the initial-value problem for an ODS considered in previous chapters. Thus, from a geometrical viewpoint, we go from prescribing values at a point (a 0-dimensional manifold) to doing the same on a hyper-surface (an $N-1$-manifold) as the number of independent variables go from 1 to N.

Keeping the same notation, let us set $a = \xi(0)$. Let $D \subset \mathbb{R}^N$ be an open neighborhood of a and let $u : D \mapsto \mathbb{R}$ be a function. It will be said that u is a solution in D of (IVP) if it is a solution in D of (9.6) and, moreover,

$$\begin{cases} u(\xi(s_1,\ldots,s_{N-1})) = u_0(s_1,\ldots,s_{N-1}) \\ \forall(s_1,\ldots,s_{N-1}) \in \mathcal{O} \quad \text{such that } \xi(s_1,\ldots,s_{N-1}) \in D. \end{cases} \tag{9.11}$$

Of course, this means that u coincides with u_0 at any point in D that belongs to S.

Example 9.3. Let S be the straight line $t = 0$ in the plane. Consider the system

$$\begin{cases} u_t + uu_x = 0 \\ u|_S = u_0(x), \end{cases} \tag{9.12}$$

where $u_0 \in C^1(\mathbb{R})$. This is a Cauchy problem for Burgers's equation (9.7). Here, $\mathcal{O} = \mathbb{R}$ and $\xi(s) \equiv (0,s)$.

Let D be the following open set in $\mathbb{R}^2$:

$$D = \{(t,x) \in \mathbb{R}^2 : 0 < t < T, \ 0 < x < L\},$$

and let $u : D \mapsto \mathbb{R}$ be a solution in D to (9.12). Then, we can interpret that u provides the velocity of a gas whose particles fill $(0,L)$ during the time interval $(0,T)$, and is equal to u_0 at initial time $t = 0$.[1] □

---

[1] In fact, (9.12) is a very simplified formulation of the so-called fundamental problem of mechanics: we want to determine the mechanical state of a medium at any time $t > 0$ using as information the physical properties of the medium and the description of its mechanical state at time $t = 0$. Note that, in the case of (9.12), the mechanical state is given by u, the physical properties are specified by the PDE, and the initial state is fixed by u_0; for more details, see for instance (Chorin and Marsden, 1990).

9.2   Resolution of the Cauchy Problem

In this section, we will solve (IVP).

   More precisely, we will prove that, under some conditions on the data, (IVP) possesses exactly one local solution (defined in a neighborhood of $a = \xi(0)$).

   The data of the problem are the following:

- $U \subset \mathbb{R}^{N+1}$ is a non-empty open set, $f_i \in C^1(U)$ for $i = 1, \ldots, N$ and $g \in C^1(U)$.
- S is a hyper-surface of class C^1 in $\mathbb{R}^N$, given by (9.10). We will set $a = \xi(0)$.
- $u_0 \in C^1(\mathcal{O})$. We will also set $u_a = u_0(0)$.

Theorem 9.1. *Assume that*

$$(\xi(s_1, \ldots, s_{N-1}), u_0(s_1, \ldots, s_{N-1})) \in U \quad \forall(s_1, \ldots, s_{N-1}) \in \mathcal{O} \qquad (9.13)$$

and

$$\det\left(f(a, u_a) \,\bigg|\, \frac{\partial \xi}{\partial s_1}(0) \,\bigg|\, \cdots \,\bigg|\, \frac{\partial \xi}{\partial s_{N-1}}(0)\right) \neq 0, \qquad (9.14)$$

where $f(a, u_0(0))$ is the vector of components $f_i(a, u_0(0))$. Then, there exist an open neighborhood D of a and a unique function $u : D \mapsto \mathbb{R}$ that solve the Cauchy problem (IVP) *in D (Figure 9.1).*

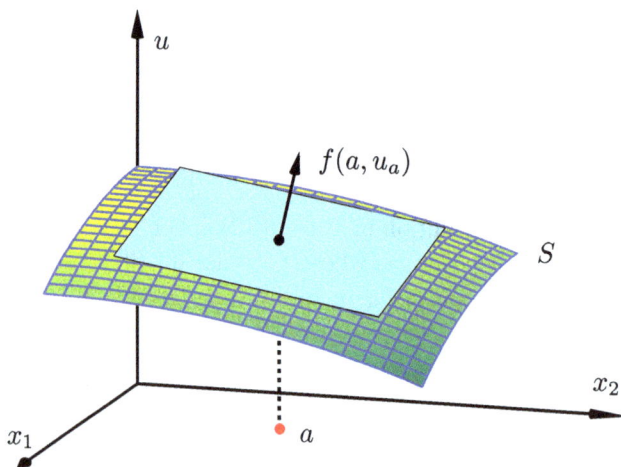

Fig. 9.1   Visualization of the assumptions in Theorem 9.1.

Remark 9.1. The assumption (9.13) says that the data ξ, u_0, and U must be compatible.

On the other hand, in order to give a sense of the problem, we must assume that $(a, u_0(0)) \in U$ and this implies that there exists $\mathcal{O}' \subset \mathcal{O}$ such that (9.13) holds. □

Remark 9.2. The assumption (9.14) is usually called a *transversality condition*. It can also be viewed as a compatibility condition on the data.

Indeed, if a solution u to (IVP) exists and we set $p_i = \partial_i u(a)$, we have

$$\sum_{i=1}^{N} f_i(a, u_a) \, p_i = g(a, u_a). \tag{9.15}$$

But differentiating (9.11) with respect to the s_j and then taking $s_1 = \cdots = s_{N-1} = 0$, we see that the identities (9.13) also give

$$\sum_{i=1}^{N} \frac{\partial \xi_i}{\partial s_j}(0) \, p_i = \frac{\partial u_0}{\partial s_j}(0). \tag{9.16}$$

If we view (9.15)–(9.16) as a system of N equations for the N unknowns $p_1, \ldots, p_N$, we find that the existence and uniqueness of solution for any g and u_0 are ensured only if the associated matrix of coefficients is non-singular. But this happens if and only if one has (9.14).

In other words, if we do not have (9.14), we can find data g and u_0 such that (9.15)–(9.16) is unsolvable. Evidently, in that case, there is no solution to (IVP).

This argument justifies the hypothesis (9.14). □

Remark 9.3. For the proof of existence, we will use the *method of characteristics*.

It relies on the introduction and properties of the so-called *characteristic system* corresponding to (9.6), that is the following ODS:

$$y' = f(y, v), \quad v' = g(y, v). \tag{9.17}$$

Let us give a sketch of the argument (the details can be found in Section 9.3):

- We will add to (9.17) the Cauchy conditions

$$y(0) = \xi(s_1, \ldots, s_{N-1}), \quad v(0) = u_0(s_1, \ldots, s_{N-1}). \tag{9.18}$$

This will make it possible to define in a neighborhood of 0 in $\mathbb{R}^N$ the continuously differentiable function $(t, s_1, \ldots, s_{N-1}) \mapsto (\overline{y}, \overline{v})(t; s_1, \ldots, s_{N-1})$

that associates to every $(t, s_1, \ldots, s_{N-1})$ the value that the solution to (9.17)–(9.18) takes at t.

- It will be seen that, thanks to (9.14), $\overline{y}$ is invertible in a neighborhood of zero.
- Let ψ be the local inverse, that is well defined and continuously differentiable in a neighborhood of $\overline{y}(0; 0, \ldots, 0) = a$, and let us set

$$u(x) = \overline{v}(\psi(x)).$$

Then, there exists a neighborhood of a where u is also well defined and solves the Cauchy problem (IVP).

The existence of a local solution to (IVP) is thus ensured. □

Remark 9.4. Thanks to the previous argument, (IVP) possesses at least one solution u, defined in a neighborhood of a.

Let us indicate how we can prove uniqueness.

Let $\hat{u}$ be another solution. In order to prove that u and $\hat{u}$ coincide in a new (eventually smaller) neighborhood of a, it suffices to consider the functions $\overline{y}$ and $\overline{v}$ introduced in the previous remark and the function $\hat{v}$, given by

$$\hat{v}(t; s_1, \ldots, s_{N-1}) = \hat{u}(\overline{y}(t; s_1, \ldots, s_{N-1}))$$

for any $(t; s_1, \ldots, s_{N-1})$ in a neighborhood of 0 in $\mathbb{R}^N$.

It is not difficult to check that, for any $(s_1, \ldots, s_{N-1})$ sufficiently close to the origin, the function

$$z := \overline{v}(\cdot; s_1, \ldots, s_{N-1}) - \hat{v}(\cdot; s_1, \ldots, s_{N-1})$$

is the unique solution to a Cauchy problem of the form

$$\begin{cases} z' = G(t, z) \\ z(0) = 0, \end{cases}$$

where G is well defined and of class C^1 and $G(t, 0) \equiv 0$.

Consequently, $z \equiv 0$.

This shows that u and $\hat{u}$ coincide in a neighborhood of a; see Section 9.3 for more details. □

Remark 9.5. In order to be applied, the arguments used to prove the existence and uniqueness of the solution need C^1 functions f_i in the PDE in (IVP) (at least) in a neighborhood of $(a, u_0(0))$.

It would be interesting to know whether or not this hypothesis can be weakened.

For example, let us assume that the f_i and g are only continuous in U. A natural attempt would be to approximate these functions by regular $f_{i,n}$ and g_n with $n = 1, 2, \ldots$, solve the corresponding approximated Cauchy problems and see if it is possible to take limits and find a solution.

However, in general, it is not clear whether this strategy works and the existence of a classical solution to (IVP) with continuous coefficients in the PDE is open. $\square$

Example 9.4. (a) Let us consider the Cauchy problem

$$\begin{cases} x_1 \partial_1 u + (x_2 + 1)\partial_2 u = u \\ u|_S = (x_1 + 1)|_S, \end{cases}$$

where S is the straight line $x_1 - x_2 = 0$. Let us see how the solution can be computed in a neighborhood of $(0, 0)$.

This is a particular case of (IVP) with $\mathcal{O} = \mathbb{R}$, $\xi(s) \equiv (s, s)$ and $u_0(s) \equiv s + 1$. The transversality condition (9.14) is fulfilled for these data, since

$$\left(f(a, u_0(0)) \Big| \frac{\partial \xi}{\partial s}(0) \right) = \begin{pmatrix} 0 & 1 \\ 1 & 1 \end{pmatrix}.$$

Consequently, the existence and uniqueness of a local solution is ensured. Let us consider the characteristic system

$$y_1' = y_1, \quad y_2' = y_2 + 1, \quad v' = v,$$

together with the initial conditions

$$y_1(0) = s, \quad y_2(0) = s, \quad v(0) = s + 1.$$

With the notation in Remark 9.3, one has

$$\overline{y}(t; s) = (se^t, (s + 1)e^t - 1), \quad \overline{v}(t; s) = (s + 1)e^t$$

for every $(t, s) \in \mathbb{R}^2$. Consequently, the change of variables $(x_1, x_2) \to (t, s)$ is given by

$$x_1 = se^t, \quad x_2 = (s + 1)e^t - 1$$

and the solution is

$$u(x_1, x_2) = \overline{v}(t; s) = x_2 + 1 \quad \forall (x_1, x_2) \in \mathbb{R}^2.$$

(b) Consider now the Cauchy problem

$$\begin{cases} u_t + uu_x = 0 \\ u|_S = u_0, \end{cases}$$

where S is the straight line $t = 0$ and $u_0 \in C^1(\mathbb{R})$ is given. Again, let us see how the solution can be found.

In this case, we must take $\xi(s) \equiv (0, s)$ and interpret the initial condition as follows:

$$u(s, 0) = u_0(s) \quad \text{in a neighborhood of } 0.$$

It is easy to check that the transversality condition is satisfied. On the other hand, the characteristic system (where we have changed the notation slightly) is the following:

$$\frac{d\sigma}{d\tau} = 1, \quad \frac{dy}{d\tau} = v, \quad \frac{dv}{d\tau} = 0.$$

The Cauchy conditions that we must take into account to complete this system are

$$\sigma(0) = 0, \quad y(0) = s, \quad v(0) = u_0(s).$$

Consequently, we get

$$\bar{\sigma}(\tau; s) = \tau, \quad \bar{y}(\tau; s) = u_0(s)\tau + s, \quad \bar{v}(\tau; s) = u_0(s) \quad \forall (\tau, s) \in \mathbb{R}^2.$$

The change of variables is now

$$t = \tau, \quad x = u_0(s)\tau + s,$$

and allows to define (τ, s) uniquely for each (t, x) near $(0, 0)$. Therefore, the solution is defined (at least) in a neighborhood of $(0, 0)$ as follows:

$$u(t, x) = u_0(s), \quad \text{with } s \text{ given by } s + u_0(s)t = x.$$

In this example, we observe that the solution is constant along the lines $x = u_0(s)t + s$. These lines are in fact the trajectories described by the functions $(\bar{\sigma}, \bar{y})(\cdot; s)$.

Accordingly, they are known as the *characteristic straight lines*.

It is also observed that, if u_0 is strictly decreasing in an interval, the solution cannot be defined in the whole $\mathbb{R}^2$.

Indeed, if this is the case, $s_1 < s_2$ and $u_0(s_1) > u_0(s_2)$, the characteristic straight lines "starting" from $(s_1, 0)$ and $(s_2, 0)$ meet at some point (t, x) with $t > 0$; at this point, u would have to take two different values, $u_0(s_1)$ and $u_0(s_2)$, which is absurd.

This example shows that, in general, a global existence result cannot b expected for (IVP), even if $U = \mathbb{R}^N$ and $\mathcal{O} = \mathbb{R}^{N-1}$. $\square$

9.3   Proof of Theorem 9.1*

As already said, we will prove that at least one solution to (9.6) exists by the method of characteristics.

Thus, for every $(s_1, \ldots, s_{N-1}) \in \mathcal{O}$, let us consider the Cauchy problem composed of the characteristic ODS

$$y' = f(y, v), \quad v' = g(y, v) \tag{9.19}$$

and the initial conditions

$$y(0) = \xi(s_1, \ldots, s_{N-1}), \quad v(0) = u_0(s_1, \ldots, s_{N-1}). \tag{9.20}$$

Since f, g, ξ, and u_0 are continuously differentiable functions, for every $(s_1, \ldots, s_{N-1}) \in \mathcal{O}$ there exists a unique maximal solution to (9.19) – (9.20), defined in the open interval $J(s_1, \ldots, s_{N-1})$.

In view of the results on continuous and differentiable dependence in Chapter 7, the set

$$\Lambda = \{(t, s_1, \ldots, s_{N-1}) \in \mathbb{R}^N : (s_1, \ldots, s_{N-1}) \in \mathcal{O}, \ t \in J(s_1, \ldots, s_{N-1})\}$$

is open, the function $(\overline{y}, \overline{v}) : \Lambda \mapsto \mathbb{R}^{N+1}$ that assigns to each $(t, s_1, \ldots, s_{N-1}) \in \Lambda$ the value at t of the associated maximal solution to (9.19) − (9.20) is continuously differentiable and, in particular, one has

$$\overline{y}'(0,0) = \left(\frac{\partial \overline{y}}{\partial t}(0,0) \middle| \frac{\partial \overline{y}}{\partial s_1}(0,0) \middle| \cdots \middle| \frac{\partial \overline{y}}{\partial s_{N-1}}(0,0) \right)$$

$$= \left(f(a, u_a) \middle| \frac{\partial \xi}{\partial s_1}(0) \middle| \cdots \middle| \frac{\partial \xi}{\partial s_{N-1}}(0) \right)$$

(for clarity, here and henceforth, 0 and $(0,0)$, respectively, denote the origins in $\mathbb{R}^{N-1}$ and $\mathbb{R}^N$).

Consequently, thanks to (9.14), $\overline{y}$ is invertible in a neighborhood of zero.

More precisely, there exist open sets Δ and D in $\mathbb{R}^N$ with $(0,0) \in \Delta$ and $a = \overline{y}(0,0) \in D$, such that $(t, s_1, \ldots, s_{N-1}) \in \Delta \mapsto \overline{y}(t, s_1, \ldots, s_{N-1}) \in D$ is a diffeomorphism.

Let ψ be the local inverse of $\overline{y}$, defined and continuously differentiable in D and set

$$u(x) = \overline{v}(\psi(x)) \quad \forall x \in D.$$

Then u is well defined, is of class C^1, and, furthermore, is a solution in D to the Cauchy problem (IVP).

Indeed, if $(s_1, \ldots, s_{N-1}) \in \mathcal{O}$ and $\xi(s_1, \ldots, s_{N-1}) \in D$, we have

$$\psi(\xi(s_1, \ldots, s_{N-1})) = (0, s_1, \ldots, s_{N-1}),$$

whence

$$u(\xi(s_1, \ldots, s_{N-1})) = \overline{v}(0, s_1, \ldots, s_{N-1}) = u_0(s_1, \ldots, s_{N-1})$$

for all $(s_1, \ldots, s_{N-1}) \in \mathcal{O}$ such that $\xi(s_1, \ldots, s_{N-1}) \in D$.

This proves that u satisfies the initial condition.

On the other hand, if we denote by $\psi_0, \psi_1, \ldots, \psi_{N-1}$ the components of ψ, in accordance with the *Chain Rule*, we have

$$\frac{\partial u}{\partial x_i}(x) = \frac{\partial \overline{v}}{\partial t}(\psi(x))\frac{\partial \psi_0}{\partial x_i}(x) + \sum_{k=1}^{N-1} \frac{\partial \overline{v}}{\partial s_k}(\psi(x))\frac{\partial \psi_k}{\partial x_i}(x) \quad \forall x \in D$$

for all $i = 1, \ldots, N$. Thus,

$$\sum_{i=1}^{N} f_i(x, u(x))\frac{\partial u}{\partial x_i}(x) = \frac{\partial \overline{v}}{\partial t}(\psi(x)) \sum_{i=1}^{N} f_i(x, u(x))\frac{\partial \psi_0}{\partial x_i}(x)$$

$$+ \sum_{k=1}^{N-1} \frac{\partial \overline{v}}{\partial s_k}(\psi(x)) \sum_{i=1}^{N} f_i(x, u(x))\frac{\partial \psi_k}{\partial x_i}(x). \quad (9.21)$$

Taking into account that, for every i,

$$f_i(x, u(x)) = \frac{\partial y_i}{\partial t}(\psi(x)) \quad \forall x \in D$$

and the matrices $\overline{y}'(\psi(x))$ and $\psi'(x)$ are the inverse to each other, we obtain

$$\sum_{i=1}^{N} f_i(x, u(x))\frac{\partial \psi_0}{\partial x_i}(x) = \sum_{i=1}^{N} \frac{\partial y_i}{\partial t}(\psi(x))\frac{\partial \psi_0}{\partial x_i}(x) = 1$$

and

$$\sum_{i=1}^{N} f_i(x, u(x))\frac{\partial \psi_k}{\partial x_i}(x) = \sum_{i=1}^{N} \frac{\partial y_i}{\partial t}(\psi(x))\frac{\partial \psi_k}{\partial x_i}(x) = 0$$

for every $x \in D$ and every $k = 1, \ldots, N - 1$.

Finally, using these identities in (9.21), we find that

$$\sum_{i=1}^{N} f_i(x, u(x))\frac{\partial u}{\partial x_i}(x) = \frac{\partial \overline{v}}{\partial t}(\psi(x)) = g(x, u(x)) \quad \forall x \in D.$$

In other words, u is a solution in D to the PDE in (IVP).

Now, let us see that the solution we have constructed is unique.

Let $\hat{D}$ be an open set containing a and let $\hat{u} : \hat{D} \mapsto \mathbb{R}$ be a solution in $\hat{D}$ to (IVP).

We will see that u and $\hat{u}$ coincide in a new, eventually smaller, neighborhood of a.

To this purpose, let $\hat{\Delta}$ be the following set:

$$\hat{\Delta} = \{(t; s_1, \ldots, s_{N-1}) \in \Delta : \overline{y}(t; s_1, \ldots, s_{N-1}) \in \hat{D}\}.$$

Obviously, $\hat{\Delta}$ is an open set in $\mathbb{R}^N$ containing $(0,0)$.

Let us denote by $\hat{O}$ the projection of $\hat{\Delta}$ over $\mathbb{R}^{N-1}$, that is,

$$\hat{O} = \{(s_1, \ldots, s_{N-1}) \in O : \exists t \text{ such that } (t; s_1, \ldots, s_{N-1}) \in \hat{\Delta}\}.$$

Then, by changing $\hat{\Delta}$ by a smaller neighborhood (if this is needed), we can assume that, if $(s_1, \ldots, s_{N-1}) \in \hat{O}$, the set of t such that $(t; s_1, \ldots, s_{N-1}) \in \hat{\Delta}$ is an interval that contains the origin.

Let us consider the functions $\hat{v}$ and $\hat{w}$, respectively, given by

$$\hat{v}(t; s_1, \ldots, s_{N-1}) = \hat{u}(\overline{y}(t; s_1, \ldots, s_{N-1}))$$

and

$$\hat{w}(t; s_1, \ldots, s_{N-1}) = \overline{v}(t; s_1, \ldots, s_{N-1}) - \hat{v}(t; s_1, \ldots, s_{N-1})$$

for all $(t; s_1, \ldots, s_{N-1}) \in \hat{\Delta}$.

Then, for every $(s_1, \ldots, s_{N-1}) \in \hat{O}$, the function $z := \hat{w}(\cdot, s_1, \ldots, s_{N-1})$ solves, in its interval of definition, the Cauchy problem

$$\begin{cases} z' = G(t, z) \\ z(0) = 0, \end{cases} \tag{9.22}$$

where

$$G(t, z) = g(\overline{y}(t; s_1, \ldots, s_{N-1}), \overline{v}(t; s_1, \ldots, s_{N-1}))$$
$$- g(\overline{y}(t; s_1, \ldots, s_{N-1}), \overline{v}(t; s_1, \ldots, s_{N-1}) - z).$$

Indeed, thanks to the definitions of $(\overline{y}, \overline{v})$ and $\hat{w}$ and (again) the Chain Rule,

$$\frac{\partial \hat{w}}{\partial t}(t; s_1, \ldots, s_{N-1}) = g(\overline{y}(t; s_1, \ldots, s_{N-1}), \overline{v}(t; s_1, \ldots, s_{N-1}))$$

$$- \sum_{i=1}^{N} f_i(\overline{y}(t; s_1, \ldots, s_{N-1}), \overline{v}(t; s_1, \ldots, s_{N-1}))$$

$$\times \frac{\partial \hat{u}}{\partial x_i}(\overline{y}(t; s_1, \ldots, s_{N-1}))$$

for all $(t; s_1, \ldots, s_{N-1}) \in \hat{\Delta}$.

Since $\hat{u}$ is by hypothesis a solution in $\hat{D}$ of the considered PDE, the following holds:

$$\frac{\partial \hat{w}}{\partial t}(t; s_1, \ldots, s_{N-1}) = g(\overline{y}(t; s_1, \ldots, s_{N-1}), \overline{v}(t; s_1, \ldots, s_{N-1}))$$

$$- g(\overline{y}(t; s_1, \ldots, s_{N-1}), \hat{v}(t; s_1, \ldots, s_{N-1}))$$

and we see that $z = \hat{w}(\cdot, s_1, \ldots, s_{N-1})$ is a solution to the ODE in (9.22).

On the other hand, since $\hat{u}$ and u satisfy the same initial condition, we also have that

$$\hat{v}(0; s_1, \ldots, s_{N-1}) = u_0(s_1, \ldots, s_{N-1}) = \overline{v}(0; s_1, \ldots, s_{N-1})$$

for all $(s_1, \ldots, s_{N-1}) \in \hat{\mathcal{O}}$. Therefore, the initial condition in (9.22) is also satisfied by z.

Taking into account that G is well defined and continuously differentiable and $G(t, 0) \equiv 0$, we find that $z \equiv 0$.

In other words, $\overline{v} = \hat{v}$ in $\hat{\Delta}$.

This proves that u and $\hat{u}$ coincide in a neighborhood of a and thus ends the proof.

Remark 9.6. The "classical" theory of first-order PDEs contains a lot of results and it is impossible to present here all the fundamental related ideas.

Anyway, let us mention that it makes perfect sense to consider Cauchy problems for nonlinear first-order PDEs. They take the form

$$\begin{cases} F(x, u, Du) = 0 \\ u|_S = u_0, \end{cases} \tag{9.23}$$

where the data S and u_0 are as in (IVP).

In this context, it is possible to prove a result similar to 9.1. The proof uses arguments like those above; see Note 11.9.2.   □

Exercises

E.9.1 Let S be the plane in $\mathbb{R}^3$ whose equation is $x_1 + x_3 = 1$ and consider the Cauchy problem

$$\begin{cases} (x_1 - 2x_3)\partial_1 u - x_3 \partial_3 u = x_2 - u \\ u|_S = 1 - x_1 + x_2 \end{cases}$$

in a neighborhood of $(0, 0, 1)$.

(1) Prove that this problem possesses exactly one local solution.

(2) Find the solution using the method of characteristics.

E.9.2 Same statement for the plane S with equation $x_1 + x_3 = 1$ and the Cauchy problem

$$\begin{cases} x_1 \partial_1 u + \partial_2 u - x_1 \partial_3 u = u - x_1 \\ u|_S = x_1 \end{cases}$$

in a neighborhood of $(0,0,0)$.

E.9.3 Same statement for the surface S with equation $x_1^2 = x_3$ and the Cauchy problem

$$\begin{cases} x_2 \dfrac{\partial u}{\partial x_1} + x_1 \dfrac{\partial u}{\partial x_2} + 2 \dfrac{\partial u}{\partial x_3} = 2(1 + x_1 x_2) \\ u|_S = \dfrac{1}{2}(x_2^2 + 3x_3) \end{cases}$$

in a neighborhood of $(0,0,0)$.

E.9.4 Same statement for the plane S with equation $x_3 = 1$ and the Cauchy problem

$$\begin{cases} u^2 \dfrac{\partial u}{\partial x_1} + x_2 \dfrac{\partial u}{\partial x_2} + x_3 \dfrac{\partial u}{\partial x_3} = u + x_2(1 + x_3) \\ u|_S = 0 \end{cases}$$

in a neighborhood of $(0,0,1)$.

E.9.5 Solve the following Cauchy problems, checking previously that a unique local solution exists:

(1) $x_1^2 \partial_1 u + x_2^2 \partial_2 u = u$, $u|_s = 1$ in a neighborhood of $(0,1)$, where S is the straight line $x_2 = 1$.

(2) $x_1 \partial_1 u + 5x_2 \partial_2 u = 3x_1 + u$, $u|_s = 2$ in a neighborhood of $(1,1)$, where S is the straight line $x_2 = x_1$.

(3) $u \partial_1 u + \partial_2 u = 1$, $u|_S = x_1^2$ in a neighborhood of $(0,0)$, where S is the straight line $x_2 = 0$.

(4) $x_1 \partial_1 u + \partial_2 u - \partial_3 u = u$, $u|_s = 1$ in a neighborhood of $(0,0,1)$, where S is the sphere $x_1^2 + x_2^2 + x_3^2 = 1$.

E.9.6 Same statement for the following Cauchy problems:

(1) $5\partial_1 u + x_1\partial_2 u + \partial_3 u = 0$, $u_{|_S} = 1$ in a neighborhood of $(0,1,1)$, where S is the plane $x_2 - x_1 = 1$.

(2) $(x_1 + x_2^2)\partial_1 u + x_2\partial_2 u + \partial_3 u = u$, $u|_S = x_2^2$ in a neighborhood of $(0,0,1)$, where S is the plane $x_3 = 1$.

(3) $x_1\partial_1 u + x_2\partial_2 u + x_3\partial_3 u = 3u$, $u_{|_S} = 0$ in a neighborhood of $(0,0,1)$, where S is the plane $x_3 = 1$.

(4) $\partial_1 u + (x_1 + x_2)\partial_2 u - \partial_3 u = u^2$, $u|_S = 1$ in a neighborhood of $(0,0,0)$, where S is the plane $x_3 = 0$.

E.9.7 Solve the following Cauchy problems by the method of characteristics in a neighborhood of a point in $\mathbb{R}^2$ that must be determined appropriately:

(1) $\begin{cases} x_2\dfrac{\partial u}{\partial x_1} - x_1\dfrac{\partial u}{\partial x_2} = 0 \\ u(x_1, 0) = f(x_1) \end{cases}$ where the function $f : \mathbb{R} \to \mathbb{R}$ is given.

(2) $\begin{cases} \dfrac{\partial u}{\partial x_1} + \dfrac{\partial u}{\partial x_2} = u^2 \\ u(x_1, 0) = f(x_1) \end{cases}$ where the function $f : \mathbb{R} \to \mathbb{R}$ is given.

(3) $\begin{cases} \dfrac{\partial u}{\partial x_2} = x_1 u\dfrac{\partial u}{\partial x_1} \\ u(x_1, 0) = x_1 \end{cases}$

(4) $\begin{cases} \dfrac{\partial u}{\partial x_1} - \dfrac{\partial u}{\partial x_2} = a(x_1 + x_2) \\ u(0, x_2) = x_2^2 \end{cases}$ where $a \in \mathbb{R}$ is given.

E.9.8 Find a surface in $\mathbb{R}^3$ containing the circumference

$$\begin{cases} x^2 + y^2 = 1 \\ z = 0 \end{cases}$$

that meets orthogonally all spheres centered at the origin.

E.9.9 Find a surface in $\mathbb{R}^3$ containing the circumference

$$\begin{cases} x^2 + y^2 = 1 \\ z = 1 \end{cases}$$

that meets orthogonally all the surfaces of the family $z(x + y) = C(3z + 1)$, with $C \in \mathbb{R}$.

E.9.10 Find a function $z = z(x, y)$ that solves the PDE

$$\left(x\frac{\partial z}{\partial x} + y\frac{\partial z}{\partial y}\right)(x^2 + y^2 - 1) = (x^2 + y^2)z$$

such that the surface it defines contains the curve

$$\begin{cases} x = 2z \\ x^2 + y^2 = 4. \end{cases}$$

E.9.11 We want to find a surface $z = u(x, y)$ described by a function u that solves the PDE

$$(x - y)y^2\frac{\partial u}{\partial x} + (y - x)x^2\frac{\partial u}{\partial y} = (x^2 + y^2)u \qquad (9.24)$$

and contains the curve $S : xz = 1, y = 0$.

(1) Formulate this search as a Cauchy problem for the PDE (9.24) in a neighborhood of a point in S.
(2) Prove that this problem possesses exactly one local solution.
(3) Determine the surface that satisfies the desired properties.

E.9.12 Same statement for the PDE

$$x(u + 2a)\frac{\partial u}{\partial x} + (xu + 2yu + 2ay)\frac{\partial u}{\partial y} = (u + a)u \quad \text{con } a > 0$$

and the curve $S : z = 2\sqrt{ax}, y = 0$.

Chapter 10

Basic Ideas in Control Theory*

This chapter is devoted to providing an introduction to the control of systems governed by ODEs and ODSs.

Roughly speaking, we consider control problems where a differential system appears and we try to choose some data (initial data, parameters, etc.) in such a way that the associated solution satisfies good properties.

For example, if the solution to the system represents the trajectory of a satellite and we want to put it into Earth's orbit at a prescribed time, it will be of major interest to decide the launch speed; if the solution furnishes the total number of individuals of a bacteria population along a time interval and our desire is to reach a prescribed quantity at final time, we will want to know what can be done at each time, etc.

There are (at least) three general strategies that serve to control or govern the behavior of the solution:

- Optimal control formulations. As shown in what follows, this amounts to finding the data and/or parameters that minimize a cost (or maximize a benefit) whose values depend at least in part on the solution(s) to the differential system.

  Related questions, arguments, results, and applications will cover most part of this chapter; see Sections 10.1–10.3.
- Controllability strategies. Here, the aim is to drive the solution to the system from an initial state (the initial data imposed to the solution) to a prescribed final state.

  We will briefly speak of controllability problems in Section 10.4.

- Stabilization techniques. These methods are devoted to getting a desired asymptotic behavior of the solution as the independent variable t grows to infinity.

  An example of the stabilization (or stabilizability) problem is the following: find the values of the parameters that convert an ODS for which the trivial solution is unstable into a new system for which stability properties are ensured. For reasons of space, we will not consider this interesting issue; see Nikitin (1994) and Sontag (1990) for some results.

In what follows, we will give a collection of examples and applications that justify the usefulness and interest of this subject.

In order to formulate and solve rigorously the control problems in this chapter, we need some elementary tools from Lebesgue theory and functional analysis. They are briefly presented in the following section.

10.1   Some Tools and Preliminary Results

It will be convenient to consider some appropriate spaces of (discontinuous) functions on $(0, T)$.

For this purpose, let us start with some definitions and properties.

We will denote by $\mathcal{M}(0, T)$ the set of real-valued Lebesgue-measurable functions defined on the interval $(0, T)$.[1]

This set is a linear space for the usual algebraic rules (the sum of functions and the product of a function by a real number).

Recall that, if $v : (0, T) \mapsto \mathbb{R}$ is measurable and nonnegative, we can always define the *integral* of v with respect to the Lebesgue measure in $(0, T)$. It is either a nonnegative real number or $+\infty$ and will be denoted by

$$\int_0^T v \, dt \quad \text{or simply} \quad \int_0^T v.$$

---

[1] Henri Léon Lebesgue (1875–1941) was a French mathematician. In 1942, he introduced the Lebesgue integrals, a new concept of integrals that generalizes and extends the classical notion and makes it possible to compute "areas" bounded by nonsmooth curves. The basic idea was to identify the sets of points where the function takes prescribed values and then measure the sizes of these sets, instead of identifying the values taken at prescribed sets (the method leading to the Riemann integral). The Lebesgue integral has played a key role in the motivation and growth of functional analysis.

By definition, one has

$$\int_0^T v := \sup\{I(z) : z \in \mathcal{S}(0,T),\ 0 \le z \le v\},$$

where $\mathcal{S}(0,T)$ denotes the set of measurable *step functions*, that is, functions of the form

$$z = \sum_{i=1}^I z_i 1_{A_i}, \quad \text{with } z_i \in \mathbb{R} \text{ and Lebesgue-measurable } A_i \subset (0,T)^2;$$

for any such z, we set $I(z) := \sum_{i=1}^I z_i |A_i|$, where $|A_i|$ is the measure of A_i.

If $v : (0,T) \mapsto \mathbb{R}$ is measurable but not necessarily nonnegative, we can consider the nonnegative measurable functions $v_+ := \max(v,0)$ and $v_- := -\min(v,0)$ and their integrals

$$\int_0^T v_+ \quad \text{and} \quad \int_0^T v_-.$$

If both are finite, it is said that v is *integrable*. In that case, the integral of v is, by definition, the quantity

$$\int_0^T v := \int_0^T v_+ - \int_0^T v_-.$$

The following basic properties are satisfied:

- Let $\mathcal{L}^1(0,T)$ be the set of measurable and integrable functions on $(0,T)$. Then, $\mathcal{L}^1(0,T)$ is a linear subspace of $\mathcal{M}(0,T)$ and $v \mapsto \int_0^T v$ is a linear form on $\mathcal{L}^1(0,T)$.
- If $u, v \in \mathcal{M}(0,T)$, the integral of one of them makes sense and

$$u = v \quad \text{almost everywhere (a.e.) in } (0,T),^3$$

  then the integral of the other one is also meaningful and both integrals coincide.
- If $u \in \mathcal{M}(0,T)$, $u \ge 0$ a.e. in $(0,T)$ and $\int_0^T u = 0$, then $u = 0$ a.e.
- If $u \in \mathcal{L}^1(0,T)$, then

$$\left| \int_0^T u \right| \le \int_0^T |u|. \tag{10.1}$$

---

[2]Recall that, for any set G, 1_G denotes the associated characteristic function.
[3]This means that $u(t) = v(t)$ at every $t \in (0,T)$ except, possibly, at a set $Z \subset (0,T)$ of measure $|Z| = 0$.

All the integrals mentioned in the sequel, unless explicitly stated, are carried out with respect to the Lebesgue measure. Usually, the integration element dt will be omitted.

In a similar way, given a non-empty open set $\Omega \subset \mathbb{R}^N$, we can introduce the spaces $\mathcal{M}(\Omega)$ and $\mathcal{L}^1(\Omega)$ containing the Lebesgue-measurable and the measurable and integrable functions on Ω, respectively.

The previous assertions can be adapted.

The following are fundamental and well-known properties of integrable functions:

- *Lebesgue's Dominated Convergence Theorem*: If $\{v_n\}$ is a sequence in the space $\mathcal{L}^1(0,T)$ that converges pointwise a.e. in $(0,T)$ to v and there exists $w \in \mathcal{L}^1(0,T)$ such that $|v_n(t)| \leq w(t)$ a.e. for all n, then $v \in \mathcal{L}^1(0,T)$ and

$$\lim_{n \to +\infty} \int_0^T |v_n - v| = 0.$$

In particular, we also have

$$\lim_{n \to +\infty} \int_0^T v_n = \int_0^T v.$$

- *Fubini's Theorem*: Assume that $T, S > 0$ and $f : (0,T) \times (0,S) \mapsto \mathbb{R}$ belongs to $\mathcal{L}^1((0,T) \times (0,S))$, that is, f is measurable and integrable with respect to the product Lebesgue measure on $(0,T) \times (0,S)$. Then, the following holds:

  (1) For t a.e. in $(0,T)$, the function $s \mapsto f(t,s)$ is measurable and integrable in $(0,S)$.
  (2) The function $t \mapsto \int_0^S f(t,s)\,ds$, that is well defined for t a.e. in $(0,T)$, is measurable and integrable.
  (3) One has

$$\int_0^T \left(\int_0^S f(t,s)\,ds \right) dt = \int_{(0,T) \times (0,S)} f(t,s)\,d(t,s).$$

  Obviously, a similar result holds exchanging the roles of t and s.

  On the other hand, the assumption $f \in \mathcal{L}^1((0,T) \times (0,S))$ can be replaced by this one: $(t,s) \mapsto f(t,s)$ is measurable and nonnegative. With this last hypothesis, the result is known as *Tonelli's Theorem*.

For the proofs, see for instance Lang (1993).

In this chapter, we will work with classes of measurable functions determined by the following equivalence relation in $\mathcal{M}(0,T)$:

$$u \sim v \Leftrightarrow u(t) = v(t) \text{ a.e. in } (0,T).$$

For brevity, we will refer to them frequently as *functions*.

Obviously, the sum of these functions and the product by a real number are well-defined operations and provide new "functions" of the same kind. Accordingly, we can speak of the linear space $M(0,T)$ of (classes of) measurable functions on $(0,T)$.

In fact, in the sequel, unless otherwise stated, when speaking of a measurable function $v : (0,T) \mapsto \mathbb{R}$, we will be implicitly making reference to the associated class in $M(0,T)$.

Let us now introduce the Lebesgue spaces. They will play a fundamental role in the formulation and solution of the control problems introduced in what follows.

We will begin by introducing the spaces of p-summable functions.

Thus, for any p satisfying $1 \leq p < +\infty$, $L^p(0,T)$ is by definition the set of (classes of) measurable functions $v : (0,T) \mapsto \mathbb{R}$ such that $|v|^p$ is integrable, i.e. satisfy

$$\|v\|_{L^p} := \left(\int_0^T |v|^p \right)^{1/p} < +\infty.$$

It is well known that $L^p(0,T)$ is a linear space for the usual algebraic rules and the real-valued mapping $v \mapsto \|v\|_{L^p}$ is a norm in $L^p(0,T)$.

For the proof of this assumption, the key tool is the so-called *Minkowski inequality*:

$$\left(\int_0^T |u+v|^p \right)^{1/p} \leq \left(\int_0^T |u|^p \right)^{1/p} + \left(\int_0^T |v|^p \right)^{1/p} \quad \forall u,v \in L^p(0,T).$$

$$(10.2)$$

For $p = 1$, this is evident. For $p > 1$, (10.2) is in turn a consequence of the *Hölder inequality*:

If $1 < p < \infty$, p' is the *conjugate* to p, i.e. the unique exponent satisfying

$$1 < p' \leq +\infty, \quad \frac{1}{p} + \frac{1}{p'} = 1$$

and the functions u and v satisfy $u \in L^p(0,T)$ and $v \in L^{p'}(0,T)$, then $uv \in L^1(0,T)$ and

$$\int_0^T |u\,v| \leq \left(\int_0^T |u|^p \right)^{1/p} \left(\int_0^T |v|^{p'} \right)^{1/p'}. \tag{10.3}$$

Obviously, the usual spaces of continuous and continuously differentiable functions in $[0,T]$ are subspaces of $L^p(0,T)$. Moreover, it is not difficult to prove that C^∞ functions are dense in $L^p(0,T)$.

More precisely, the following holds:

For any $v \in L^p(0,T)$ and any $\varepsilon > 0$, there exists $v_\varepsilon \in C^\infty([0,T])$ such that

$$\|v - v_\varepsilon\|_{L^p} \leq \varepsilon.$$

In fact, v_ε can be chosen with compact support in $(0,T)$, that is, satisfying $v_\varepsilon(t) = 0$ for $0 \leq t \leq \delta$ and $T - \delta \leq t \leq T$ for some δ (that depends on ε).

For an elementary proof relying on truncation and regularization by convolution, see for instance Lang (1993).

For all $p \in [1, +\infty)$, endowed with the norm $\|\cdot\|_{L^p}$, $L^p(0,T)$ is a *separable Banach space*.

Recall that this means the following:

- As a metric space, $L^p(0,T)$ is *complete*, i.e. any Cauchy sequence in $L^p(0,T)$ converges in this space.

  This comes from *Lebesgue's Dominated Convergence Theorem;* see for instance Lang (1993) for a proof and more details.
- As a topological space, it is *separable*, i.e. there exists a set $E \subset L^p(0,T)$ that is *dense* and *countable*.

  For example, if $(0,T)$ is bounded, a set E with these properties is the family of polynomial functions on $(0,T)$ with rational coefficients.

In particular, when $p = 2$, $L^2(0,T)$ is a separable Hilbert space, since $\|\cdot\|_{L^2}$ is induced by the scalar product

$$(u,v)_{L^2} := \int_0^T u\,v.$$

Henceforth, it will be understood that the notations $L^p(0,T)$ and in particular $L^2(0,T)$ refer not only to the sets of p-summable and square-summable functions, but also to the associated vector and normed structures.

Let us now define the space of essentially bounded functions.

Thus, if the function $v : (0,T) \mapsto \mathbb{R}$ is given, the *essential supremum* of v is by definition the quantity

$$\text{ess sup}_{(0,T)} v = \inf\{A : v(t) \leq A \text{ a.e. in } (0,T)\}.$$

Note that, if u and v are measurable and $u \sim v$, their essential suprema necessarily coincide. Accordingly, we will denote by $L^\infty(0,T)$ the set of (classes of) measurable functions bounded a.e. in $(0,T)$, that is, satisfying

$$\|v\|_{L^\infty} := \text{ess sup}_{(0,T)} |v| < +\infty.$$

Again, $L^\infty(0,T)$ is a linear space and $v \mapsto \|v\|_{L^\infty}$ is a norm in $L^\infty(0,T)$. Moreover, endowed with the norm $\|\cdot\|_{L^\infty}$, $L^\infty(0,T)$ is a (non-separable) Banach space.

An obvious consequence of the definition of $L^\infty(0,T)$ and Hölder inequalities is that the spaces $L^p(0,T)$ are ordered.

More precisely, if $1 \leq p < q \leq +\infty$, one has $L^q(0,T) \subset L^p(0,T)$ and the identity mapping from $L^q(0,T)$ into $L^p(0,T)$ is continuous, since

$$\|v\|_{L^p} \leq T^a \|v\|_{L^q} \quad \forall v \in L^p(0,T), \text{ with } a = \frac{1}{p} - \frac{1}{q}.$$

Example 10.1. Let us set $u_{a,b}(t) := t^{-a} (\log 2/t)^{-b}$ for $0 < a, b \leq 1$. Then, if $a < 1$, $u_{a,b} \in L^p(0,1)$ for all $p < 1/a$. Contrarily, $u_{1,b} \notin L^1(0,1)$. $\square$

A relevant property concerning convergent sequences in $L^p(0,T)$ is the following:

Let $\{v_n\}$ be a sequence in $L^1(0,T)$ and assume that v_n converges in this space to a function v; then there exists a subsequence of $\{v_n\}$ that converges pointwise a.e. to v.

We will end this section with some important results from functional analysis.

Let H be a Hilbert space with norm $\|\cdot\|_H$ and scalar product $(\cdot,\cdot)_H$.

Recall that the norm induces a distance and, consequently, generates a metric topology in H that permits to speak of open, closed, and compact sets, Cauchy and convergent sequences, etc.

In particular, a sequence $\{v_k\}$ in H converges to $v \in H$ if

$$\lim_{k \to +\infty} \|v_k - v\|_H = 0.$$

For reasons that will become clear very soon, it will be said in that case that v_k *converges strongly to v in H*.

It is possible to define other interesting topologies in H.

Here, we call "interesting" any topology that ensures the continuity of the vector mappings $(v, w) \mapsto v + w$ and $(\lambda, v) \mapsto \lambda v$, respectively, on $H \times H$ and $\mathbb{R} \times H$.

One of these is the so-called *weak topology*.

By definition, a set $G \subset H$ is *open* for the weak topology (or weakly open) if, for any $v_0 \in G$, there exist finitely many $w_i \in H$ with $1 \leq i \leq m$ and $\varepsilon > 0$ such that

$$v \in H, \quad |(w_i, v - v_0)_H| \leq \varepsilon \quad \text{for all} \quad i = 1, \ldots, m \;\Rightarrow\; v \in G.$$

It is easy to check that this is an "interesting" topology and, also, that any weakly open set is open in the usual sense in H.

In H, a sequence $\{v_k\}$ converges to v for the weak topology if and only if the following is satisfied:

$$\lim_{k \to +\infty} (w, v_k - v)_H = 0 \quad \forall w \in H.$$

In that case, it is usual to say that v_k *converges weakly to* v (or $v_k \to v$ weakly in H).

Of course, if $v_k \to v$ in the usual sense, we also have $v_k \to v$ weakly in H.

However, in an infinite dimensional Hilbert space, weak convergence is strictly weaker than the usual convergence.

In other words, there always exist sequences that converge weakly and do not converge for the norm topology (see for instance Exercise E.10.1).

The following fundamental result holds:

Theorem 10.1. *Let H be a Hilbert space and let $\{v_k\}$ be a sequence in H. Assume that $\{v_k\}$ is bounded, that is*

$$\exists C > 0 \quad \text{such that} \quad \|v_k\|_H \leq C \;\; \forall k \geq 1.$$

Then $\{v_k\}$ possesses weakly convergent subsequences.

The proof can be found for instance in Brézis (1983). It relies on the following two ideas:

(1) There exists an orthonormal system $\{e_1, e_2, \ldots\}$ such that a subsequence of $\{v_k\}$ converges weakly to v in H if and only if the corresponding subsequence of $\{(v_k, e_n)_H\}$ converges to $(v, e_n)_H$ for all n.

(2) After a diagonal extraction process, it is possible to find a subsequence of $\{v_k\}$ and a point $v \in H$ with this property.

This result indicates that, in any Hilbert space, bounded sets are sequentially weakly relatively compact.

In lots of problems concerning differential equations, this allows to extract from a sequence of approximations subsequences that converge (at least in the weak sense) and, in the limit, furnish a solution.

Example 10.2. Let us consider the Hilbert space $L^2(0,1)$, the sets

$$A_{i,n} := \{t \in [0,1] : i2^{-n} \leq t < (i+1)2^{-n}\} \quad \text{for } n \geq 1,\ 0 \leq i \leq 2^n - 1$$

and the functions

$$v_n = \sum_{i=0}^{2^n-1} (-1)^i 1_{A_{i,n}} .$$

Then $v_n \to 0$ weakly in $L^2(0,T)$.

However, we cannot have strong convergence to zero, since this would imply pointwise convergence a.e. for a subsequence, which is impossible. $\square$

In any Hilbert space H, the following important property is satisfied:

Let $\{v_n\}$ be a sequence in H and assume that $v_n \to v$ weakly in H and, furthermore, $\|v_n\|_H \to \|v\|_H$. Then $v_n \to v$ strongly.

The proof is very easy.

Indeed, one has

$$\|v_n - v\|_H^2 = \|v_n\|_H^2 - 2(v_n, v)_H + \|v\|^2$$

for all n. But, by assumption, this goes to zero as $n \to +\infty$.

Consequently, the assertion holds.

Let us now recall some properties of convex sets and convex functions:

Proposition 10.1. *Let H be a Hilbert space and let $K \subset H$ be a non-empty convex set. Then K is closed if and only if it is weakly closed. In fact, this happens if and only if it is sequentially weakly closed, that is, if and only if one has the following:*

If $v_k \in K$ for all k and $v_k \to v$ weakly in H, then $v \in K$.

Proposition 10.2. *Let H be a Hilbert space, let $K \subset H$ be a non-empty closed convex set, and let $F : K \mapsto \mathbb{R}$ be a continuous convex function. Then F is sequentially weakly lower semi-continuous (s.w.l.s.c.), that is:*

If $v_k \in K$ for all k and $v_k \to v$ weakly in H, then $\liminf_{k \to +\infty} F(v_k) \geq F(v)$.

The proofs of these results can be found for instance in Ekeland and Témam (1999).

In the control problems considered in the following sections, they will be essential to prove the existence of a solution.

Example 10.3. Any set of the form

$$K := \{v \in L^2(0,T) : v(t) \in I \text{ a.e. in } (0,T)\},$$

where $I \subset \mathbb{R}$ is a closed interval is a closed convex set in $L^2(0,T)$.

This is a consequence of the fact that strong convergence in $L^2(0,T)$ implies a.e. pointwise convergence for a subsequence.

Also, if Y is a Banach space, $M \in \mathcal{L}(L^2(0,T);Y)$, $G \subset Y$ is a non-empty closed convex set and

$$U := \{v \in L^2(0,T) : Mv \in G\},$$

we again have that U is closed and convex.   □

Let H be a Hilbert space and let $F : H \mapsto \mathbb{R}$ be a function. Recall that it is said that F is Gâteaux-differentiable at $u \in H$ in the direction $v \in H$ if the following quantity exists:

$$DF(u;v) := \lim_{s \to 0} \frac{1}{s} \left(F(u+sv) - F(u) \right).$$

If this happens for any $u, v \in H$, it will be said that F is Gâteaux-differentiable.

If, furthermore, $DF(u; \cdot) : H \mapsto \mathbb{R}$ is a linear continuous form for every $u \in H$ and the mapping $u \mapsto DF(u; \cdot)$ is continuous, it will be said that F is continuously differentiable (or of class C^1) on H.[4]

In the sequel, for any C^1 function $F : H \mapsto \mathbb{R}$, $F'(u)$ will stand for the *gradient* of F at u, that is, the unique element in H satisfying

$$(F'(u), v)_H = DF(u;v) \quad \forall v \in H.$$

---

[4]In that case, F is also Fréchet-differentiable at every $u \in H$ and its Fréchet-differential at u is $DF(u; \cdot)$, that is,

$$\lim_{v \to 0} \frac{\|F(u+v) - F(u) - DF(u;v)\|_H}{\|v\|_H} = 0 \quad \forall u \in H.$$

Let us end this section with some properties satisfied by differentiable convex functions:

Proposition 10.3. *Let H be a Hilbert space and let $F : H \mapsto \mathbb{R}$ be a C^1 convex function. Then*

$$F(v) - F(u) \geq (F'(u), v - u)_H \quad \forall u, v \in H.$$

Furthermore, $F' : H \mapsto H$ is a monotone mapping, that is,

$$(F'(v) - F'(u), v - u)_H \geq 0 \quad \forall u, v \in H.$$

The proof is easy; see Exercise E.10.2.

10.2 A Very Simple Optimal Control Problem: A Linear ODE, No Constraints

In the optimal control problems considered in this book, we will dispose of the following elements:

- An initial or boundary-value problem for an ODE or an ODS where at least one of the data (a coefficient, a right-hand side variable, etc.) must be chosen.
- A family $\mathcal{U}_{\mathrm{ad}}$ of *admissible* data.
- A function $J : \mathcal{U}_{\mathrm{ad}} \mapsto \mathbb{R}$ whose values depend on the chosen data and the associated solution(s) to the ODE or ODS.

The initial (or boundary)-value problem is usually called the *state equation,* the datum we must choose is called the *control* and the function J is the *cost* or *objective* functional.

In all cases, the problem we want to solve reduces to minimizing the cost J in the set $\mathcal{U}_{\mathrm{ad}}$ of admissible controls.

In this section, it will be assumed that the state equation is

$$\begin{cases} y' = ay + v, \quad t \in (0, T) \\ y(0) = y_0 \end{cases} \tag{10.4}$$

and we will consider the following basic control problem:

$$\begin{cases} \text{Minimize } J(v) := \dfrac{\alpha}{2} \int_0^T |y - y_d|^2 \, \mathrm{d}t + \dfrac{\mu}{2} \int_0^T |v|^2 \, \mathrm{d}t \\ \text{Subject to } v \in L^2(0, T), \ (y, v) \text{ satisfies (10.4).} \end{cases} \tag{10.5}$$

The data are the following:

$\alpha, \mu \in \mathbb{R}_+$ with $\alpha + \mu = 2$, $T > 0$, $a \in \mathbb{R}$, $y_0 \in \mathbb{R}$, and $y_d \in L^2(0, T)$.

Let us give an interpretation.

For example, $y : [0, T] \mapsto \mathbb{R}$ can be the population density of an isolated species that is nourished by an external agent at a rate indicated by the function $v = v(t)$.

Of course, y is the state, v is the control, the whole space $L^2(0, T)$ is the set of admissible controls, and J is the cost functional.

Our interest is to find a control v that makes $J(v)$ as small as possible. In view of the terms that appear in $J(v)$, this can be interpreted as the search of a reasonably cheap feeding strategy that leads to a state close to the desired population y_d along $(0, T)$.

The relevance that we give to these objectives (not too expensive control and proximity of the state to y_d) is determined by the choice of α and μ. For instance, small μ means that we do not mind applying expensive strategies too much.

We will have to work with states associated to controls in the Hilbert space $L^2(0, T)$.

Consequently, we will have to extend the concept of the solution to (10.4).

Definition 10.1. Let $v \in L^2(0, T)$ be given. It will be said that y is a generalized solution to (10.4) in $(0, T)$ if $y \in C^0([0, T])$, y is differentiable for almost all $t \in [0, T]$,

$$y'(t) = ay(t) + v(t) \quad \text{a.e. in} \quad (0, T)$$

and $y(0) = y_0$.

For every $v \in L^2(0, T)$ and every $y_0 \in \mathbb{R}$, there exists exactly one generalized solution y to (10.4) in $(0, T)$, given by

$$y(t) = e^{at} y_0 + \int_0^t e^{a(t-s)} v(s) \, ds \quad \forall t \in [0, T]. \tag{10.6}$$

Indeed, it is easy to check that the function furnished by (10.6) satisfies all the required properties.

On the other hand, if y and z are maximal solutions to (10.4), then $y - z$ is a "classical" solution to a homogeneous linear ODE of the first order and satisfies $(y - z)(0) = 0$, whence $y(t) = z(t)$ for all t.

For an optimal control problem like (10.5), the main related questions are the following:

(1) The existence and uniqueness of a solution.

If it exists, it is usually called an *optimal control* and the associated state is called an *optimal state*.
(2) The optimality characterization.

Let us accept the existence of one or several optimal controls.

It is then interesting to deduce properties that permit to distinguish them from non-optimal controls. They will be called *optimality conditions*.

In practice, the role played by these conditions is similar to the role played by the necessary and/or sufficient conditions of minima for standard extremal problems.
(3) The computation of the optimal control(s).

We will see that there exist (at least) two essentially different ways to perform this task: either using the optimality conditions (the so-called indirect method) or directly applying descent algorithms (the direct method).

In both cases, the objective is to generate a sequence of admissible controls that converge (in an appropriate sense) to an optimal control.
(4) The sensibility or dependence of the optimal control(s) with respect to parameters.

It is also relevant to know whether (for instance) the optimal control depends continuously on the initial data, the coefficient a, etc.

In the particular case of (10.5), since the state equation and the cost functional are very simple, the answers to these questions are not too involved.

The first result is the following:

Theorem 10.2. *The optimal control problem (10.5) possesses exactly one solution $\hat{v}$.*

Proof. First, let us see that there exists at least one optimal control.

Let us set $m := \inf_{v \in L^2(0,T)} J(v)$ (m must be a nonnegative real number) and let $\{v_n\}$ be a minimizing sequence, that is, such that

$$J(v_n) \to m \quad \text{as} \quad n \to +\infty.$$

By definition of infimum, there exist sequences of this kind.

The sequence $\{v_n\}$ is bounded; indeed, if this were not the case, in view of the definition of J, we would have $\limsup_{n\to+\infty} J(v_n) \to +\infty$, which is obviously an absurdity.

Since $L^2(0,T)$ is a Hilbert space, there exists a subsequence (again denoted $\{v_n\}$) that converges weakly in $L^2(0,T)$ toward a control $\hat{v}$, i.e. such that

$$(v_n, v)_{L^2} \to (\hat{v}, v)_{L^2} \text{ as } n \to +\infty \ \forall v \in L^2(0,T).$$

Note that, since J is convex and continuous, in view of Proposition 10.1, one has

$$m = \lim_{n\to+\infty} J(v_n) = \liminf_{n\to+\infty} J(v_n) \geq J(\hat{v}).$$

Consequently, $J(\hat{v}) = m$ and the existence of an optimal control is established.

However, for completeness, we will give now a different (direct) proof of this fact.

Let us denote by y_n (resp. $\hat{y}$) the solution to (10.4) with v replaced by v_n (resp. $\hat{v}$).

Thanks to (10.6), we have $y_n \to \hat{y}$ strongly in $L^2(0,T)$, that is,

$$\|y_n - \hat{y}\|_{L^2} \to 0 \text{ as } n \to +\infty. \tag{10.7}$$

Consequently,

$$\int_0^T |y_n - y_d|^2 \, dt \to \int_0^T |\hat{y} - y_d|^2 \, dt \text{ as } n \to +\infty. \tag{10.8}$$

On the other hand, the sequence $\{v_n\}$ converges strongly in $L^2(0,T)$. Indeed, for all $n \geq 1$, one has

$$\frac{\mu}{2}\|v_n - \hat{v}\|_{L^2}^2 = \frac{\mu}{2}\|v_n\|_{L^2}^2 - \mu(v_n,\hat{v})_{L^2}^2 + \frac{\mu}{2}\|\hat{v}\|_{L^2}^2$$

$$= J(v_n) - J(\hat{v}) + \frac{\alpha}{2}\left(\int_0^T |\hat{y} - y_d|^2 \, dt - \int_0^T |y_n - y_d|^2 \, dt\right)$$

$$+ \mu\left(\|\hat{v}\|_{L^2}^2 - (v_n,\hat{v})_{L^2}^2\right).$$

Taking limits in the right-hand side as $n \to +\infty$, we get $m - J(\hat{v})$, that must be ≤ 0. Therefore,

$$\|v_n - \hat{v}\|_{L^2} \to 0 \text{ as } n \to +\infty.$$

In particular,

$$\int_0^T |v_n|^2 \, dt \to \int_0^T |\hat{v}|^2 \, dt \quad \text{as} \quad n \to +\infty \tag{10.9}$$

and, from (10.8) and (10.9), we see again that $J(v_n) \to J(\hat{v})$, which proves that $\hat{v}$ is an optimal control.

In order to prove uniqueness, let us assume that there exist two optimal controls $\hat{v}$ and $\hat{w}$ with $\hat{v} \neq \hat{w}$.

This means that $\hat{v}, \hat{w} \in L^2(0,T)$, and $J(\hat{v}) = J(\hat{w}) = m :=$ $\inf_{v \in L^2(0,T)} J(v)$.

Let us introduce $v := \frac{1}{2}(\hat{v} + \hat{w})$. Since J is the sum of a convex and a strictly convex function (the square of the norm in $L^2(0,T)$), we have that J is also strictly convex, whence $J(v) < m$, which is an absurdity.

This ends the proof. $\qquad\square$

The characterization of the (unique) optimal control is furnished by the following result:

Theorem 10.3. *Let $\hat{v}$ be the unique solution to* (10.5) *(the unique optimal control) and let $\hat{y}$ be the associated solution to* (10.4) *(the associated optimal state). Then, there exists $\hat{p}$ such that the triplet $(\hat{y}, \hat{p}, \hat{v})$ satisfies the system*

$$\begin{cases} \hat{y}' = a\hat{y} + \hat{v}, \quad t \in (0,T) \\ \hat{y}(0) = y_0 \\ -\hat{p}' = a\hat{p} + \hat{y} - y_d, \quad t \in (0,T) \\ \hat{p}(T) = 0 \\ \hat{v} = -\dfrac{\alpha}{\mu}\hat{p}, \end{cases} \tag{10.10}$$

which is called the optimality system for (10.5). *The function $\hat{p}$ is called the adjoint state corresponding to $\hat{v}$.*

Proof. Let w be an arbitrary function in $L^2(0,T)$. First, we observe that

$$\exists \lim_{s\to 0} \frac{J(\hat{v}+sw) - J(\hat{v})}{s} = \alpha \int_0^T (\hat{y} - y_d) \cdot z \, dt + \mu \int_0^T \hat{v} \cdot w \, dt,$$

where z solves

$$\begin{cases} z' = az + w, \quad t \in (0,T) \\ z(0) = 0 \end{cases} \tag{10.11}$$

in the sense of Definition 10.1.

This means that J is differentiable (in the Gâteaux sense) at $\hat{v}$ and, also, that its Gâteaux-derivative in the direction w is given by

$$DJ(\hat{v}; w) = \alpha \int_0^T (\hat{y} - y_d) \cdot z \, dt + \mu \int_0^T \hat{v} \cdot w \, dt.$$

Consequently, since $\hat{v}$ is the optimal control,

$$\alpha \int_0^T (\hat{y} - y_d) \cdot z \, dt + \mu \int_0^T \hat{v} \cdot w \, dt = 0.$$

Let us introduce the function $\hat{p}$, with

$$\begin{cases} -\hat{p}' = a\hat{p} + \hat{y} - y_d, & t \in (0, T) \\ \hat{p}(T) = 0. \end{cases} \tag{10.12}$$

Then

$$\int_0^T (\hat{y} - y_d) \cdot z \, dt = \int_0^T (-\hat{p}' - a\hat{p}) \cdot w \, dt = \int_0^T \hat{p} \cdot w \, dt$$

and, consequently,

$$\int_0^T (\alpha \hat{p} + \mu \hat{v}) \cdot w \, dt = 0 \quad \forall w \in L^2(0, T),$$

that is, the last equality in (10.10) is satisfied.

Hence, the optimal control $\hat{v}$ is, together with the optimal state $\hat{y}$ and the corresponding solution to the adjoint state equation (the optimal adjoint state) a solution to (10.10).   □

With the help of a function called the *Hamiltonian*, (10.10) can be rewritten in a simple way, that is easy to remember.

Thus, let us introduce H, with

$$H(y, p, v) := \frac{1}{2}|y - y_d|^2 + \frac{\mu}{2\alpha}|v|^2 + p \cdot (ay + v). \tag{10.13}$$

Then, the optimality system is the following:

$$\begin{cases} \hat{y}' = \dfrac{\partial H}{\partial p}(\hat{y}, \hat{p}, \hat{v}), & t \in (0, T) \\[2mm] \hat{y}(0) = y_0 \\[2mm] -\hat{p}' = \dfrac{\partial H}{\partial y}(\hat{y}, \hat{p}, \hat{v}), & t \in (0, T) \\[2mm] \hat{p}(T) = 0 \\[2mm] H(\hat{y}(t), \hat{p}(t), \hat{v}(t)) \leq H(\hat{y}(t), \hat{p}(t), v) \quad \forall v \in \mathbb{R}, \ t \in (0, T). \end{cases} \tag{10.14}$$

The result that affirms that the optimal control $\hat{v}$, together with $\hat{y}$ and $\hat{p}$, satisfies (10.14) is frequently known as *Pontryaguin's Maximal Principle*.

In the case of the optimal control problem (10.5), the existence of a function $\hat{p}$ such that $(\hat{y}, \hat{p}, \hat{v})$ satisfies (10.10) is not only a necessary condition of optimality; it is also sufficient.

This is because the ODE is linear and the cost functional is quadratic. In this respect, see Exercise E.10.19.

Indeed, from the argument used in the proof of Theorem 10.3, one has

$$DJ(\hat{v}; w) = \int_0^T (\alpha \hat{p} + \mu \hat{v}) \cdot w \, dt \quad \forall w \in L^2(0,T)$$

and, consequently, $\hat{v}$ is the optimal control if and only if

$$\int_0^T (\alpha \hat{p} + \mu \hat{v}) \cdot w \, dt = 0 \quad \forall w \in L^2(0,T).$$

Hence, $\hat{v}$ is the optimal control if and only if (10.10) holds.

The system (10.10) furnishes in a very natural way a first iterative method for the computation of the optimal control.

It is the following:

ALG 1:

(a) Choose $v^0 \in L^2(0,T)$.
(b) Then, for given $\ell \geq 0$ and $v^\ell \in L^2(0,T)$, compute y^ℓ, p^ℓ, and $v^{\ell+1}$ as follows:

$$\begin{cases} (y^\ell)' = a y^\ell + v^\ell, & t \in (0,T) \\ y^\ell(0) = y_0, \end{cases}$$

$$\begin{cases} -(p^\ell)' = a p^\ell + y^\ell - y_d, & t \in (0,T) \\ p^\ell(T) = 0, \end{cases}$$

and

$$v^{\ell+1} = -\frac{\alpha}{\mu} p^\ell.$$

These iterates can be viewed as the result of applying a *method of successive approximations*.

More precisely, let us consider the mapping $\Lambda : L^2(0,T) \mapsto L^2(0,T)$ that associates to each $v \in L^2(0,T)$, first, the corresponding solution to (10.4); then, the solution to

$$\begin{cases} -p' = ap + y - y_d, & t \in (0,T) \\ p(T) = 0 \end{cases}$$

and, finally, the control

$$\Lambda(v) := -\frac{\alpha}{\mu}p. \qquad (10.15)$$

Then, **ALG 1** amounts to first choosing $v^0 \in L^2(0,T)$ and then, for every $\ell \geq 0$, computing

$$v^{\ell+1} = \Lambda(v^\ell).$$

The following convergence result holds:

Theorem 10.4. *Let $\hat{v}$ be the unique solution to (10.5). There exists $\epsilon_0 > 0$ such that, if $0 < \alpha/\mu \leq \epsilon_0$, the sequence $\{v^\ell\}$ produced by* **ALG 1** *starting from an arbitrarily chosen $v^0 \in L^2(0,T)$ converges strongly in $L^2(0,T)$ to $\hat{v}$ as $\ell \to +\infty$.*

For the proof, see Exercise E.10.11.

Other iterative methods can be deduced.

In particular, since the derivative of J at any v is easy to compute, we can consider gradient descent methods.

Thus, for any $v \in L^2(0,T)$, let us denote by $J'(v)$ the gradient of J at v, i.e. the unique function in $L^2(0,T)$ satisfying

$$(J'(v),w)_{L^2} = DJ(v;w) \quad \forall w \in L^2(0,T).$$

One has $J'(v) = \alpha p + \mu v$, for all $v \in L^2(0,T)$, where p is the adjoint state associated to v.

In a gradient method, the strategy relies on the construction of controls $v^1, v^2, \ldots$ such that, for every $\ell \geq 0$, $v^{\ell+1}$ is of the form $v^\ell - \rho^\ell d^\ell$, where d^ℓ is either $J'(v^\ell)$ or something not too different from $J'(v^\ell)$ and ρ^ℓ is an adequate positive number.

The use of $J'(v^\ell)$ (or something close to $J'(v^\ell)$) as descent direction is explained by the fact that, if $J'(v^\ell) \neq 0$ and $M > 0$ is fixed, the problem

$$\begin{cases} \text{Minimize} \quad J(v^\ell - w) \\ \text{Subject to: } w \in L^2(0,T), \quad \|w\|_{L^2} \leq M \end{cases}$$

is solved for

$$w = \frac{M}{\|J'(v^\ell)\|_{L^2}} J'(v^\ell).$$

In other words, $J'(v^\ell)$ is the steepest descent direction at v^ℓ.

Among all gradient descent methods, the best known and most used is the *optimal step gradient method*. In the context of (10.5), it is the following:

ALG 2:

(a) Choose $v^0 \in L^2(0,T)$.
(b) Then, for any given $\ell \geq 0$ and $v^\ell \in L^2(0,T)$, compute y^ℓ, p^ℓ, d^ℓ (the gradient of J at v^ℓ), ρ^ℓ (the optimal step), and $v^{\ell+1}$ as follows:

$$\begin{cases} (y^\ell)' = ay^\ell + v^\ell, & t \in (0,T) \\ y^\ell(0) = y_0, \end{cases}$$

$$\begin{cases} -(p^\ell)' = ap^\ell + y^\ell - y_d, & t \in (0,T) \\ p^\ell(T) = 0, \end{cases}$$

$$d^\ell = \alpha p^\ell + \mu v^\ell,$$

$$\rho^\ell \text{ solution to} \quad \begin{cases} \text{Minimize} \quad J(v^\ell - \rho d^\ell) \\ \text{Subject to: } \rho \geq 0, \end{cases}$$

$$v^{\ell+1} = v^\ell - \rho^\ell d^\ell.$$

We see that, at each step, the task is reduced to solving two Cauchy problems and a one-dimensional extremal problem.

In this case, this is very easy, because we can find the solutions to these Cauchy problems explicitly and, furthermore, the function to minimize is polynomial of degree two in ρ.

For other more complicated problems, what we have to do at each step can be more complex but, as shown in what follows, corresponds to a similar scheme.

For **ALG 2**, a convergence result can also be established. More precisely, one has:

Theorem 10.5. *Let $\hat{v}$ be the unique solution to (10.5) and let $v^0 \in L^2(0,T)$ be given. Then, the sequence $\{v^\ell\}$ furnished by **ALG 2** starting from v^0 converges strongly in $L^2(0,T)$ to $\hat{v}$ as $\ell \to +\infty$.*

Proof. The proof is as follows. First, we observe that any couple of consecutive gradients $J'(v^\ell)$ and $J'(v^{\ell+1})$ are orthogonal as follows:

$$(J'(v^{\ell+1}), J'(v^\ell))_{L^2} = 0 \quad \forall \ell \geq 0.$$

This is implied by the identities

$$\frac{\mathrm{d}}{\mathrm{d}t} J(v^\ell - \rho d^\ell)\bigg|_{\rho=\rho^\ell} = 0,$$

that hold for all $\ell \geq 0$.

As a consequence, for any $k \geq 0$ one has

$$J(v^k) - J(v^{k+1}) \geq \frac{\mu}{2}\|v^k - v^{k+1}\|_{L^2}^2$$

and, after addition in k,

$$J(v^0) - J(v^{\ell+1}) \geq \frac{\mu}{2}\sum_{k=0}^{\ell}\|v^k - v^{k+1}\|_{L^2}^2.$$

This shows that the sequence $\{J(v^\ell)\}$ is decreasing and $v^\ell - v^{\ell+1}$ goes to zero in $L^2(0,T)$ as $\ell \to +\infty$.

Using the expression of $J'(v)$, we easily deduce that $J'(v^\ell) - J'(v^{\ell+1})$ also goes to zero. Moreover, since

$$\|J'(v^\ell)\|_{L^2}^2 = (J'(v^\ell), J'(v^\ell) - J'(v^{\ell+1}))_{L^2}$$
$$\leq \|J'(v^\ell)\|_{L^2}\|J'(v^\ell) - J'(v^{\ell+1})\|_{L^2},$$

we deduce the same for $J'(v^\ell)$.

The controls v^ℓ are uniformly bounded in $L^2(0,T)$. Consequently, we can extract a sequence $\{v^{\ell'}\}$ that converges weakly in $L^2(0,T)$ to some $\hat{v}$.

Using again the expression of $J'(v)$, we see that $J'(v^{\ell'}) \to J'(\hat{v})$ weakly in $L^2(0,T)$.

Hence, $J'(\hat{v}) = 0$. But this proves that $\hat{v}$ is the optimal control.

We have just seen that there exists a subsequence weakly converging to the optimal control.

Since the optimal control is unique and the previous argument can be applied to any subsequence of $\{v^\ell\}$, we find that, in fact, the whole sequence converges.

Finally, we observe that the convergence is strong. Indeed, one has

$$v^\ell = \frac{1}{\mu}\left(J'(v^\ell) - \alpha p^\ell\right).$$

Both terms on the right-hand side converge strongly. Consequently, the same can be said for v^ℓ.

This ends the proof. $\qquad\qquad\qquad\qquad\qquad\qquad\qquad\qquad\qquad\qquad\square$

A simplified version of **ALG 2** is the following algorithm, usually called the *fixed step gradient method*:

ALG 2':

(a) Fix $\bar p > 0$ and choose $v^0 \in L^2(0,T)$.
(b) Then, for any given $\ell \geq 0$ and $v^\ell \in L^2(0,T)$, compute y^ℓ, p^ℓ, d^ℓ (the gradient of J at v^ℓ) and $v^{\ell+1}$ as follows:

$$\begin{cases} (y^\ell)' = ay^\ell + v^\ell, \quad t \in (0,T) \\ y^\ell(0) = y_0, \end{cases}$$

$$\begin{cases} -(p^\ell)' = ap^\ell + y^\ell - y_d, \quad t \in (0,T) \\ p^\ell(T) = 0, \end{cases}$$

$$d^\ell = \alpha p^\ell + \mu v^\ell,$$

$$v^{\ell+1} = v^\ell - \bar p d^\ell.$$

It can be proved that, if $\bar p$ is sufficiently small and $v^0 \in L^2(0,T)$ is given, the sequence furnished by **ALG 2'** starting from v^0 converges strongly in $L^2(0,T)$ to the optimal control; see for instance Ciarlet (1989) for more details.

For completeness, we will present now another method, whose behavior and performance are considerably better. This is the so-called *optimal step conjugate gradient method,* introduced by Hestenes and Stiefel in Hestenes and Stiefel (1952–1953).

The motivation is as follows. At the ℓth step, for given $v^0, v^1, \ldots, v^\ell$, we look for a new control $v^{\ell+1}$ of the form

$$v^{\ell+1} = v^\ell + \sum_{k=0}^\ell \alpha_{k,\ell} J'(v^k),$$

with the $\alpha_{k,\ell} \in \mathbb{R}$ such that

$$J\left(v^\ell + \sum_{k=0}^\ell \alpha_{k,\ell} J'(v^k)\right) \leq J\left(v^\ell + \sum_{k=0}^\ell \alpha_k J'(v^k)\right) \quad \forall \alpha_0, \ldots, \alpha_k \in \mathbb{R}.$$

This would lead to gradients that are orthogonal to all previous gradients, that is, such that

$$(J'(v^{\ell+1}), J'(v^k))_{L^2} = 0 \quad \forall \ell \geq 0, \quad \forall k = 0, 1, \dots, \ell \qquad (10.16)$$

(this is why one usually speaks of "conjugate" gradients).

In fact, the computation of each $v^{\ell+1}$ can be performed by minimizing the function $\rho \mapsto J(v^\ell - \rho d^\ell)$ for some appropriate d^ℓ. More precisely, what we have to do is the following:

ALG 3:

(a) Choose $v^0 \in L^2(0, T)$.

(b) Compute y^0, p^0, g^0 (the gradient of J at v^0), ρ^0 (the optimal step), and v^1 as in **ALG 2** and take $d^0 = g^0$, that is:

$$\begin{cases} (y^0)' = ay^0 + v^0, & t \in (0, T) \\ y^0(0) = y_0, \end{cases}$$

$$\begin{cases} -(\varphi^0)' = a\varphi^0 + y^0 - y_d, & t \in (0, T) \\ \varphi^0(T) = 0, \end{cases}$$

$$d^0 = g^0 = \alpha\varphi^0 + \mu v^0,$$

$$\rho^0 \text{ solution to } \quad \begin{cases} \text{Minimize } J(v^0 - \rho d^0) \\ \text{Subject to: } \rho \geq 0, \end{cases}$$

$$v^1 = v^0 - \rho^0 d^0.$$

(c) Then, for given $\ell \geq 1$, $v^\ell \in L^2(0, T)$, and $g^{\ell-1}, d^{\ell-1} \in L^2(0, T)$, compute y^ℓ, p^ℓ, g^ℓ (the gradient of J at v^ℓ), d^ℓ (the new descent direction), ρ^ℓ (the optimal step), and $v^{\ell+1}$ as follows:

$$\begin{cases} (y^\ell)' = ay^\ell + v^\ell, & t \in (0, T) \\ y^\ell(0) = y_0, \end{cases}$$

$$\begin{cases} -(p^\ell)' = ap^\ell + y^\ell - y_d, & t \in (0, T) \\ p^\ell(T) = 0, \end{cases}$$

$$g^\ell = \alpha p^\ell + \mu v^\ell,$$

$$\gamma^\ell = \frac{\|g^\ell\|_{L^2}^2}{\|g^{\ell-1}\|_{L^2}^2}, \quad d^\ell = g^\ell + \gamma^\ell d^{\ell-1},$$

$$\rho^\ell \text{ solution to } \begin{cases} \text{Minimize} \quad J(v^\ell - \rho d^\ell) \\ \text{Subject to: } \rho \geq 0, \end{cases}$$

$$v^{\ell+1} = v^\ell - \rho^\ell d^\ell.$$

The advantage of **ALG 3** with respect to **ALG 2** is that it involves essentially the same amount of computational work and is however considerably faster.

In fact, if it were applied to a similar problem in finite dimension equal to q and no rounding errors were produced in the computations, in view of (10.16), it would converge after q iterates.

Let us state a convergence result for **ALG 3**:

Theorem 10.6. *Let $\hat{v}$ be the unique solution to (10.5) and let $v^0 \in L^2(0,T)$ be given. Assume that there exists $\beta > 0$ such that*

$$(J'(v^\ell), d^\ell)_{L^2} \geq \beta \|J'(v^\ell)\|_{L^2} \|d^\ell\|_{L^2} \quad \forall \ell \geq 0. \tag{10.17}$$

*Then the sequence $\{v^\ell\}$ furnished by **ALG 3** starting from v^0 converges strongly in $L^2(0,T)$ to $\hat{v}$ as $\ell \to +\infty$.*

The proof can be found in Polak (1997).

In fact, the assumption (10.17) can be weakened: it is sufficient to have

$$(J'(u^n), d^n)_{L^2(\omega)} \geq \beta_n \|J'(u^n)\|_{L^2(\omega)} \|d^n\|_{L^2(\omega)}$$

for some $\beta_n > 0$ satisfying $\sum_{n \geq 0} \beta_n^2 = +\infty$.

The conjugate gradient algorithm has been very popular.

A lot of versions and variants have been and are being used for solving many extremal problems; see more details for instance in Andrei (2020).

The optimality system (10.10) is also useful to analyze the way the optimal control depends on data and parameters.

Thus, we can put

$$\hat{v} = h(t; \alpha, \mu, T, a, y_0, y_d)$$

for some h that is, among other things, of class C^1.

This is a direct consequence of the continuity and differentiability of the maximal solution of a Cauchy problem with respect to initial data and parameters, recall the results in Chapter 7.

Furthermore, we can easily compute the partial derivatives of $\hat{v}$ with respect to any of the arguments of h.

For example, let us denote, respectively, by z, ψ, and ω the partial derivatives of $\hat{y}$, $\hat{p}$, and $\hat{v}$ with respect to y_0. Then, a simple calculus shows that (z, ψ, ω) is the unique solution to

$$
\begin{cases}
z' = az + k, \quad t \in (0, T) \\
z(0) = 1 \\
-\psi' = a\psi + z, \quad t \in (0, T) \\
\psi(T) = 0 \\
\omega = -\dfrac{\alpha}{\mu}\psi.
\end{cases}
$$

Many other similar assertions can be deduced.

Thus, for instance, we can analyze the behavior of $\hat{y}$ and $\hat{v}$ with respect to α or μ, etc.; see Exercises E.10.5 and E.10.6.

10.3 Other Optimal Control Problems for Equations and Systems

We will present in the following paragraphs several optimal control problems corresponding to different ODEs and ODSs, cost functionals, and admissible control sets. Some other more complex problems can be found in Note 11.10.

It will be seen that results similar to Theorems 10.2 and 10.3, iterative algorithms similar to **ALG 1** to **ALG 3**, and convergence properties like those above can be established.

10.3.1 *A linear ODS with no constraint*

Let us consider now the non-scalar Cauchy problem

$$
\begin{cases}
y' = Ay + Bv, \quad t \in (0, T) \\
y(0) = y_0,
\end{cases}
\tag{10.18}
$$

where $A \in \mathcal{L}(\mathbb{R}^N; \mathbb{R}^N)$, $B \in \mathcal{L}(\mathbb{R}^m; \mathbb{R}^N)$, and $y_0 \in \mathbb{R}^N$.

In this case, the control $v = (v_1, \dots, v_m)^t$ belongs to $L^2(0, T)^m$ and $y = (y_1, \dots, y_N)^t$, with N and m being two integers ≥ 1.

Of course, the most interesting situation corresponds to a large N and a small m since, in that case, we are trying to control "many components" with "few scalar controls".

Definition 10.2. Let $v \in L^2(0,T)^m$ be given. It will be said that y is a generalized solution to (10.4) if $y \in C^0([0,T]; \mathbb{R}^N)$, y is differentiable at almost every $t \in [0,T]$,

$$y'(t) = Ay(t) + Bv(t) \quad \text{a.e. in } (0,T)$$

and $y(0) = y_0$.

For every $v \in L^2(0,T)^m$, there exists a unique generalized solution y to (10.4), given by

$$y(t) = e^{At}y_0 + \int_0^t e^{A(t-s)}Bv(s)\,ds \quad \forall t \in [0,T]. \tag{10.19}$$

The argument that leads to (10.19) is similar to the argument used in Section 10.2: first, the function given by (10.6) satisfies the appropriate properties; second, if y and z are two maximal solutions to (10.18), then $y - z$ is a "classical" solution to a homogeneous linear ODS with zero initial conditions, whence $y = z$.

The optimal control problem is the following:

$$\begin{cases} \text{Minimize } J(v) := \dfrac{\alpha}{2} \displaystyle\int_0^T |y - y_d|^2\,dt + \dfrac{\mu}{2} \displaystyle\int_0^T |v|^2\,dt \\ \text{Subject to } v \in L^2(0,T)^m, \quad (y,v) \text{ satisfies (10.18)}, \end{cases} \tag{10.20}$$

where α, μ, T are as before $y_0 \in R^N$ and $y_d \in L^2(0,T)^N$.

We can prove several results similar to those in Section 10.2:

• Again, we get an existence and uniqueness result:

Theorem 10.7. *The optimal control problem* (10.20) *possesses exactly one solution* $\hat{v}$.

The proof is identical to the proof of Theorem 10.2.

• The optimal control is characterized by an optimality system. More precisely, the following holds:

Theorem 10.8. *Let* $\hat{v}$ *be the unique solution to* (10.20) (*the unique optimal control*) *and let* $\hat{y}$ *be the corresponding solution to* (10.18) (*the associated state*). *Then, there exists* $\hat{p}$ *such that the triplet* $(\hat{y}, \hat{p}, \hat{v})$ *satisfies* *the optimality system*

$$\begin{cases} \hat{y}' = A\hat{y} + B\hat{v}, \quad t \in (0,T) \\ \hat{y}(0) = y_0 \\ -\hat{p}' = A^t\hat{p} + \hat{y} - y_d, \quad t \in (0,T) \\ \hat{p}(T) = 0 \\ \hat{v} = -\dfrac{\alpha}{\mu}B^t\hat{p}. \end{cases} \tag{10.21}$$

Again, the proof is essentially the same as the proof of Theorem 10.3. With the help of the Hamiltonian

$$H(y,p,v) := \frac{1}{2}|y - y_d|^2 + \frac{\mu}{2\alpha}|v|^2 + p \cdot (Ay + Bv),$$

the system (10.21) can be written in the following form:

$$\begin{cases} \hat{y}'_i = \dfrac{\partial H}{\partial p_i}(\hat{y},\hat{p},\hat{v}), \quad 1 \le i \le N, \quad t \in (0,T) \\ \hat{y}(0) = y_0 \\ -\hat{p}'_i = \dfrac{\partial H}{\partial y_i}(\hat{y},\hat{p},\hat{v}), \quad 1 \le i \le N, \quad t \in (0,T) \\ \hat{p}(T) = 0 \\ H(\hat{y}(t),\hat{p}(t),\hat{v}(t)) \le H(\hat{y}(t),\hat{p}(t),v) \quad \forall v \in \mathbb{R}^m, \quad t \in (0,T). \end{cases} \tag{10.22}$$

For this control problem, we again have that the existence of $\hat{p}$ such that $(\hat{y},\hat{p},\hat{v})$ satisfies (10.10) is a necessary and sufficient condition for optimality.

- Once more, with the help of (10.21), we have a first iterative algorithm for the computation of the optimal control $\hat{v}$ as follows:

$$v^\ell \mapsto y^\ell \mapsto p^\ell \mapsto v^{\ell+1},$$

with

$$\begin{cases} (y^\ell)' = Ay^\ell + Bv^\ell, \quad t \in (0,T) \\ y^\ell(0) = y_0, \end{cases}$$

$$\begin{cases} -(p^\ell)' = A^t p^\ell + y^\ell - y_d, \quad t \in (0,T) \\ p^\ell(T) = 0 \end{cases}$$

and

$$v^{\ell+1} = -\frac{\alpha}{\mu}B^t p^\ell.$$

As in Section 10.2, we can also formulate gradient and conjugate gradient algorithms for 10.20.

Again, convergence results can be established; see Exercise E.10.7.

- Obviously, arguing as in Section 10.2, the way the optimal control depends on the parameters and data arising in (10.18)–(10.20) can be analyzed.

10.3.2 A nonlinear ODE with no constraint

In this section, the considered state equation is

$$\begin{cases} y' = g(t, y) + v, & t \in (0, T) \\ y(0) = y_0, \end{cases} \tag{10.23}$$

where $y_0 \in \mathbb{R}$ and the function g belongs to $C^1(\mathbb{R}^2)$ and satisfies

$$\left| \frac{\partial g}{\partial t} \right| + \left| \frac{\partial g}{\partial y} \right| \leq C \text{ in } \mathbb{R}^2.$$

Definition 10.3. Let $v \in L^2(0, T)$ be given. A generalized solution to (10.23) (in $(0, T)$) is any function $y \in C^0([0, T])$ that is differentiable at almost every $t \in (0, T)$ and satisfies

$$y'(t) = g(t, y(t)) + v(t) \quad \text{a.e. in } (0, T)$$

and the initial condition $y(0) = y_0$.

Under the previous conditions on g, it can be proved that, for every $y_0 \in \mathbb{R}$ and every $v \in L^2(0, T)$, there exists exactly one generalized solution y to (10.23).

The proof of this assertion is a little more complex than in the case of a linear equation or system; see in this respect Exercise E.10.20.

Consider the control problem

$$\begin{cases} \text{Minimize } J(v) := \dfrac{\alpha}{2} \displaystyle\int_0^T |y - y_d|^2 \, dt + \dfrac{\mu}{2} \displaystyle\int_0^T |v|^2 \, dt \\ \text{Subject to } v \in L^2(0, T), \ (y, v) \text{ satisfies } (10.23). \end{cases} \tag{10.24}$$

- This time, the arguments leading to the existence of a solution to (10.24) are more involved.

Moreover, it is much more difficult to get uniqueness (in principle, there can be more than one optimal control).

Specifically, the following is obtained:

Theorem 10.9. *The optimal control problem* (10.24) *possesses at least one solution.*

For the proof, see Exercise E.10.21.

• In what concerns optimality, the following can be established:

Theorem 10.10. *Let $\hat{v}$ be a solution to* (10.24) *(an optimal control) and let $\hat{y}$ be the solution to* (10.23) *(the corresponding state). Then, there exists $\hat{p}$ such that the triplet $(\hat{y}, \hat{p}, \hat{v})$ satisfies the optimality system*

$$\begin{cases} \hat{y}' = g(t, \hat{y}) + \hat{v}, \quad t \in (0, T) \\ \hat{y}(0) = y_0 \\ -\hat{p}' = \dfrac{\partial g}{\partial y}(t, \hat{y})\,\hat{p} + \hat{y} - y_d, \quad t \in (0, T) \\ \hat{p}(T) = 0 \\ \hat{v} = -\dfrac{\alpha}{\mu}\hat{p}. \end{cases} \tag{10.25}$$

See Exercise E.10.22 for the proof.

• As in previous cases, we are led to several iterative algorithms that furnish sequences of controls that, under some appropriate conditions, converge to a solution.

The details are left to the reader; see Exercises E.10.16 and E.10.17.

• Finally, we can also analyze the dependence of the solution(s) with respect to initial data and parameters.

Of course, the situation is more complex; see in this respect Exercise E.10.18.

10.3.3 *A linear ODS with inequality constraints*

We consider now a more realistic situation where v is constrained to belong to the following family of admissible controls:

$$\mathcal{U}_{\mathrm{ad}} := \{v \in L^2(0, T)^m : v_{*,i} \le v_i(t) \le v_i^* \ \text{ a.e. for } \ 1 \le i \le m\}.$$

The optimal control is the following:

$$\begin{cases} \text{Minimize } J(v) := \dfrac{\alpha}{2} \int_0^T |y - y_d|^2 \, \mathrm{d}t + \dfrac{\mu}{2} \int_0^T |v|^2 \, \mathrm{d}t \\ \text{Subject to } v \in \mathcal{U}_{\mathrm{ad}}, \ (y,v) \text{ satisfies (10.18).} \end{cases} \tag{10.26}$$

With regards to existence and uniqueness of solution, we have:

Theorem 10.11. *The optimal control problem* (10.26) *possesses exactly one solution* $\hat{v}$.

Proof. For the proof of existence, we can argue as in the proof of Theorem 10.2. Only one assertion needs a more delicate proof:

If the $v_n \in \mathcal{U}_{\mathrm{ad}}$ and the sequence $\{v_n\}$ converges weakly in $L^2(0,T)^m$ toward a control $\hat{v}$, then $\hat{v} \in \mathcal{U}_{\mathrm{ad}}$.

This is true because $\mathcal{U}_{\mathrm{ad}}$ is closed and convex (recall Example 10.3).

On the other hand, the proof of uniqueness does not have any additional difficulty. $\square$

- For the characterization of the optimal control, it is convenient to introduce the projectors $P_{\mathrm{ad},i} : \mathbb{R} \mapsto [v_{*,i}, v_i^*]$ with $i = 1, \ldots, m$, given as follows:

$$P_{\mathrm{ad},i}(z) = \begin{cases} v_{*,i} & \text{if } z < v_{*,i} \\ z & \text{if } z \in [v_{*,i}, v_i^*] \\ v_i^* & \text{if } z > v_i^*. \end{cases}$$

The following result holds:

Theorem 10.12. *Let* $\hat{v}$ *be the unique solution to* (10.26) *and let* $\hat{y}$ *be the corresponding solution to* (10.18). *Then, there exists* $\hat{p}$ *such that the triplet* $(\hat{y}, \hat{p}, \hat{v})$ *satisfies the optimality system*

$$\begin{cases} \hat{y}' = A\hat{y} + B\hat{v}, \ \ t \in (0,T) \\ \hat{y}(0) = y_0 \\ -\hat{p}' = A^t \hat{p} + \hat{y} - y_d, \ \ t \in (0,T) \\ \hat{p}(T) = 0 \\ \hat{v} = (\hat{v}_1, \ldots, \hat{v}_m), \text{ with } \hat{v}_i = P_{\mathrm{ad},i}\left(-\dfrac{\alpha}{\mu}(B^t\hat{p})_i\right) \text{ for } 1 \le i \le m. \end{cases} \tag{10.27}$$

Proof. The proof is similar to the proof of Theorem 10.3.

The function J is differentiable in the Gâteaux sense at $\hat{v}$ and its Gâteaux-derivative in any admissible direction $v - \hat{v}$ must satisfy

$$DJ(\hat{v}; v - \hat{v}) \geq 0.$$

Now, taking into account the definition of $J(v)$, we see that, if $\hat{p}$ solves (10.12), we necessarily have

$$\int_0^T (\alpha B^t \hat{p} + \mu \hat{v}) \cdot (v - \hat{v})\, dt \geq 0 \quad \forall v \in \mathcal{U}_{\text{ad}}, \quad \hat{v} \in \mathcal{U}_{\text{ad}}.$$

It is not difficult to see that this is equivalent to (10.27).

This ends the proof. □

In this case, just as when there is no constraint, thanks to the linearity of the ODS and the quadratic structure of the cost functional, the existence of $\hat{p}$ such that $(\hat{y}, \hat{p}, \hat{v})$ satisfies (10.27) is necessary and sufficient for the optimality of $\hat{v}$.

• With regards to iterative algorithms, we can again use the optimality system. Indeed, the following fixed-point iterates can be considered:

$$v^\ell \mapsto y^\ell \mapsto p^\ell \mapsto v^{\ell+1},$$

with

$$\begin{cases} (y^\ell)' = Ay^\ell + Bv^\ell, & t \in (0, T) \\ y^\ell(0) = y_0, \end{cases}$$

$$\begin{cases} -(p^\ell)' = A^t p^\ell + y^\ell - y_d, & t \in (0, T) \\ p^\ell(T) = 0 \end{cases}$$

and

$$v^{\ell+1} = (v_1^{\ell+1}, \dots, v_m^{\ell+1}), \text{ with } v_i^{\ell+1} = P_{\text{ad},i}\left(-\frac{\alpha}{\mu}(B^t p^\ell)_i\right) \text{ for } 1 \leq i \leq m.$$

We can also introduce algorithms of the gradient and conjugate gradient kinds for the computation of the optimal control.

This time, constraints must be taken into account, which brings some additional difficulties.

In particular, determining optimal steps ρ^ℓ is not so simple if we want to respect the fact that controls belong to $\mathcal{U}_{\text{ad}}$; see Exercise E.10.8.

- Arguing as in Section 10.2, the way the optimal control depends on the parameters and data arising in (10.18)–(10.20) can be analyzed.

10.4 Controllability Problems

This section is devoted to explaining what is controllability.

In order to describe the situation, let us consider the simplest case of a controlled linear ODE of the first order

$$y' = ay + v(t), \quad t \in (0, T), \tag{10.28}$$

where $a \in \mathbb{R}$.

Equation (10.28) is (exactly) controllable in the following sense: for any $y_0, y_T \in \mathbb{R}$, there always exist controls $v \in L^2(0, T)$ such that the generalized solution to this ODE together with the initial condition $y(0) = y_0$ satisfies

$$y(T) = y_T.$$

Indeed, the generalized solution to the Cauchy problem (10.4) is given by (10.6):

$$y(t) = e^{at} y_0 + \int_0^t e^{a(t-s)} v(s) \, \mathrm{d}s \quad \forall t \in [0, T].$$

Thus, the desired controls are those satisfying

$$\int_0^T e^{a(t-s)} v(s) \, \mathrm{d}s = y_T - e^{aT} y_0,$$

an equation for v that is always solvable.

Obviously, this means that we can drive the solutions to (10.28) from an arbitrary state y_0 to another arbitrary state y_T. It is also obvious that the control that performs this task is not unique at all.

As shown in what follows, when we deal with a non-scalar linear ODS, the situation can be completely different.

Thus, let us consider the system

$$y' = Ay + Bv(t), \quad t \in (0, T), \tag{10.29}$$

where $A \in \mathcal{L}(\mathbb{R}^N; \mathbb{R}^N)$, $B \in \mathcal{L}(\mathbb{R}^m; \mathbb{R}^N)$, and $N \geq m$.

As already mentioned, it is particularly interesting the case where N is large and m is small.

Definition 10.4. It will be said that (10.29) is exactly controllable if, for any $y_0, y_T \in \mathbb{R}^N$, there always exist controls $v \in L^2(0,T)^m$ such that the generalized solution to this ODS together with the initial condition $y(0) = y_0$ satisfies

$$y(T) = y_T. \tag{10.30}$$

Now, the generalized solution to the Cauchy problem (10.18) is given by (10.19):

$$y(t) = e^{tA}y_0 + \int_0^t e^{(t-s)A} Bv(s)\, ds \quad \forall t \in [0,T]$$

and we want to get controls satisfying

$$\int_0^T e^{(T-s)A} Bv(s)\, ds = y_T - e^{TA}y_0. \tag{10.31}$$

But it is not in principle clear whether this equation in $\mathbb{R}^N$ is solvable.

In the sequel, we will use the notation

$$[B;A] := [B|AB|\cdots|A^{N-1}B].$$

The following holds:

Theorem 10.13. *The linear system* (10.29) *is exactly controllable if and only if*

$$\operatorname{rank}[B;A] = N. \tag{10.32}$$

Proof. First, note that exact controllability holds if and only if, for any $z \in \mathbb{R}^N$, there exists $v \in L^2(0,T)$ such that

$$\int_0^T v(s) \cdot B^t e^{(T-s)A^t} \varphi_T\, ds = (z, \varphi_T) \quad \forall \varphi_T \in \mathbb{R}^N. \tag{10.33}$$

Let us assume that (10.32) holds and let us set

$$[\varphi_T] := \left(\int_0^T |B^t e^{(T-s)A^t} \varphi_T|^2\, ds \right)^{1/2} \quad \forall \varphi_T \in \mathbb{R}^N. \tag{10.34}$$

Then $[\cdot]$ is a norm in $\mathbb{R}^N$.

Indeed, the unique nontrivial property to check is that $[\varphi_T] > 0$ for $\varphi_T \neq 0$. This can be proved as follows.

Let $\varphi_T \in \mathbb{R}^N$ be given and let us set

$$\varphi(s) := e^{(T-s)A^t} \varphi_T \quad \forall s \in [0,T]. \tag{10.35}$$

If $[\varphi_T] = 0$, then $B^t \varphi \equiv 0$ and, in particular, $0 = \varphi(T) = B^t \varphi_T$.

We also have $\frac{\mathrm{d}^j}{\mathrm{d}t^j}B^t\varphi \equiv 0$ for all $j \geq 1$, whence

$$0 = \frac{\mathrm{d}^j}{\mathrm{d}t^j}B^t\varphi\Big|_{t=T} = B^t(A^t)^j\varphi_T \quad \forall j \geq 1.$$

This yields

$$\begin{pmatrix} B^t \\ B^t A^t \\ \vdots \\ B^t(A^t)^{N-1} \end{pmatrix} \varphi_T = 0.$$

But this is a linear system where the coefficient matrix has maximal rank, since it is the transpose of $[B; A]$.

Consequently, $\varphi_T = 0$ and we have proved that $[\,\cdot\,]$ is a norm.

Now, let us fix $z \in \mathbb{R}^N$ and set

$$I_0(\varphi_T) := \frac{1}{2}[\varphi_T]^2 - (z, \varphi_T) \quad \forall \varphi_T \in \mathbb{R}^N.$$

Since $[\,\cdot\,]$ is a norm in $\mathbb{R}^N$, the function I_0 possesses exactly one minimizer $\hat{\varphi}_T \in \mathbb{R}^N$. One has $DI_0(\hat{\varphi}_T; \varphi_T) = 0$ for all $\varphi_T \in \mathbb{R}^N$, that is,

$$\int_0^T B^t e^{(T-s)A^t}\hat{\varphi}_T \cdot B^t e^{(T-s)A^t}\varphi_T \, \mathrm{d}s = (z, \varphi_T) \quad \forall \varphi_T \in \mathbb{R}^N.$$

Consequently, if we set $v(s) := B^t e^{(T-s)A^t}\hat{\varphi}_T$ for all $s \in [0, T]$, we find a control in $L^2(0, T)^m$ satisfying (10.33).[5] This proves that (10.32) implies exact controllability.

By the way, note that, for each $\varphi_T \in \mathbb{R}^N$, the function φ introduced in (10.35) is the unique solution to the backwards in time system

$$\begin{cases} -\varphi' = A^t\varphi, & t \in (0, T) \\ \varphi(T) = \varphi_T. \end{cases}$$

Since $[\,\cdot\,]$ is a norm in the finite dimensional space $\mathbb{R}^N$, there exists a constant $K_T > 0$ such that

$$|\varphi_T|^2 \leq K_T \int_0^T |B^t\varphi(s)|^2 \, \mathrm{d}s \quad \forall \varphi_T \in \mathbb{R}^N. \tag{10.36}$$

Conversely, let us see that, if rank $[B; A] < N$, (10.29) is not exactly controllable.

---

[5]Actually, we have found a control in $C^\infty([0, T])^m$ that does the work.

Thus, let V be the space

$$V := \left\{ \sum_{j=0}^{N-1} \alpha_j A^j B z : z \in \mathbb{R}^m, \quad \alpha_0, \dots, \alpha_{N-1} \in \mathbb{R} \right\}.$$

Then, the final state corresponding to any initial data y_0 and any control $v \in L^2(0,T)^m$ satisfies $y(T) \in e^{TA} y_0 + V$.

Indeed, it is clear that v can be written as the strong limit in $L^2(0,T)^m$ of a sequence of step functions

$$v_k = \sum_{i=1}^{I_k} z_{ik} 1_{A(i,k)}, \quad \text{for } k \geq 1;$$

see to this purpose Exercise E.10.25. Consequently,

$$\begin{aligned} y(T) &= e^{TA} y_0 + \int_0^T e^{(T-s)A} B v(s) \, ds \\ &= e^{TA} y_0 + \lim_{k \to +\infty} \int_0^T e^{(T-s)A} B v_k(s) \, ds \\ &= e^{TA} y_0 + \lim_{k \to +\infty} \sum_{i=1}^{I_k} \left(\int_{A(i,k)} e^{(T-s)A} \, ds \right) B z_{ik}. \end{aligned}$$

Taking into account the definition of the exponential matrix, it is clear that this last term can be written in the form

$$e^{TA} y_0 + \lim_{k \to +\infty} \left(\lim_{\ell \to +\infty} \sum_{j=1}^{\ell} \alpha_{jk\ell} A^j B w_{jk\ell} \right)$$

for some $\alpha_{jk\ell} \in \mathbb{R}$ and some $w_{jk\ell} \in \mathbb{R}^m$.

However, in view of Caley–Hamilton's Theorem, we know that all terms of the form $A^j B w_{jk\ell}$ with $j \geq N$ can be rewritten as sums of products of powers of A of exponent $\leq N-1$ times B times vectors in $\mathbb{R}^m$.

This proves that $y(T)$ is the limit (in $\mathbb{R}^N$) of a sequence of vectors in $e^{TA} y_0 + V$ and, consequently, belongs to this linear manifold.

But V is a proper subspace of $\mathbb{R}^N$, since $\dim V = \text{rank}\,[B;A] < N$. Consequently, not all vectors in $\mathbb{R}^N$ are attainable and (10.29) is not exactly controllable. $\qquad\square$

It is usual to say that (10.32) is the *Kalman condition* for (10.29).

A lot of comments can be made on this result.

First, note that the necessary and sufficient condition (10.32) is independent of T. In other words, either (10.29) is exactly controllable for all $T > 0$ or it is not for any T.

Consequently, if it is satisfied, it is completely meaningful to consider the function $k = k(T; y_0)$, with

$$k(T; y_0) := \inf_{v \in \mathcal{U}_T(y_0)} \|v\|_{L^2(0,T)^m} \quad \forall T > 0, \quad \forall y_0 \in \mathbb{R}^N,$$

where $\mathcal{U}_T(y_0)$ stands for the set of controls $v \in L^2(0, T)^m$ satisfying (10.31).

Obviously, this quantity can be interpreted as the *cost of controllability*, that is, the minimal price we have to pay to drive the system from y_0 exactly to y_T.

It would be interesting to get sharp estimates of $k(T; y_0)$ and, in particular, see how this quantity grows as $T \to 0$.

Another comment is that the fact that the dimension of $\mathbb{R}^N$ is finite is crucial (since (10.34) is a norm, we immediately have (10.36)). For a similar controllability problem where the state "lives" in an infinite dimensional Hilbert space, the situation is much more complex; see some additional comments in Coron (2007).

There exists a partial generalization of Theorem 10.13 that works for linear ODSs with variable coefficients.

More precisely, let us consider the system

$$y' = A(t)y + Bv(t), \quad t \in (0, T), \tag{10.37}$$

where $A \in C^0([0, T]; \mathcal{L}(\mathbb{R}^N; \mathbb{R}^N))$ and again $B \in \mathcal{L}(\mathbb{R}^m; \mathbb{R}^N)$ and $N \geq m$.

Assume that $v \in L^2(0, T)^m$. As before, a generalized solution in $[0, T]$ to (10.37) is any function $y \in C^0([0, T]; \mathbb{R}^N)$ that is differentiable at almost any $t \in (0, T)$ and satisfies

$$y'(t) = A(t)y(t) + Bv(t) \quad \text{a.e. in} \quad (0, T).$$

Let $F \in C^1([0, T]; \mathcal{L}(\mathbb{R}^N; \mathbb{R}^N))$ be a fundamental matrix for the homogeneous system associated to (10.37). Then, for every $y_0 \in \mathbb{R}^n$ and every $v \in L^2(0, T)^m$, there exists exactly one generalized solution to (10.37) satisfying $y(0) = y_0$. It is given by

$$y(t) = F(t)F(0)^{-1}y_0 + \int_0^t F(t)F(s)^{-1}Bv(s) \, ds \quad \forall t \in [0, T].$$

As before, it will be said that (10.37) is exactly controllable at time T if, for any $y_0, y_T \in \mathbb{R}^N$, there always exist controls $v \in L^2(0, T)^m$ such that

the generalized solution corresponding to the initial condition $y(0) = y_0$
satisfies (10.30).

The following result holds:

Theorem 10.14. *Let* $B_i = B_i(t)$ *be given by*

$$B_0(t) \equiv B; \quad B_i(t) = B'_{i-1}(t) - A(t)B_{i-1}(t) \quad for \ i \geq 1$$

and let us denote by $V(t)$ *the space spanned by* $B_i(t)z$ *with* $i \geq 0$ *and* $z \in \mathbb{R}^m$. *Assume that*

$$V(\bar{t}) = \mathbb{R}^N \quad \text{for some } \bar{t} \in [0, T]. \tag{10.38}$$

Then (10.37) *is exactly controllable at time* T.

For the proof, see Coron (2007).

Unfortunately, (10.38) is a sufficient but not necessary condition for the controllability of (10.37). A counter example, as well as other comments and controllability results, are given in Coron (2007).

Exercises

E.10.1 Let H be an infinite dimensional Hilbert space and let $\{e_1, e_2, \dots\}$ be an orthonormal family.

    (1) Prove that, for any $w \in H$, the series $\sum_{n \geq 1} |(w, e_n)_H|^2$ converges in H and satisfies the so-called *Bessel inequality*:

$$\sum_{n \geq 1} |(w, e_n)_H|^2 \leq \|w\|_H^2 .$$

    (2) Prove that, for any $w \in H$, the series $\sum_{n \geq 1} (w, e_n)_H e_n$ converges in H.

    (3) Prove that the sequence $\{e_n\}$ converges to 0 in H weakly but not strongly.

E.10.2 Let H be a Hilbert space and assume that $u, v \in H$.

    (1) Prove that

$$\frac{1}{s} \left(F(u + s(v - u)) - F(u) \right) \leq F(v) - F(u) \quad \forall s \in (0, 1].$$

    (2) Deduce the first inequality in Proposition 10.3.

    (3) Then, using this inequality twice, deduce the second one.

E.10.3 Prove that, if α/μ is sufficiently small, the mapping Λ in (10.15) is a contraction. Deduce that the corresponding (fixed-point) iterative process converges in $L^2(0,T)$ to the optimal control $\hat{v}$.

E.10.4 Prove (10.7).

E.10.5 Consider the optimality system (10.10). Prove that the solution is continuously differentiable with respect to α and find a system satisfied by the corresponding partial derivatives of $\hat{y}$, $\hat{p}$, and $\hat{v}$.

E.10.6 Consider again the optimality system (10.10). Prove that the solution is continuously differentiable with respect to μ and find a system satisfied by the partial derivatives of $\hat{y}$, $\hat{p}$, and $\hat{v}$.

E.10.7 Consider the optimal control problem (10.20) and the associated optimality system (10.21).

(1) Formulate the optimal step gradient algorithm for this problem and prove a convergence result.

(2) Formulate the related optimal step conjugate gradient algorithm.

E.10.8 Consider the optimal control problem (10.26) and the associated optimality system (10.27).

(1) Formulate an optimal step gradient algorithm with projection for this problem.

(2) Formulate a similar algorithm of the conjugate gradient kind.

E.10.9 Consider the optimal control problem

$$\begin{cases} \text{Minimize } I(v) := \dfrac{\alpha}{2}|y(T)|^2 + \dfrac{\mu}{2}\int_0^T |v|^2 \, dt \\ \text{Subject to } v \in L^2(0,T),\ (y,v) \text{ satisfies } (10.4). \end{cases} \qquad (10.39)$$

Prove that this problem possesses exactly one solution $\hat{v}$ (the optimal control), deduce the corresponding optimality system, and find a related iterative algorithm for the computation of $\hat{v}$ that converges when α/μ is sufficiently small.

E.10.10 Same statement for the optimal control problem

$$\begin{cases} \text{Minimize } I(v) := \dfrac{\alpha}{2}|y(T)|^2 + \dfrac{\mu}{2}\int_0^T |v|^2 \, dt \\ \text{Subject to } v \in L^2(0,T)^m,\ (y,v) \text{ satisfies } (10.18). \end{cases}$$

E.10.11 Prove Theorem 10.4. To this purpose, prove that there exists $\epsilon_0 > 0$ such that, if $0 < \alpha/\mu \le \epsilon_0$, the mapping Λ defined in (10.15) is a contraction.

E.10.12 Same statement for the optimal control problem
$$\begin{cases} \text{Minimize } I(v) := \dfrac{\alpha}{2}|y(T)|^2 + \dfrac{\mu}{2}\displaystyle\int_0^T |v|^2 \, dt \\ \text{Subject to } v \in L^2(0,T)^m, \ (y,v) \text{ satisfies (10.18).} \end{cases}$$

E.10.13 Prove Theorem 10.7.

    Hint: Adapt the argument in the proof of Theorem 10.2.

E.10.14 Prove Theorem 10.8.

    Hint: Adapt the argument in the proof of Theorem 10.3.

E.10.15 Prove that, in (10.20), if α/μ is sufficiently small, the mapping that leads from v_ℓ to $v_{\ell+1}$ is a contraction and, consequently, the v^ℓ converge in $L^2(0,T)^m$ to the optimal control $\hat{v}$.

E.10.16 Consider the optimal control problem (10.24) and the associated optimality system (10.25). Formulate a fixed-point algorithm for the computation of a solution to this problem and prove a convergence result.

E.10.17 Consider again the optimal control problem (10.24) and the optimality system (10.25). Formulate the optimal step gradient and the optimal step conjugate gradient algorithm for the computation of a solution.

E.10.18 Let $(\hat{y}, \hat{p}, \hat{v})$ be a solution to the optimality system (10.25). Find conditions on g that ensure the differentiability of $\hat{y}$, $\hat{p}$, and $\hat{v}$ with respect to y_0, α, and μ. Assuming these conditions, find the systems satisfied by the corresponding derivatives.

E.10.19 Prove that, in the case of the optimal control problems (10.5), (10.20), and (10.26), the existence of (an adjoint state) $\hat{p}$ such that the triplet $(\hat{v}, \hat{y}, \hat{p})$ solves the optimality system is a necessary and sufficient condition for $\hat{v}$ to be optimal.

E.10.20 Consider the Cauchy problem (10.23), where $y_0 \in \mathbb{R}$, g belongs to $C^1(\mathbb{R}^2)$ and satisfies
$$\left|\frac{\partial g}{\partial t}\right| + \left|\frac{\partial g}{\partial y}\right| \le C \ \text{ in } \ \mathbb{R}^2,$$
and $v \in L^2(0,T)$. Prove that there exists a unique generalized solution.

    Hint: First, prove the result when $v \in C^0([0,T])$. Then, proceed by approximation.

E.10.21 Prove Theorem 10.9 following these steps:

(1) Consider a minimizing sequence $\{v_n\}$ and check that the v_n are uniformly bounded and the set of the corresponding states y_n is bounded and equi-continuous in $C^0([0,T])$.

(2) Deduce that there exists a subsequence (again indexed by n) that converges weakly in $L^2(0,T)$ to a control v_* such that the corresponding states y_n converge strongly to the state associated to v_*.

(3) Deduce that v_* is optimal, that is, $J(v_*) \leq J(v)$ for all $v \in L^2(0,T)$.

E.10.22 Prove Theorem 10.10 following these steps:

(1) Prove that the function $v \mapsto J(v)$ is differentiable at $\hat{v}$ and, also, that its Gâteaux-derivative at $\hat{v}$ in the direction w is

$$DJ(\hat{v}; w) = \alpha \int_0^T (\hat{y} - y_d)\, \hat{z}\, dt + \mu \int_0^T \hat{v} w\, dt$$

for any $w \in L^2(0,T)$, where $\hat{z}$ is the solution to

$$\begin{cases} \hat{z}' = \dfrac{\partial g}{\partial y}(t, \hat{y})\, \hat{z} + w, \quad t \in (0,T) \\ \hat{z}(0) = 0. \end{cases}$$

(2) Introduce appropriately the adjoint state $\hat{p}$ associated to $\hat{v}$ and, using that $DJ(\hat{v}; w) = 0$ for all w, deduce (10.25).

E.10.23 Consider the optimal control problem

$$\begin{cases} \text{Minimize } I(v) := \dfrac{\alpha}{2}|y(T)|^2 + \dfrac{\mu}{2}\int_0^T |v|^2\, dt \\ \text{Subject to } v \in L^2(0,T),\ (y,v) \text{ satisfies (10.23),} \end{cases} \tag{10.40}$$

where y_0 and g are as in Exercise E.10.20. Prove that this problem possesses at least one solution $\hat{v}$ (an optimal control) and deduce the corresponding optimality system.

E.10.24 Formulate a fixed-point algorithm for the computation of a solution to the optimal control problem (10.40) and prove a convergence result.

E.10.25 Prove that any function $w \in L^2(0,T)$ can be written as the strong limit in $L^2(0,T)$ of a sequence of step functions

$$v_k = \sum_{i=1}^{I_k} z_{ik} \mathbf{1}_{A(i,k)},$$

where the $z_{ik} \in \mathbb{R}^m$ and, for each $k \geq 1$, the $A(i,k)$ with $1 \leq i \leq I_k$ form a collection of disjoint subintervals of $(0,T)$.

Chapter 11

Additional Notes*

In this chapter, we will frequently find normed spaces (denoted $X, Y, Z, \ldots$).

Recall that $\mathcal{L}(X; Y)$ stands for the vector space of linear continuous mappings $A : X \mapsto Y$.

Also, recall that $\mathcal{L}(X; Y)$ is a normed space for the norm

$$\|A\|_{\mathcal{L}(X;Y)} = \sup_{x \in X \setminus \{0\}} \frac{\|Ax\|_Y}{\|x\|_X} = \sup_{\substack{x \in X \\ \|x\|_X = 1}} \|Ax\|_Y .$$

When $X = Y$, we will simply write $\mathcal{L}(X)$ instead of $\mathcal{L}(X; X)$.

11.1 Generalities

11.1.1 *Historical origin of differential equations*

It is accepted nowadays that differential equations constitute one of the most adequate tools to identify a function satisfying some specific properties or describe a real-life phenomenon of any kind.

In general terms, we are speaking of properties and phenomena governed by laws or statements that must be satisfied by a physical variable (a position, a temperature, a population density, etc.).

This goal has been pursued since the ancient times. Thus, there are reasons to view the following achievements as antecedents of the analysis of differential equations:

- The works by Archimedes[1] on integration of functions and computation of areas.
- The contributions of Copernicus[2] and Galileo[3] to the deduction of the laws of mechanics.
- The calculus of quotients, tangent lines, and critical points and the connection of differentiation and integration, apparently suggested for the first time by Pierre de Fermat.[4]

---

[1]Archimedes (287–212 B.C.) was a Greek physician, mathematician, and engineer. He is considered one of the most important scientists of classical antiquity. Among many other achievements, he postulated the laws of buoyancy and applied them to floating and submerged objects. According to *Archimedes's Principle,* on a body immersed in a fluid the exerted buoyant force is equal to the weight of the fluid that the body displaces and acts in the upward direction at the center of mass of the displaced fluid. A well-known legend says that he used this law to detect whether a crown was made of impure gold. After achieving satisfactorily the experiment, he exclaimed "Eureka."
[2]Nicolaus Copernicus (1473–1543) was a Polish mathematician, astronomer, physician, and economist. He formulated a model of the universe that placed the Sun and not the Earth at the center. This had been done much earlier by Aristarchus of Samos (310–230 B.C.), but very probably Copernicus was not aware of it. Copernicus's model was published in his book *De revolutionibus orbium coelestium,* a key work in the history of science, frequently viewed as a starting point of the Scientific Revolution. In 1517, he formulated (apparently for the first time) a *quantity theory of money,* a law that states that the general price level of goods and services is directly proportional to the money supply, i.e. the amount of money in circulation. A little later, in 1519 he formulated an economic principle, later called *Gresham's law,* that essentially says that "bad money" drives out "good money" (for instance, coins made of cheap metal may gradually drive out from circulation expensive metal objects).
[3]Galileo di Vincenzo Bonaiuti de' Galilei (1564–1642) was an Italian astronomer, physicist, and engineer. Many science historians refer to him as the "father" of observational astronomy, scientific method, and modern science. He studied and described a lot of physical phenomena, contributed to applied science and technology, and used the telescope for fruitful and transcendental observations. Galileo was a supporter of Copernican heliocentrism. For this reason, he was investigated by the Roman Inquisition in 1615, which concluded that "heliocentrism was foolish, absurd, and heretical since it contradicted Holy Scripture." After the publication of his *Dialogue Concerning the Two Chief World Systems* in 1632, he was prosecuted by the Inquisition, found "vehemently suspect of heresy," sentenced to formal imprisonment, and forced to recant. One day after, the sentence was commuted to house arrest, and he had to suffer for the rest of his life. A popular legend tells that, after recanting, Galileo said quietly "E pur si muove" (and yet it moves).
[4]Pierre de Fermat (1601–1665) was a French mathematician. He made some discoveries that can be viewed as seminal or starting points in infinitesimal calculus. In this respect,

If we go through the achievements strictly corresponding to the study of differential equations obtained along History, we find three well-defined periods:

11.1.1.1 *The algebraic-resolution period*

Here, the contributions mainly consist in determining explicit solutions to some simple ODEs with elementary methods and, then, interpreting geometrically and/or physically the results.

The period begins with the introduction and foundation of "calculus." This is attributed to Leibniz[5] and Newton (Figures 11.1 and 11.2).[6]

---

let us mention his technique of *adequality,* that permits to compute maxima and minima of many functions, tangents to curves, areas, centers of mass, etc. For instance, in order to find the maximizer of the function $f(x) = -x^2 + 2x$, he wrote that $f(x)$ is "very similar" to $f(x + \varepsilon)$ for any small ε, then simplified the corresponding expressions, divided by ε, and he neglected the remaining small quantity, to deduce that $x = 1$; a "handmade" method actually equivalent to what is usual in differential calculus. He also made notable contributions to analytic geometry, probability, and optics. His most famous achievements were *Fermat's Principle* for light propagation (also deduced using his adequality technique) and the statement of *Fermat's Last Theorem* in number theory, written without proof in a note at the margin of a copy of Diophantus's *Arithmetica.* The first successful proof of this theorem was announced in 1994 by Andrew Wiles and formally published in 1995, 358 years later.

[5]Gottfried Wilhelm Leibniz (1646–1716) was a German mathematician, philosopher, and diplomat. He was a prominent figure in both the history of philosophy and the history of mathematics. In mathematics, his greatest achievement was the development of the main ideas of differential and integral calculus. He also made contributions in other fields: linear algebra, geometry, and even seminal computation science. As a philosopher, he was representative of rationalism and idealism of the 17th century and a supporter of optimism, i.e. the conclusion that our world is the best possible world that God could have created (sometimes, this viewpoint was satirized by other philosophers, like Voltaire).

[6]Sir Isaac Newton (1642–1727) was an English physicist, theologian, inventor, alchemist, and mathematician. He has been widely recognized as one of the greatest scientists of all time. As summarized by Joseph Louis Lagrange, "Newton was the greatest genius that ever existed and also the most fortunate, since a system that governs the world can be found only once." He made relevant contributions to many areas of physics and mathematics: he formulated the laws of motion and universal gravitation; then, he used his mathematical description of gravity to derive Kepler's laws of planetary motion, the trajectories of comets, the precession of the equinoxes, and other phenomena; he demonstrated that the motion of objects on Earth and celestial bodies could be accounted for by the same principles; he built the first practical reflecting telescope; he also formulated an empirical law of cooling, etc. In the framework of fluid mechanics, Newton was particularly interested in the law that governs the force of resistance experimented by a body immersed in a fluid. He also postulated that, in a real-world fluid, the internal efforts among particles are proportional to the velocity spatial variations. In addition

Fig. 11.1   Gottfried Wilhelm von Leibniz, 1646–1716.

Among other things, Leibniz devised the notations x, $y(x)$, $\mathrm{d}x$, $\mathrm{d}y$, etc., and discovered the identity

$$\int y\,\mathrm{d}y = \frac{y^2}{2}.$$

Simultaneously, thanks to the work of Newton, which managed to use ODEs to describe the planetary motion, the relevance of this tool for the description of natural phenomena was revealed.

Newton was able to give a formulation and even solve this problem.

He also achieved a first classification of differential equations.

It is well known that he disputed Leibniz's authorship of the fundamental concepts of calculus.

However, it is usually accepted nowadays that Newton and Leibniz developed their ideas independently, following different methods: Leibniz was mainly motivated by the description and computation of tangents and areas. On the other hand, Newton's main interests were to model the motion of bodies and to compute trajectories and velocities.

---

to his work on calculus, as a mathematician, Newton contributed to the study of power series, generalized the *Binomial Theorem* to non-integer exponents, developed a method for approximating the roots of a function, etc. In a letter to Robert Hooke in February 1676, he wrote the celebrated sentence "If I have seen further, it is by standing on the shoulders of giants."

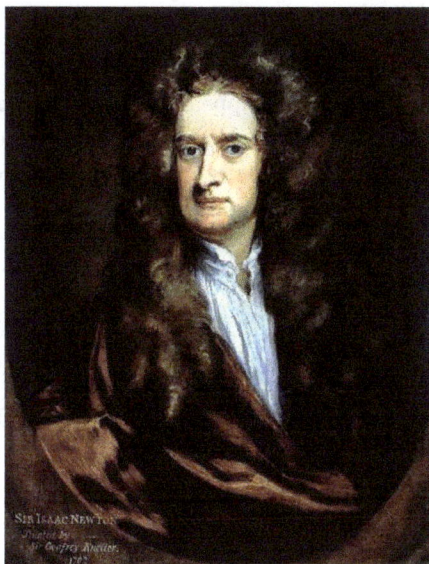

Fig. 11.2   Sir Isaac Newton, 1643–1727.

Later, we find some remarkable contributions:

- New integration methods, relying on variable separation and integrating factors, thanks to Jakob, Johann, and Daniel Bernoulli.[7]
- The clarification of integrability and the role played by elementary functions, thanks to Riccati.[8]
- A systematic use of methods, assisted by an appropriate ODE classification, thanks to Euler.[9]

---

[7] Jakob, Johann, and Daniel Bernoulli (resp. 1654–1705, 1667–1748, and 1700–1782) were Swiss mathematicians. Daniel Bernoulli (1700–1782) was the most distinguished of the Bernoulli family. He made contributions not only in mathematics but also in many other fields: medicine, biology, mechanics, physics, astronomy, etc. *Bernoulli's Theorem* in fluid dynamics states that, if the speed of a fluid increases, we simultaneously observe a decrease of the pressure or a decrease of the potential energy.

[8] Jacopo Francesco Riccati (1676–1754) was a mathematician and jurist, born in Venice. He is known above all for having studied the ODE that bears his name.

[9] Leonhard Euler (1707–1783) was a Swiss mathematician, physicist, and engineer. He made a lot of relevant contributions in many areas of mathematics: differential calculus, graph theory, topology, analytic number theory, etc. Pierre-Simon de Laplace described Euler's great influence in mathematics: "Read Euler, read Euler, he is the master of us all." The Euler equation for an ideal fluid is a relation between the velocity and pressure of a fluid. It is based on *Newton's Second Law of Motion* and seems to be the

11.1.1.2   *The analytical period*

This starts with the work of D'Alembert (Figure 11.3)[10] and Lagrange.[11]

The first significative advance appears to be the method of variation of constants, which leads to the resolution of non-homogeneous linear ODEs.

The period continues with Clairaut,[12] which extends some previous works by Euler on the integrating factor.

The main contributions of these times deal with the following:

- New basic concepts, properties, and resolution techniques for linear ODEs.
- New applications in physics and other sciences.
- The rigorous formulation of the Cauchy problem and the first general existence and uniqueness results.

In what concerns the theoretical analysis of initial-value problems, the first step is accomplished in the context of analytical functions, mainly

---

second PDE found in History; more details are given in what follows. This equation first appeared in the paper "Principes généraux du mouvement des fluides" published in *Mémoires de l'Académie des Sciences de Berlin* in 1757. Only 5 years before, D'Alembert had introduced the one-dimensional wave equation (also called the string equation); let us complete these historical data by recalling that Euler and Daniel Bernoulli formulated, respectively, in 1759 and 1762 the two-dimensional and three-dimensional wave equations.

[10] Jean le Rond D'Alembert (1717–1813), considered a genuine representative of the *Enlightened Movement,* conceived science as a whole and supported enthusiastically its utility as a tool for human progress. He was first a physicist and then a mathematician and was particularly interested in the *three-body problem,* the *precession of equinoccia,* and the behavior of *vibrating strings.* Together with Taylor, he seems to be one of the first mathematicians interested in PDEs.

[11] Joseph-Louis Lagrange (1738–1813) was a mathematician and astronomer, born in Italy and later became a naturalized French. He made very significant contributions to the fields of analysis, number theory, and both classical and celestial mechanics: he was one of the pioneers in calculus of variations, introducing the Lagrange multipliers to solve constrained problems. Then, he was able to get a fundamental connection of mechanical principles and variational calculus. He also introduced the method of variation of parameters (or constants). He studied the three-body problem and found some particular solutions (the so-called Lagrangian points). Another field where he made advances is interpolation. He also made contributions to number theory and algebra, etc. In 1794, Lagrange became the first professor of analysis at the École Polytechnique.

[12] Alexis Claude Clairaut (1713–1765) was a French mathematician and astronomer. His work helped to establish the validity of the principles and results outlined by Newton in the *Principia.* He is known for the ODE that bears his name.

Fig. 11.3   Jean le Rond D'Alembert, 1717–1813.

thanks to the work of Cauchy.[13] Here, we find for the first time results of existence, uniqueness, and resolution, see Note 11.1.3.

Later, these results were extended, by Picard[14] and Lindelöf,[15] to ODEs involving functions with bounded derivatives. Subsequently, approximate solutions were introduced, as well as the *equi-continuity* concept.

---

[13] Augustin-Louis Cauchy (1780–1857) was a French mathematician, engineer, and physicist. He made pioneering contributions in mathematics analysis, continuum mechanics, and other areas. He was probably the first to state and rigorously prove theorems of calculus. Possibly, more concepts and theorems have been named after Cauchy than after any other mathematician: *Cauchy's Theorem* for the existence and uniqueness of solution to initial-value problems; Cauchy condition; Cauchy sequences; *Cauchy's Theorem* in complex variable analysis; 16 concepts and theorems are named for Cauchy in elasticity, etc. In his book *Cours d'Analyse*, Cauchy emphasized the importance of rigor in analysis. Thus, he gave an explicit definition of the *infinitesimal* in terms of a sequence tending to zero and proved *Taylor's Theorem* rigorously, establishing the form of the remainder. Cauchy had a great influence over his contemporaries and successors. He wrote approximately 800 research articles and five textbooks on a variety of topics.

[14] Charles Émile Picard (1856–1941) was a French mathematician. In addition to his theoretical work, Picard made contributions to applied mathematics, including the theories of telegraphy and elasticity.

[15] Ernst Leonard Lindelöf (1870–1946) was a Finnish mathematician. He made contributions in real analysis, complex analysis, and topology.

This allowed to prove existence in more general conditions, in view of the contributions of Peano.[16]

11.1.1.3 *Qualitative theory and connection to numerical analysis*

This period starts with Poincaré[17] and Liapunov,[18] who were interested in the initial steps by some questions with origins in *celestial mechanics* and the stability properties of fluids in motion, respectively.

The initial-value problems for linear and nonlinear ODSs were studied from the viewpoint of stability.

For a solution to be stable and/or attractive, a lot of necessary and sufficient criteria were investigated. Also, a lot of consequences of the deduced properties for systems modeling real-world problems were deduced.

The *attractor* concept was introduced. This way, through the definition and properties of fractal sets, the field was connected to *chaos theory*.

A very relevant characteristic of this period is its great connection with the theory to functional analysis, with lots of contributions and advances.

On the other hand, the achievements of the theory were increasingly complemented with the numerical analysis of the problems. This was

---

[16] Giuseppe Peano (1858–1932) was an Italian mathematician. He wrote over 200 books and papers on mathematical analysis, logic, and set theory. He also invented an international auxiliary language, *Latino sine flexione,* which is a simplified version of Classical Latin.

[17] Jules Henri Poincaré (1854–1912) was a French mathematician, physicist, and engineer. He is often described as "The Last Universalist," since he shone in all existing fields in his time. He made many original fundamental contributions to pure and applied mathematics, mathematical physics, and celestial mechanics. More precisely, he made relevant advances in algebraic topology, the theory of analytic functions of several complex variables, the theory of Abelian functions, algebraic geometry, hyperbolic geometry, number theory, electromagnetism, the special theory of relativity, etc. It is worth mentioning, his contributions to the three-body problem (leading to the foundations of modern chaos theory, the special theory of relativity (presenting the Lorentz transformations in their modern symmetrical form), and his formulation of the so-called *Poincaré conjecture,* one of the most famous problems in mathematics, solved in 2002–2003 by Perelman. In 1882, he initiated the *qualitative theory of differential equations,* on the basis that, although the solutions to an ODE cannot be found in general, a lot of information can be deduced.

[18] Aleksandr Mikhailovich Liapunov (1857–1918) was a Russian mathematician and physicist. He is known for his development of the stability theory of a dynamical system, as well as for his many contributions to mathematical physics and probability theory. His work had a very significant impact and a lot of concepts are named after him: Liapunov equation, exponents, function, stability, central limit theorem, etc.

favored since the 1950's by the emergence and development of computers and the advances in scientific informatics.

In this specific aspect of the history of differential equations, it is adequate to emphasize the work of von Neumann,[19] with seminal and, at the same time, specially enriching inputs.

At present, the advances in informatics have allowed us to understand the behavior of the solutions to ODSs even in cases of high complexity.

Indeed, for systems with origins in many realistic phenomena, the numerical resolution is possible and produces, with a reasonable computational effort, very accurate approximations of the solutions.

More details on the historical and contemporary successes of the theory of differential equations can be found, for example, in Archibalda *et al.* (2004); Nápoles and Negrón (2002); and Nápoles (2008) and the references therein.

11.1.2   Population models

The best-known and most-used models of population growth are the *exponential* and the *logistic model*. The first was introduced by Malthus[20] in 1798 and the second one by Verhülst[21] in 1838. In both cases, we are considering *closed* models, that is, migrations are discarded.

---

[19] John von Neumann (1903–1957) was a Hungarian-American mathematician, physicist, and engineer. He made major contributions to many fields, including mathematics (foundations of mathematics, measure theory, functional analysis, ergodic theory, group theory, operator algebras, numerical analysis, geometry, etc.), physics (quantum mechanics, hydrodynamics, nuclear physics, etc.), economics (game theory and general equilibrium theory), computing (von Neumann architecture, linear programming, numerical meteorology, scientific computing, self-replicating machines, etc.), and statistics. He is considered to be possibly the most influential researcher in scientific computing of all time. He is the author of about 60 articles in pure mathematics, 60 in applied mathematics, 20 in physics, and about 10 on other subjects. He prematurely died at 54, maybe at a moment where he had reached a knowledge and maturity level that would have led him to produce outstanding results. His last work, an unfinished manuscript written while he was in the hospital, was later published in book form with the title *The Computer and The Brain*.

[20] Thomas Robert Malthus (1776–1834) was an English economist in the fields of political economy and demography. He is known for the population growth law that bears his name.

[21] Pierre François Verhülst (1804–1849) was a Belgian mathematician. He is known for the logistic law, where he assumed that the growth of an isolated species increases proportionally to the population size and is limited by competition, that is proportional to the square of the population size.

In general terms, if we assume that the function $p = p(t)$ denotes the density (or relative number of individuals) of a population, it is completely natural to assume that p must satisfy an ODE of the form

$$\dot{p} = f(t, p)$$

(or $\dot{p} = f(p)$), with appropriate properties of f.

An important aspect of population dynamics is *species interaction.*

Indeed, when dealing with eco-systems, it is usual to find (for instance) two species that interact and the right model is not a single ODE but a non-scalar system of the form

$$\begin{cases} \dot{p} = f(t, p, q) \\ \dot{q} = g(t, p, q), \end{cases}$$

again for appropriate f and g.

Thus, it is usual to find ecological systems where a species, the *predator,* depends on another, the *prey,* for surviving. In turn, the growth of the prey is limited by the action (and growth) of the predator.

In a simplified situation, we can assume that f and g are independent of t and polynomial of degree 2. We obtain the ODS

$$\begin{cases} \dot{p} = ap - bpq \\ \dot{q} = -cq + dpq, \end{cases} \tag{11.1}$$

where a, b, c, and d are positive constants and $p = p(t)$ and $q = q(t)$ are, respectively, assumed to represent the prey and predator populations.

With $q \equiv 0$, that is, in the absence of a predator, we accept that the prey possesses limited resources and therefore it is governed by the Malthus law.

On the other hand, if $p \equiv 0$, the predator has no means to survive and q is governed by the linear ODE $\dot{q} = -cq$, whose solutions go exponentially to zero as $t \to +\infty$.

It can be proved that any solution (p, q) for which both p and q are positive is periodic.

It is also frequent to find situations where several species coexist and compete with each other for the same resources.

A particular case is described by the system

$$\begin{cases} \dot{p} = ap - bpq \\ \dot{q} = cq - dpq, \end{cases} \tag{11.2}$$

where the number of species is two; compared to (11.1).

Now, except for a reduced number of very particular situations (unrealizable in practice), one of the species grows to infinity and the other one tends to extinction.

Other relevant aspects to consider are territorial occupation and migration: if resources run out, populations have to migrate to other available regions; contrarily, if resources are abundant, other similar species come from neighboring locations.

In fact, for the evolution of human population, the most decisive elements along History have been spatial changes and transitions.

In order to model the behavior of one or several species where territorial characteristics are important, it is convenient to use PDEs and not ODEs.

A relatively simple example is as follows: we accept that the population is located in the open set $D \subset \mathbb{R}^2$ (a fixed geographical region) during the time interval $(0, T)$ and we denote by $u : D \times (0, T) \mapsto \mathbb{R}$ the corresponding relative number of individuals; if resources at hand are sufficient, it is reasonable to assume that u satisfies the PDE

$$\frac{\partial u}{\partial t} - k\left(\frac{\partial^2 u}{\partial x_1^2} + \frac{\partial^2 u}{\partial x_2^2}\right) = f(x, t, u), \quad (x_1, x_2, t) \in D \times (0, T),$$

that of course will have to be complemented with suitable additional conditions.

The terms where we find second-order spatial partial derivatives are called *diffusion terms* (k is a positive constant). They indicate that the population has a tendency to spread or diffuse in D and, also, that this diffusion is carried out "migrating" from regions where u is large to regions where u is small.

The right-hand side is a (positive or negative) growth term that indicates the way the population evolves. As before, in a simplified situation, we can take $f(x, t, u) = \pm au$ for some $a > 0$.

11.1.3 *Initial-value problems*

The appellation "initial-value problem" is clearly motivated by the role of differential equations as models in physics (Figure 11.4).

It allows to designate those problems where we want to find or compute the solution to a differential system for which the behavior is known at a fixed value of the independent variable (which very frequently plays the role of time):

(CP)
$$\begin{cases} y' = f(t, y) \\ y(t_0) = y_0 . \end{cases}$$

Fig. 11.4   Henri Poincaré, 1854–1912.

The initial-value problems were studied in a systematic way for the first time by Cauchy, who made two fundamental contributions:

- He clarified the need and convenience of adding initial-value conditions to ODEs. More precisely, he showed that initial-value conditions make it possible to select a function within the whole family of solutions.
  From the viewpoint of modeling in physics and other sciences, this was crucial. It became evident that the task to describe the behavior of many phenomena that evolve with time can be achieved by establishing the law to be satisfied, fixing the "initial" status, and then computing the corresponding solution, that is, finding out the status at any other time.
- Under some conditions, he was able to prove the existence of a solution and even provide a constructive method of resolution.

In the context of Cauchy problems (CP), the method applied by Cauchy is valid when f is analytical near (t_0, y_0) and uses as working hypothesis that the solution φ is also analytical in a neighborhood of t_0.

The fundamental idea is to use the information provided by y_0 and the coefficients of the power expansion of f near (t_0, y_0) to find the coefficients of the power expansion of φ near t_0 and then check that this expansion converges.

To this second part of the proof, Cauchy introduced the so-called *method of majorization.*

Cauchy was also able to formulate for the first time the initial-value problem in the context of PDEs.[22]

He tried to adapt the method of majorization, but the complete successful extension only came later, in 1875, with the contribution of Kowalevskaya.[23]

The existence and uniqueness result, again with analytical data and leading to an analytical solution, is usually known as the *Cauchy–Kowalevskaya Theorem.*

11.1.4 *Impossibility of the explicit resolution of some elementary ODEs*

Let us see that, even in the case of the Riccati ODE (Figure 11.5)

$$y' = a(t)\, y^2 + b(t)\, y + c(t), \tag{11.3}$$

where a, b, and c are rational functions, it is impossible in general to obtain solutions amenable to be expressed by explicit formulas.

In order to fix ideas, let us first recall what is an *algebraic function* and what is an *elementary function.*

It will be said that $\varphi = \varphi(t)$, defined in a finite union of intervals in $\mathbb{R}$, is an *algebraic function* if there exists a polynomial in the variables t and y, $P = P(t, y)$, such that

$$P(t, \varphi(t)) \equiv 0.$$

The algebraic functions are also called 0-*rank elementary functions.* In short, we will say that φ is 0-EF.

---

[22]An additional Cauchys contribution is the so-called method of characteristics for first-order PDEs, introduced in 1819, together with Johan F. Pfaff (1765–1825).

[23]Sofia Vasilyevna Kowalevskaya (1850–1891) was a Russian mathematician. She was a pioneer for women in mathematics around the world, with important contributions to analysis, partial differential equations, and mechanics. She was the first woman to obtain a doctorate in mathematics, the first female full professor in northern Europe, and one of the first female editors of a scientific journal (*Acta Matematica*) in an almost exclusively male-dominated world.

Fig. 11.5   Alexander Liapunov, 1857–1918.

We will say that a 1-*rank monomial* is a function of the form

$$q = e^\varphi \quad \text{or} \quad q = \int \varphi \, dt,$$

where φ is 0-EF; then, we will call 1-*rank elementary function* (in short 1-EF) to any function φ such that, for some polynomial $Q = Q(v_1, \ldots, v_\ell, z_1, \ldots, z_m, y)$ in the variables v_i, z_j and y, one has

$$Q(p_1(t), \ldots, p_\ell(t), q_1(t), \ldots, q_m(t), \varphi(t)) \equiv 0,$$

where the p_i are 0-EFs and the q_j are 1-rank monomials.

Obviously, these definitions can be extended by induction. Thus, we can speak of n-*rank monomials* and, more generally, of n-*rank elementary functions* (that is, n-EFs) for every $n \geq 1$.

The family of all these functions is, by definition, the family of elementary functions.

Accordingly, it can be said that the elementary functions are those that can be obtained by operating algebraically with the exponential and integral operators and algebraic functions of the variable t arbitrarily many times in arbitrary order.

Let us consider the Riccati ODE

$$y' = -y^2 + c(t), \tag{11.4}$$

where c is an EF. We will say that it is possible "to compute explicitly" the solutions to (11.4) if all the solutions to this ODE are elementary functions. We have the following result, due to Liouville[24]:

Theorem 11.1. *If* (11.4) *possesses n-EF solutions, then it also possesses* $(n-1)$-*EF solutions. Consequently, if there exists a solution that is an elementary function, there exists a solution that is an algebraic function.*

Now, let us consider in particular the ODE

$$y' = -y^2 + 1 + \frac{s(s+1)}{t^2}, \tag{11.5}$$

where $s \in \mathbb{R}$.

Using appropriate results from complex analysis, it can be proved that (11.5) possesses solutions in the class 0-EF if and only if s is an integer.

Consequently, in general, the solutions to (11.5) are not elementary functions. This confirms that, in practice, we cannot expect to solve explicitly ODEs, even if they are as simple as (11.4).

For more details on this subject, see Kolmogorov *et al.* (1988). On the other hand, for a more complete information on the origin and usefulness of Riccati ODEs, see Bittanti (1992) and Amigó (2008).

11.1.5 *Linear equations and systems*

Linear ODEs appear in a natural way when we are looking for "perturbations" of solutions to more or less complicated ODEs with a real-world motivation.

A precise situation that serves to give an example is the following.

Consider the equation of the ideal pendulum

$$\theta'' = -k \sin \theta, \tag{11.6}$$

where k is a positive constant.

Here, we are denoting by $\theta(t)$ the angle determined by the vertical axis (downwards directed) and the pendulum at time t; $\ell = g/k$ is the length of the pendulum and g is the Gravity acceleration constant.

---

[24] Joseph Liouville (1809–1882) was a French mathematician and engineer. In mathematics, he worked in several different fields: number theory, complex analysis, differential geometry, and topology. He is remembered particularly for *Liouville's Theorem*: a bounded entire function in the complex plane must necessarily be constant.

This ODE possesses two constant solutions, φ^0 and $\overline{\varphi}$, respectively, given by

$$\varphi^0(t) = 0, \quad \overline{\varphi}(t) = \pi, \quad \forall t \in \mathbb{R}. \tag{11.7}$$

If we look for solutions φ "close" to φ^0, it is reasonable to assume that $\varphi \simeq \varphi^0 + \psi \equiv \psi$, where ψ solves the *linearized* equation

$$x'' = -kx. \tag{11.8}$$

In a similar way, if we are interested in solutions to (11.6) that are "close" to $\overline{\varphi}$, we must put $\varphi \simeq \overline{\varphi} + \zeta$, where ζ is now a solution to the following linear equation:

$$x'' = kx. \tag{11.9}$$

In both cases, what we did is to replace the right-hand side of the original ODE $(-k\sin\theta)$ by the terms that appear in the Taylor expansion of the first order:

$$-k(\sin\theta) \sim -k(\sin\theta)|_{\theta=0} - k\,\frac{\mathrm{d}}{\mathrm{d}\theta}(\sin\theta)\bigg|_{\theta=0} \cdot \theta = -k\theta$$

for (11.8) and

$$-k(\sin\theta) \sim -k(\sin\theta)|_{\theta=\pi} - k\,\frac{\mathrm{d}}{\mathrm{d}\theta}(\sin\theta)\bigg|_{\theta=\pi} \cdot \theta = k\theta$$

in the case of (11.9).

In view of the results in Chapter 8, it is clear that φ_0 is an asymptotically stable solution to (11.6), while $\overline{\varphi}$ is unstable.

The particular forms of (11.8) and (11.9) confirm this.

Let us present a second example. Consider the nonlinear ODS

$$\begin{cases} u' = f_0(u) - \sigma v \\ v' = u - v, \end{cases} \tag{11.10}$$

where $f_0(u) \equiv \lambda u - u^3 - \kappa$ (σ, λ, and κ are positive constants).

This ODS serves to model the evolution of many excitable systems (for instance, the behavior of a neuron); it is usual to say that it is of the *FitzHugh–Nagumo* kind.[25]

---

[25]It is also common to say that this is a system of the *activator–inhibitor* kind. One component stimulates production, while the other one tends to inhibit or even "forbid" the growth. The original motivation for the FitzHugh–Nagumo model was to isolate conceptually the essentially mathematical properties of excitation and propagation from the electrochemical properties of sodium and potassium ion flow; see FitzHugh (1969).

The stationary (i.e. constant) solutions are the functions

$$u(t) \equiv \overline{u}, \quad v(t) \equiv \overline{v},$$

where $\overline{u} = \overline{v} = f_0(\overline{u})/\sigma$, that is,

$$(\lambda - \sigma)\overline{u} - \overline{u}^3 - \kappa = 0, \quad \overline{v} = \overline{u}. \tag{11.11}$$

Consequently, it makes sense to linearize (11.10) near a stationary solution and look for

$$u = \overline{u} + \varphi, \quad v = \overline{v} + \psi,$$

where $(\overline{u}, \overline{v})$ satisfies (11.11) and φ and ψ solve the linear ODS

$$\begin{cases} \varphi' = (\lambda - 3\overline{u}^2)\varphi - \sigma\psi \\ \psi' = \varphi - \psi. \end{cases} \tag{11.12}$$

Linear ODEs and ODSs also appear as a direct consequence of assuming (exactly or approximately) linear physical laws.

A typical and well-known example is the *harmonic oscillator* (with or without friction).

This system models the motion of a body subject to the action of a force proportional to its distance to a fixed point (the origin), eventually damped by a second force, this time proportional to its velocity.

Accordingly, the ODE is

$$mx'' = -kx - ax',$$

where m and k are positive constants (m is the mass of the body and k is the elastic constant of the system) and $a \geq 0$ (a is the friction constant).

From the mathematical viewpoint, it is very convenient to analyze linear ODSs with details, both theoretically and numerically.

First, because for them many more things can be said than in a general case; in particular, many explicit formulas can be found for the solutions. Also, because the arguments that lead to existence, uniqueness, regularity, and other properties of the solutions are much more clear and simple.

On the other hand, for linear systems it is possible to design and analyze much more in-depth numerical methods.

11.2 Local Analysis of the Cauchy Problem

11.2.1 *Other fixed-point theorems*

As we already said, apart from Banach's fixed-point theorem (Theorem 2.2), there exist many other results that assert the existence of fixed points.

We will mention here the well-known Brouwer, Schauder, and Leray-Schauder Theorems[26, 27, 28]:

Theorem 11.2 (Brouwer). *Let $K \subset \mathbb{R}^m$ be a non-empty compact convex set and let $\mathcal{T} : K \mapsto K$ be a continuous mapping. Then, $\mathcal{T}$ possesses at least one fixed point in K.*

Theorem 11.3 (Schauder). *Let X be a Banach space, let $K \subset X$ be a non-empty convex closed bounded set, and let $\mathcal{T} : K \mapsto K$ be a continuous mapping such that $\mathcal{T}(K)$ is relatively compact. Then, $\mathcal{T}$ possesses at least one fixed point in K.*

Theorem 11.4 (Leray–Schauder; Fixed-Point). *Let X be a Banach space and let $\mathcal{T} : X \times [0,1] \mapsto X$ be a continuous mapping such that, for all bounded sets $B \subset X$, $\mathcal{T}(B \times [0,1])$ is relatively compact. Assume that*

$$\mathcal{T}(x,0) = 0 \quad \forall x \in X$$

and, also, that there exists $C > 0$ such that

$$x \in X, \quad \lambda \in [0,1], \quad x = \mathcal{T}(x,\lambda) \Rightarrow \|x\|_X \leq C.$$

Then, $\mathcal{T}(\cdot, 1)$ possesses at least one fixed point in X, that is, there exists $\hat{x} \in X$ such that $\hat{x} = \mathcal{T}(\hat{x}, 1)$.

For the proofs, consequences, and applications to mathematical physics, see for instance Zeidler (1986, 1988).

---

[26]Luitzen Egbertus Jan Brouwer (1881–1966) was a Dutch mathematician and philosopher. He worked in topology, set theory, measure theory, and complex analysis. He is known as the founder of modern topology. As a philosopher, Brouwer was a defender of *intuitionism,* a constructivist school of mathematics opposed to Hilbert's *formalism.*

[27]Juliusz Pawel Schauder (1899–1943) was a Polish mathematician of Jewish origin. He is known for his work in functional analysis, partial differential equations, and mathematical physics. He belonged to a large Polish group of mathematicians, the *Lwów School of Mathematics,* renowned for its exceptional contributions to subjects such as point-set topology, set theory, and functional analysis. Other mathematicians in the group were Banach, Mazur, Orlicz, Steinhaus, and Ulam. Schauder is well known for the *Schauder Fixed-Point Theorem,* the Schauder bases, and the *Leray–Schauder Principle,* the main outcome of a very successful collaboration with Jean Leray in Paris. In 1935, he got a position of a senior assistant in the University of Lwów. In 1943, after the invasion of Germany, he was forbidden to work. He wrote to Ludwig Bieberbach asking for support but Bieberbach passed his letter to the Gestapo and Schauder was arrested and then executed.

[28]Jean Leray (1906–1998) was a French mathematician. He worked on partial differential equations and algebraic topology. Among his many contributions, one of them, produced in 1934, initiated the study of weak solutions to the Navier–Stokes equations.

Let us make some comments.

First, note that the analog of Brouwer's Theorem in a normed space of infinite dimension is false. Indeed, in the Hilbert space ℓ^2 of the square-summable sequences in $\mathbb{R},$[29] the mapping $\mathcal{T}$, given by

$$\mathcal{T}(\{x_1, x_2, \ldots\}) = \left\{ \left(1 - \sum_{i \geq 1} |x_i|^2 \right)^{1/2}, x_1, x_2, \ldots \right\} \quad \forall \{x_n\} \in \ell^2,$$

is well defined, continuous, and maps the closed unit ball $\overline{B}(0; 1)$ into itself, having no fixed point in $\overline{B}(0; 1)$.

Secondly, observe that none of the previous theorems asserts uniqueness. It is in fact very easy to exhibit examples of each of them where uniqueness fails.

Thirdly, observe that the conditions in Theorems 11.2 and 11.3 do not guarantee at all convergence to a fixed point of $\mathcal{T}$ for the algorithm of successive approximations.

Indeed, these conditions allow us to introduce successive approximations by iterating as we did in Chapter 2: we can choose x^0 in K and then set

$$x^{n+1} := \mathcal{T}(x^n)$$

for all $n \geq 0$. From the properties of $\mathcal{T}$ and K, it can be shown that the resulting sequence $\{x^n\}$ possesses convergent subsequences. However, this is not sufficient to prove the convergence to a fixed point.

Finally, let us mention that Theorem 11.3 leads to a very simple proof of the following existence result:

Consider the CP, where $(t_0, y_0) \in \mathbb{R}^{N+1}$ and (for example) $f : \mathbb{R}^{N+1} \mapsto \mathbb{R}^N$ is continuous and bounded:

$$|f(t, y)| \leq M \quad \forall (t, y) \in \mathbb{R}^{N+1}.$$

Then (CP) possesses at least one global solution, that is defined in the whole $\mathbb{R}$.

11.2.2 *Implicit and inverse functions*

As a complement to Remark 2.7, part (a), we will present here some comments on the implicit function theorem and other related results (Figure 11.6).

---

[29]See Note 11.5.1 for the definition of ℓ^2, its scalar product and the associated norm.

Fig. 11.6   Augustin-Louis Cauchy, 1780–1857.

First, let us recall a "classical" version of this theorem in spaces of infinite dimension:

Theorem 11.5 (Implicit Function). *Let X, Z, and W be Banach spaces, let $U \subset X \times Z$ be a non-empty open set and assume that $(x_0, z_0) \in U$. Let $F : U \mapsto W$ be given and suppose that $F(x_0, z_0) = w_0$. It is assumed that F is continuous and continuously differentiable with respect to z in U, that is:*

- *For every $(x, z) \in U$ there exists $D_2 F(x, z) \in \mathcal{L}(Z; W)$ such that*

$$\lim_{k \to 0} \frac{\|F(x, z + k) - F(x, z) - D_2 F(x, z)\, k\|_W}{\|k\|_Z} = 0.$$

- *The mapping $(x, z) \mapsto D_2 F(x, z)$, that is well defined from U into $\mathcal{L}(Z; W)$, is continuous.*

It is also assumed that $D_2 F(x_0, z_0)$ is an isomorphism from Z onto W. Then, there exist open neighborhoods A and B, respectively, of x_0 and z_0 and there exists a function $f \in C^0(A; Z)$ with the following properties:

(1) *$A \times B \subset U$ and $f(x) \in B$ for all $x \in A$.*
(2) *$f \in C^0(A; Z)$.*

(3) $F(x, f(x)) = w_0$ *for all* $x \in A$.
(4) f *is "unique" in the following sense: if* $(x, z) \in A \times B$ *and* $F(x, z) = p_0$, *then* $z = f(x)$.

Corollary 11.1. *Under the conditions of Theorem 11.5, if we also have* $F \in C^1(U; W)$ *and* $D_2 F(x, z)$ *is an isomorphism from* Z *onto* W *for every* $(x, z) \in U$, *then* $f \in C^1(A; Z)$ *and*

$$Df(x) = -D_2 F(x, f(x))^{-1} \circ D_1 F(x, f(x))$$

in a neighborhood of x_0.

For the proofs, see for instance Dontchev and Rockafellar (2014) and Krantz and Parks (2013).

Secondly, let us recall, in a similar context, the inverse function theorem:

Theorem 11.6 (Inverse Function). *Let* X *and* Z *be Banach spaces, let* $D \subset X$ *be a non-empty open set, and suppose that* $x_0 \in D$. *Let* $\Phi : D \mapsto Z$ *be given and let us set* $\Phi(x_0) = z_0$. *It is assumed that* Φ *is continuously differentiable in* D *and* $D\Phi(x_0)$ *is an isomorphism from* X *onto* Z. *Then, there exist open neighborhoods* A *and* B, *respectively, of* x_0 *and* z_0 *and there exists a unique function* $\Psi \in C^1(B; X)$ *with the following properties:*

(1) $\Phi(x) \in B$ *for all* $x \in A$ *and* $\Psi(z) \in A$ *for all* $z \in B$.
(2) Φ *is a diffeomorphism from* A *onto* B *and* $\Phi^{-1} = \Psi$, *that is,* $\Psi(\Phi(x)) = x$ *for all* $x \in A$ *and* $\Phi(\Psi(z)) = z$ *for all* $z \in B$.
(3) *For every* $x \in A$, *the linear mappings* $D\Phi(x)$ *and* $D\Psi(\Phi(x))$ *are the inverse to each other.*

Theorem 11.6 and Corollary 11.1 are two equivalent versions of the same property of differential calculus. Each of them can be proved starting from the other one.

A generalized version of Theorem 11.6 is *Liusternik's Theorem* Liusternik and Sobolev (1961), which says the following:

Theorem 11.7 (Liusternik). *Let* X *and* Z *be Banach spaces, let* $D \subset X$ *be a non-empty open set, and suppose that* $x_0 \in D$. *Let* $\Phi : D \mapsto Z$ *be given and let us set* $\Phi(x_0) = z_0$. *It is assumed that* Φ *is strictly differentiable at* x_0, *that is, for some* $D\Phi(x_0) \in \mathcal{L}(X; Z)$, *one has*

$$\lim_{x, y \to x_0} \frac{\|\Phi(y) - \Phi(x) - D\Phi(x_0)(y - x)\|_Z}{\|y - x\|_X} = 0.$$

282   *Ordinary Differential Equations and Applications*

It is also assumed that $D\Phi(x_0)$ is surjective. Then, there exist open neighborhoods A and B, respectively, of x_0 and z_0 and there exists a function $\Psi \in C^1(B; X)$ with the following properties:

(1) $\Phi(x) \in B$ for all $x \in A$ and $\Psi(z) \in A$ for all $z \in B$.
(2) $\Phi(\Psi(z)) = z$ for all $z \in B$.

All these results are *local:* the conclusion holds good in a neighborhood of the considered point $((x_0, z_0)$ in the case of Theorem 11.5 and x_0 in the case of Theorems 11.6 and 11.7).

However, there exists a global version of Theorem 11.6 that is less known:

Theorem 11.8 (Global Inverse Theorem). *Let X and Z be Banach spaces and let $\Phi : X \mapsto Z$ be continuously differentiable in X and such that*

(1) *$D\Phi(x)$ is an isomorphism from X onto Z for all $x \in X$.*
(2) *There exists $C > 0$ such that*

$$\|(D\Phi(x))^{-1}\|_{\mathcal{L}(Z;X)} \leq C \quad \forall x \in X.$$

Then, Φ is globally invertible in X, that is, there exists a unique function $\Psi \in C^1(Z; X)$ such that

$$\Psi(\Phi(x)) = x \quad \forall x \in X, \quad \Phi(\Psi(z)) = z, \quad \forall z \in Z.$$

Furthermore, for every $x \in X$, the linear mappings $D\Phi(x)$ and $D\Psi(\Phi(x))$ are the inverse to each other.

For the proof, possible extensions, additional comments, and applications, see for instance Chow and Hale (1982); see also Dontchev (1996).

11.3 Uniqueness of Solution in a Cauchy Problem

Let us recall a uniqueness result taken from Okamura (1941) (Figure 11.7):

Theorem 11.9. *Let us assume that*

$$\Omega = \{(t, y) \in \mathbb{R}^{N+1} : |t| < a, \ |y_i| \leq b, \ i = 1, \ldots, N\}$$

and let $f : \Omega \mapsto \mathbb{R}^N$ be a continuous function such that $f(t, 0) \equiv 0$. Then, the following assertions are equivalent:

Fig. 11.7   Sofia Vasilyevna Kowalevskaya, 1850–1891.

(1) *In any neighborhood of* 0, *the Cauchy problem*

$$\begin{cases} y' = f(t,y) \\ y(0) = 0 \end{cases} \qquad (11.13)$$

possesses at most one solution.

(2) *There exists a function* $V : \Omega \mapsto \mathbb{R}$, *locally Lipschitz-continuous with respect to the variable* y, *such that* $V(t,0) \equiv 0$, $V(t,y) \neq 0$ *for* $y \neq 0$, *and*

$$\limsup_{s \to 0} \frac{1}{s} [V(t+s, y_1 + sf_1(t,y), \ldots, y_N + sf_N(t,y)) - V(t,y)] \leq 0$$

for all $(t,y) \in \Omega$.

In particular, if the considered ODS is autonomous and there exists an associated Liapunov function,[30] the Cauchy problem (11.13) possesses at most one solution.

The proof that the first assertion implies the second one relies on the following construction of the function V:

• For any given $(t,y), (\bar{t}, \bar{y}) \in \Omega$ and any polygonal Π connecting (t,y) to $(\bar{t}, \bar{y})$ formed with segments alternatively with slope f and on a

---

[30] For the definition of a Liapunov function, see Chapter 8.

hyperplane $t = \text{Const.}$, we denote by $|\Pi|$ the sum of the lengths of the segments for which t is a constant. Then, we denote by $\ell((t, y), (\bar{t}, \bar{y}))$ the infimum of all these possible $|\Pi|$.

• We set

$$V(t, y) = \ell((0, 0), (t, y)) \quad \forall (t, y) \in \Omega.$$

11.4 Global Analysis of the Cauchy Problem

11.4.1 *Existence of a maximal solution*

Let us analyze the existence of maximal solutions to the CP when f is continuous.

To this purpose, it will be convenient to recall a result, usually known as *Zorn's Lemma,* that is a consequence of the *Axiom of Choice*:

Theorem 11.10. *Let $\mathcal{M}$ be a non-empty set, partially ordered by "$\leq$."* *Assume that $\mathcal{M}$ is inductive, that is, every totally ordered subset $\mathcal{M}_0$ possesses an upper bound. Then, $\mathcal{M}$ possesses a maximal element. In other words, there exists $m^* \in \mathcal{M}$ such that, if $m^* \leq m$ for some $m \in \mathcal{M}$, necessarily $m = m^*$.*

Let us assume that, in (CP), $f : \Omega \mapsto \mathbb{R}^N$ is (only) continuous and $(t_0, y_0) \in \Omega$. Using Theorem 11.10, it can be shown that (CP) possesses at least one maximal solution (neither extendable to the right nor to the left).

To this end, it will suffice to redo the proof of Theorem 4.1 with appropriate changes.

Thus, let us consider the set $\mathcal{S}(t_0, y_0)$, formed by the couples (I, φ), where I is an interval whose interior contains t_0 and $\varphi : I \mapsto \mathbb{R}^N$ is a solution to (CP) in I. In view of Theorem 2.6, this set is non-empty. On the other hand, it is not difficult to check that the binary relation

$$(I, \varphi) \leq (J, \psi) \iff I \subset J, \ \psi \text{ extends } \varphi$$

is a partial order in $\mathcal{S}(t_0, y_0)$ that makes this set inductive. Consequently, there exists $(I^*, \varphi^*) \in \mathcal{S}(t_0, y_0)$ that is maximal for this order. Obviously, this means that $\varphi^* : I^* \mapsto \mathbb{R}^N$ is a maximal solution to (CP).

11.4.2 *Characterization of the maximal solution*

That a solution to (CP) is maximal can be characterized by observing the properties of the corresponding right and left *terminals*.

Definition 11.1. Let $(t_0, y_0) \in \Omega$ and $(I, \varphi) \in \mathcal{S}(t_0, y_0)$, with $I = (\alpha, \beta)$. The right terminal of φ is the set

$$T^+(\varphi) := \begin{cases} \emptyset & \text{if } \beta = +\infty \\ \overline{\tau^+(\varphi)} \cap \{(\beta, y) : y \in \mathbb{R}^N\} & \text{otherwise.} \end{cases}$$

In a similar way, we can introduce the left terminal $T^-(\varphi)$. □

Thus, if $(\beta, \overline{y}) \in T^+(\varphi)$, there exists a sequence $\{t_n\}$ in I such that

$$t_n \to \beta, \quad \varphi(t_n) \to \overline{y}.$$

But this does not imply at all that $\varphi(t)$ has a limit as $t \to \beta^-$.

It does not imply either that this limit must be equal to $\overline{y}$, as the following example illustrates:

Example 11.1. Consider the CP

$$\begin{cases} y' = y - \sin(\pi/t) - (\pi/t^2)\cos(\pi/t) \\ y(-1) = 0. \end{cases} \tag{11.14}$$

The unique maximal solution to (11.14) is $\varphi : (-\infty, 0) \mapsto \mathbb{R}$, with

$$\varphi(t) = \sin(\pi/t) \quad \forall t < 0.$$

In this case,

$$T^-(\varphi) = \emptyset \text{ and } T^+(\varphi) = \{(0, y) : y \in [-1, 1]\}. \qquad □$$

The role played by terminals becomes clear in the following result:

Theorem 11.11. *Suppose that f is continuous and locally Lipschitz-continuous with respect to y in Ω, $(t_0, y_0) \in \Omega$, and $(I, \varphi) \in \mathcal{S}(t_0, y_0)$, with $I = (\alpha, \beta)$. The following four assertions are equivalent:*

(1) *φ is extendable to the right.*
(2) *$\tau^+(\varphi)$ is bounded and one has (4.4).*
(3) *$T^+(\varphi) \cap \Omega \neq \emptyset$.*
(4) *$T^+(\varphi)$ is a singleton and reduces to a point in Ω.*

The proof is not difficult and is left as an exercise.

A direct consequence of this result is that, if f fulfills the usual existence and uniqueness conditions and $(t_0, y_0) \in \Omega$, then $\varphi : (\alpha, \beta) \mapsto \mathbb{R}^N$ is the maximal solution to (CP) if and only if

$$T^+(\varphi) \cap \Omega = T^-(\varphi) \cap \Omega = \emptyset.$$

11.5   More on Linear Equations and Systems

11.5.1   *Lagrange's formula*

The idea of the method of variation of constants can be applied not only in the framework of CPs for ODSs but also in many other contexts.

Let X be a Banach space. Let us accept that, for any given $a, b \in \mathbb{R}$ with $a < b$ and any continuous function $f : [a, b] \mapsto X$, we can speak of the integral

$$\int_a^b f(s)\,\mathrm{d}s$$

and, also, that under these conditions the new function $\tilde{f}$, given by

$$\tilde{f}(t) = \int_a^t f(s)\,\mathrm{d}s \quad \forall t \in [a, b],$$

is continuously differentiable and satisfies $\tilde{f}'(t) = f(t)$ for all $t \in [a, b]$.

Let $A \in \mathcal{L}(X)$ be a linear continuous mapping. For every $x_0 \in X$, we will first consider the homogeneous linear CP

$$\begin{cases} x' = Ax \\ x(0) = x_0 . \end{cases} \tag{11.15}$$

For any given interval I that contains 0, it will be said that $\xi : I \mapsto X$ is a solution in I to (11.15) if ξ is continuously differentiable in I, satisfies

$$\xi'(t) = A\xi(t) \quad \forall t \in I$$

and, moreover, $\xi(0) = x_0$.

For every $t \in \mathbb{R}$, it makes sense to speak of e^{tA}: by definition, e^{tA} is a new linear continuous mapping in $\mathcal{L}(X)$, given by

$$\mathrm{e}^{tA} = \sum_{n \geq 0} \frac{t^n}{n!} A^n$$

(observe that this series converges in $\mathcal{L}(X)$ for all $t \in \mathbb{R}$).

Let us set

$$F(t) := \mathrm{e}^{tA} \quad \forall t \in \mathbb{R}.$$

Then, with arguments similar to those used in the proof of Theorem 5.4, it is not difficult to check that $F : \mathbb{R} \mapsto \mathcal{L}(X)$ is well defined, continuously differentiable, and satisfies

$$F'(t) = A\,F(t) \quad \forall t \in \mathbb{R}.$$

We also have that $F(t)$ is an automorphism on X and

$$F(t)^{-1} = \left(e^{tA}\right)^{-1} = e^{-tA} \quad \forall t \in \mathbb{R}.$$

Hence, the function ξ, given by

$$\xi(t) = F(t)x_0 = e^{tA} x_0 \quad \forall t \in \mathbb{R}, \tag{11.16}$$

is a solution in $\mathbb{R}$ to (11.15).

In fact, this is the unique solution. Indeed, if J is an interval containing 0 and $\eta : J \mapsto X$ solves (11.15) in J, it is clear that $t \mapsto e^{-tA} \eta(t)$ is a continuously differentiable function that coincides at 0 with x_0 and possesses a derivative equal to zero in J. Consequently, $e^{-tA} \eta(t) \equiv x_0$, that is,

$$\eta(t) = \xi(t) \quad \forall t \in J.$$

This argument shows that (11.15) possesses a unique global solution, actually defined in the whole $\mathbb{R}$, given by (11.16).

Now, let $b : I \mapsto X$ be a continuous function (again, I is an interval containing 0) and let us consider the *non-homogeneous* CP

$$\begin{cases} x' = Ax + b(t) \\ x(0) = x_0. \end{cases} \tag{11.17}$$

Again, it will be said that a solution to (11.17) in I is a continuously differentiable function $\xi : J \mapsto X$ satisfying

$$\xi'(t) = A\xi(t) + b(t) \quad \forall t \in I$$

and $\xi(0) = x_0$.

The following generalization of Theorem 5.3 holds:

Theorem 11.12. *There exists a unique solution in I to (11.17). It is given by*

$$\begin{aligned} \xi(t) &= F(t)x_0 + F(t) \int_0^t F(s)^{-1} b(s) \, ds \\ &= e^{tA} x_0 + \int_0^t e^{(t-s)A} b(s) \, ds \end{aligned} \tag{11.18}$$

for all $t \in I$.

This result can be applied, for example, in the following situation: $X = \ell^2$, $A \in \mathcal{L}(\ell^2)$, $b \in C^0(I; \ell^2)$, and $\{x_{0,n}\} \in \ell^2$.

By definition, ℓ^2 is the linear space formed by the sequences $\{x_n\}$ with $x_n \in \mathbb{R}$ and $\sum_{n \geq 1} |x_n|^2 < +\infty$; this is a Hilbert space for the scalar product $(\cdot, \cdot)_{\ell^2}$, with

$$(\{x_n\}, \{y_n\})_{\ell^2} = \sum_{n \geq 1} x_n y_n \quad \forall \{x_n\}, \{y_n\} \in \ell^2.$$

Let $A \in \mathcal{L}(\ell^2)$ be given. It can be proved that there exist unique sequences

$$\{a_{1,n}\}, \{a_{2,n}\}, \cdots \in \ell^2$$

such that, for each $\{x_n\} \in \ell^2$, one has $\{y_n\} = A\{x_n\}$ if and only if

$$y_n = \sum_{m \geq 1} a_{n,m} x_m \quad \forall n \geq 1.$$

On the other hand, for any given $b \in C^0(I; \ell^2)$, one has

$$b(t) = \{b_n(t)\} \quad \forall t \in I$$

for some functions $b_n \in C^0(I)$.

Therefore, in this case, (11.17) can be rewritten as a first-order ODS of infinite dimension as follows:

$$\begin{cases} x_n' = \displaystyle\sum_{m \geq 1} a_{n,m} x_m + b_n(t), & n \geq 1 \\ x_n(0) = x_{0,n}, & n \geq 1. \end{cases}$$

Nonetheless, the most interesting problems of the kind (11.17) appear when A is an unbounded linear operator with proper domain $D(A) \subset X$. More details will be given in Note 11.5.2.

11.5.2 *Semigroups of operators and applications*

The theory of semigroups of operators allows to generalize the results in Section 5.5 to the case of infinite dimensions, with applications in many areas, among them PDE theory.

We will present now some fundamental ideas; for a more detailed exposition, see for example Brézis (1983).

Again, let X be a Banach space and let $A : D(A) \subset X \mapsto X$ be a linear operator, not necessarily bounded, defined in the linear space $D(A)$ (the domain of A). It will be assumed that $D(A)$ is dense in X and A is

m-dissipative. This means that, for every $\lambda > 0$, the linear operator $\mathrm{Id} - \lambda A$ possesses a continuous inverse in X, denoted by J_λ, with

$$\|J_\lambda\|_{\mathcal{L}(X)} = \|(\mathrm{Id} - \lambda A)^{-1}\|_{\mathcal{L}(X)} \leq 1 \quad \forall \lambda > 0. \tag{11.19}$$

In other words, for every $\lambda > 0$ and every $f \in X$, the equation

$$u - \lambda A u = f, \quad u \in D(A)$$

possesses exactly one solution u, with $\|u\|_X \leq \|f\|_X$ (it is also usual to say that $-A$ is m-accretive).

Example 11.2. Let $B \in \mathcal{L}(X)$ and $\alpha > \|B\|_{\mathcal{L}(X)}$ be given. Set $D(A) := X$ and $A := -\alpha \mathrm{Id} + B$. Then, A is m-dissipative. □

Example 11.3. Let us take $X = C^0([0,1])$ (endowed with the usual norm),

$$D(A) := \{v \in C^1([0,1]) : v(0) = 0\}$$

and

$$Av := -v_x - c(x)v \quad \forall v \in D(A),$$

where $c \in L^\infty(0,1)$ and $c \geq 0$. Then, $D(A)$ is a dense subspace of $C^0([0,1])$ and, again, $A : D(A) \mapsto C^0([0,1])$ is m-dissipative. □

For every $x_0 \in X$, we will consider the CP

$$\begin{cases} x' = Ax \\ x(0) = x_0. \end{cases} \tag{11.20}$$

Given a function $\xi : [0, +\infty) \mapsto X$, it will be said that ξ is a solution in $[0, +\infty)$ to (11.20) if ξ is continuous in $[0, +\infty)$, continuously differentiable in $(0, +\infty)$, and satisfies

$$\xi(t) \in D(A) \quad \text{and} \quad \xi'(t) = A\xi(t) \quad \forall t > 0$$

and also $\xi(0) = x_0$.

It will be said that $A : D(A) \subset X \mapsto X$ is the *generator of a semigroup of contractions* if, for every $x_0 \in X$, the corresponding CP (11.20) possesses exactly one solution ξ in $[0, +\infty)$, with

$$\|\xi(t)\|_X \leq \|x_0\|_X \quad \forall t \geq 0.$$

The semigroup of contractions associated to A is, by definition, the associated family of linear continuous operators $\{S(t)\}_{t \geq 0}$ where, for every t, $S(t)$ is the mapping that assigns to each x_0 the unique solution to (11.20) at time t as follows:

$$S(t)x_0 = \xi(t) \quad \forall x_0 \in X, \quad \forall t \geq 0.$$

Note that $\{S(t)\}_{t \geq 0}$ is a true *semigroup* (in the algebraic sense) for the usual composition law. Indeed, it suffices to note that

$$S(t_1) \circ S(t_2) = S(t_1 + t_2) \quad \forall t_1, t_2 \geq 0.$$

The fundamental result of the theory of semigroups of operators is the following:

Theorem 11.13 (Hille–Yosida, Banach framework). *Let X be a Banach space and let $A : D(A) \subset X \mapsto X$ be a linear operator. Then, A is the generator of a semigroup of contractions if and only if $D(A)$ is dense in X and A is m-dissipative.*

Thanks to this result, the operators A for which (11.20) is uniquely solvable for all initial data, with a solution whose norm is non-increasing with t, are characterized: they have to be densely defined and must "dissipate," in the sense of (11.19).

A particular interesting situation appears when X is a Hilbert space with scalar product $(\cdot, \cdot)_X$. In this case, A is m-dissipative if and only if $-A$ is monotone, that is,

$$(-Au, u)_X \geq 0 \quad \forall u \in D(A).$$

Therefore, the following holds:

Corollary 11.2 (Hille–Yosida, Hilbert framework). *Let X be a Hilbert space and let the operator $A : D(A) \subset X \mapsto X$ be linear. The following three assertions are equivalent:*

(1) *A is the generator of a semigroup of contractions.*
(2) *$D(A)$ is dense in X and A is m-dissipative.*
(3) *$-A$ is maximal monotone, that is, $-A$ is monotone and maximal, i.e. there is no operator $B : D(B) \subset X \mapsto X$ satisfying $D(B) \supsetneq D(A)$ and $Bu = Au$ for all $u \in D(A)$.*

See Brézis (1973) and Chicone and Latushkin (1999) for the proofs of these results.

Of course, if $B \in \mathcal{L}(X)$ and $A = -\alpha \mathrm{Id} + B$ is m-dissipative, the semigroup of contractions generated by A is given by

$$S(t) = \mathrm{e}^{t(-\alpha \mathrm{Id}+B)} = \mathrm{e}^{-\alpha t}\,\mathrm{e}^{tB} \quad \forall t \geq 0;$$

see Note 11.5.1. In other words, the unique solution to (11.20) in this case is

$$\xi(t) = \mathrm{e}^{-\alpha t}\,\mathrm{e}^{tB} x_0 \quad \forall t \geq 0.$$

Let us consider again Example 11.3. Thanks to Theorem 11.13, for every $u_0 \in C^0([0,1])$, there exists exactly one function

$$u : [0,+\infty) \mapsto C^0([0,1])$$

with the following properties:

- $u \in C^0([0,+\infty); C^0([0,1])) \cap C^1((0,+\infty); C^0([0,1]))$.
- $u(t) \in D(A)$ and $u'(t) = Au(t)$ for all $t \geq 0$; in particular, one has $u(t)|_{x=0} = 0$ for all t.
- $u(0) = u_0$.

Consequently, we have solved (in the sense of semigroup theory) the following problem:

$$\begin{cases} u_t + u_x + c(x)u = 0, & (x,t) \in (0,1) \times (0,+\infty) \\ u(0,t) = 0, & t \in (0,+\infty) \\ u(x,0) = u_0(x), & x \in (0,1). \end{cases}$$

Despite being elementary, this example illustrates the utility of the theory of semigroups in the context of PDEs.

Now, let us consider the non-homogeneous problem

$$\begin{cases} x' = Ax + b(t) \\ x(0) = x_0, \end{cases} \tag{11.21}$$

where (again) X is a Banach space, $x_0 \in X$, $0 < T < +\infty$, and $b : [0,T] \mapsto X$ is a continuous function.

It is assumed that $A : D(A) \subset X \mapsto X$ is an m-dissipative linear operator with dense domain and we denote by $\{S(t)\}$ the contractions semigroup generated by A.

By analogy with (5.17) and (11.18), let us introduce the function $\xi : [0,T] \mapsto X$, with

$$\xi(t) = S(t) x_0 + \int_0^t S(t-s)\,b(s)\,\mathrm{d}s \quad \forall t \in [0,T]. \tag{11.22}$$

It will be said that ξ is the *weak solution* to (11.21).

It can be proved that, when b satisfies appropriate regularity properties, the formula (11.22) furnishes a true solution to (11.21). More precisely, one has the following:

Theorem 11.14. *Under the previous conditions, let us assume that* $b :$ $[0, T] \mapsto X$ *is continuously differentiable. Then, the weak solution* ξ *to* (11.21) *is the unique function satisfying the following:*

(1) $\xi \in C^0([0, T]; X) \cap C^1((0, T); X)$.
(2) $\xi(t) \in D(A)$ *and* $\xi'(t) = A\xi(t) + b(t)$ *for all* $t \in (0, T)$.
(3) $\xi(0) = x_0$.

When this happens, it is customary to say that ξ *is a strong solution to* (11.21).

Of course, this result can be applied to the particular problem considered in Example 11.3. Thus, under appropriate conditions on f, we can solve the non-homogeneous problem

$$\begin{cases} u_t + u_x + c(x)u = f(x, t), & (x, t) \in (0, 1) \times (0, +\infty) \\ u(0, t) = 0, & t \in (0, +\infty) \\ u(x, 0) = u_0(x), & x \in (0, 1). \end{cases}$$

11.6   More Results on Boundary-Value Problems

11.6.1   *A nonlinear boundary-value problem*

Motivated by applications in physics and other sciences, there exist a large variety of boundary-value problems for nonlinear ODSs.

They have led in turn to a lot of theoretical and numerical techniques whose description is impossible to summarize here.

See for instance Graef *et al.* (2018b) and Korman (2019) for some of them.

By way of illustration, let us consider an example.

Example 11.4. We will consider a nonlinear second-order ODE complemented with linear boundary conditions. More precisely, let $f : [\alpha, \beta] \times \mathbb{R} \mapsto \mathbb{R}$ be a continuously differentiable function and let us consider the problem

$$\begin{cases} y'' = f(x, y), & x \in [\alpha, \beta] \\ y(\alpha) = A, & y(\beta) = B, \end{cases} \tag{11.23}$$

where $A, B \in \mathbb{R}$.

Let us assume that f satisfies the *sublinear growth condition*

$$|f(x,y)| \leq C_1 + C_2|y| \quad \forall(x,y) \in [\alpha, \beta] \times \mathbb{R} \qquad (11.24)$$

(where the $C_i \in \mathbb{R}$) and the *monotonicity condition*

$$f(x, \cdot) \quad \text{is non-decreasing } \forall x \in [\alpha, \beta]. \qquad (11.25)$$

Under these assumptions, it can be proved that (11.23) possesses exactly one solution for every $A, B \in \mathbb{R}$.

This result can be proved in several ways.

As far as existence is concerned, we can (for instance) rewrite (11.23) as a fixed-point equation in the Banach space $C^0([\alpha, \beta])$ and then prove that this equation has at least one solution.

To this purpose, a useful strategy is to check that the hypotheses of Theorem 11.4 are satisfied.

Of course, the most delicate point is to prove a uniform estimate in $C^0([\alpha, \beta])$ of all possible solutions to

$$\begin{cases} y'' = \lambda f(x,y), & x \in [\alpha, \beta] \\ y(\alpha) = A, & y(\beta) = B, \end{cases}$$

where $\lambda \in [0, 1]$.

This can be deduced from (11.24) and (11.25) after some appropriate calculations. The details are left to the reader. $\qquad \square$

11.6.2 *Green's operators*

There are many reasons that make the study of Green's operators interesting. In the conditions of Remark 6.1 and keeping the same notation, let us set

$$\mathcal{G}(b) = \varphi_b \quad \forall b \in C^0(I; \mathbb{R}^N).$$

Recall that φ_b is by definition the unique solution to (6.6).

If a fundamental matrix is known for the associated homogeneous linear ODS, then Green's operator can be made explicit.

More precisely, the following result holds:

Theorem 11.15. *Assume that the homogeneous linear boundary-value problem* (6.2) *only possesses the trivial solution. Let F be a fundamental* *matrix of the ODS $y' = A(x)y$ and let us set*

$$J(s) := -(BF(\alpha) + CF(\beta))^{-1}CF(\beta)F(s)^{-1} \quad \forall s \in I$$

(note that $BF(\alpha) + CF(\beta)$ is non-singular) and

$$G(x,s) := \begin{cases} F(x)\,J(s) & \text{if } x \leq s \\ F(x)\,(J(s) + F(s)^{-1}) & \text{otherwise.} \end{cases}$$

Then, the corresponding Green operator $\mathcal{G}$ is given as follows:

$$\mathcal{G}(b) = \int_\alpha^\beta G(x,s)\,b(s)\,ds \quad \forall x \in I, \quad \forall b \in C^0(I;\mathbb{R}^N). \tag{11.26}$$

In the particular case of a second-order linear ODE, there is a still simpler way to indicate the way $\mathcal{G}$ operates.

Thus, let us consider (for example) the boundary-value problem

$$\begin{cases} y'' = a_0(x)\,y + b(x), \quad x \in [\alpha, \beta] \\ b_0 y(\alpha) + b_1 y'(\alpha) = c_0 y(\beta) + c_1 y'(\beta) = 0, \end{cases} \tag{11.27}$$

where $a_0, b \in C^0(I)$ and $b_i, c_i \in \mathbb{R}$, with $|b_0| + |b_1| > 0$ and $|c_0| + |c_1| > 0$. Of course, this is a particular case of (6.11) and we have the following:

Theorem 11.16. *Let us assume that the homogeneous boundary-value problem associated to (11.27) only possesses the trivial solution. Then:*

(1) *There exist functions $\varphi_1, \varphi_2 \in C^2(I)$ such that*

$$\varphi_i'' = a_0\varphi_i \quad \text{and} \quad \varphi_1\varphi_2' - \varphi_1'\varphi_2 \neq 0 \ \text{in } I,$$

$$b_0\varphi_1(\alpha) + b_1\varphi_1'(\alpha) = c_0\varphi_2(\beta) + c_1\varphi_2'(\beta) = 0.$$

(2) *Let us set*

$$g(x,s) = \begin{cases} \varphi_1(x)\varphi_2(s) & \text{if } x \leq s \\ \varphi_1(s)\varphi_2(x) & \text{otherwise.} \end{cases}$$

For every $b \in C^0(I)$, the unique solution to (11.27) is the function φ_b, with

$$\varphi_b(x) = \int_\alpha^\beta g(x,s)b(s)\,ds \quad \forall x \in I.$$

The proofs of these results are not difficult. It suffices to check that the indicated functions solve the corresponding boundary-value problems.

11.6.3 *Sturm–Liouville problems*

Let us consider the following

Sturm–Liouville problem:
Find pairs $(\lambda, \varphi) \in \mathbb{R} \times C^2(I)$ such that φ is a nontrivial solution to the boundary-value problem

$$\begin{cases} -y'' + a_0(x)\, y = \lambda y, & x \in [\alpha, \beta] \\ b_0\, y(\alpha) + b_1\, y'(\alpha) = c_0\, y(\beta) + c_1\, y'(\beta) = 0. \end{cases} \tag{11.28}$$

Here, we will assume again that $a_0 \in C^0(I)$ and $b_i, c_i \in \mathbb{R}$, with $|b_0| + |b_1| > 0$ and $|c_0| + |c_1| > 0$.

In order to solve adequately (11.28), it is convenient to work in the Hilbert space $L^2(I)$ of the (classes of) measurable and square-integrable functions; recall the definition and basic properties given in Section 10.1.

If (λ, φ) is a solution to (11.28), it is usually said that λ is an *eigenvalue* of the differential operator

$$-\frac{\mathrm{d}^2}{\mathrm{d}x^2} + a_0(x) \tag{11.29}$$

for the boundary conditions

$$b_0\, y(\alpha) + b_1\, y'(\alpha) = c_0\, y(\beta) + c_1\, y'(\beta) = 0 \tag{11.30}$$

and, also, that φ is an associated eigenfunction.

It can be proved that the eigenvalues of (11.29)–(11.30) are *simple;* that is, if (λ, φ) and (λ, ψ) solve the previous Sturm–Liouville problem, then φ and ψ are necessarily linearly dependent.

It can also be proved that the eigenfunctions associated to different eigenvalues are *orthogonal* in the usual sense of the scalar product in $L^2(I)$; in other words, if (λ, φ) and (μ, ψ) solve the Sturm–Liouville problem and $\lambda \neq \mu$, then

$$(\varphi, \psi)_{L^2} := \int_\alpha^\beta \varphi(x)\, \psi(x)\, \mathrm{d}x = 0.$$

The most important result that can be established in this context is the following:

Theorem 11.17. *Let us denote by Λ the set of eigenvalues of the previous Sturm–Liouville problem. Then:*

(1) *Λ is a countable infinite set. The points in Λ can be written as the terms of a strictly increasing sequence $\{\lambda_n\}$, with*

$$\lim_{n \to \infty} \lambda_n = +\infty.$$

(2) *For every $n \geq 1$, let φ_n be an eigenfunction associated to λ_n, with*

$$\|\varphi_n\|_{L^2} := \left(\int_\alpha^\beta |\varphi(x)|^2 \, dx \right)^{1/2} = 1.$$

Then, $\{\varphi_n\}$ is a Hilbertian basis of $L^2(I)$. That is, it is an orthonormal family in $L^2(I)$ such that, for every $f \in L^2(I)$, one has

$$f = \sum_{n \geq 1} (f, \varphi_n)_{L^2} \, \varphi_n$$

in the sense of the convergence in $L^2(I)$.

As a consequence of this result, we can find expressions of the solutions to a large family of boundary-value problems for second-order ODEs.

Thus, let us assume for simplicity that the homogeneous problem

$$\begin{cases} -y'' + a_0(x)\, y = 0, & x \in [\alpha, \beta] \\ b_0\, y(\alpha) + b_1\, y'(\alpha) = c_0\, y(\beta) + c_1\, y'(\beta) = 0 \end{cases}$$

only possesses the trivial solution.

Then, the corresponding eigenvalues satisfy $\lambda_n \neq 0$ for all n and, for every $b \in C^0(I)$, the unique solution to

$$\begin{cases} -y'' + a_0(x)\, y = b(x), & x \in [\alpha, \beta] \\ b_0\, y(\alpha) + b_1\, y'(\alpha) = c_0\, y(\beta) + c_1\, y'(\beta) = 0 \end{cases}$$

is given by

$$\varphi_b = \sum_{n \geq 1} \frac{1}{\lambda_n} \, (b, \varphi_n)_{L^2} \, \varphi_n \,.$$

These results rely on the spectral theory of compact operators in Hilbert spaces. This can be found for instance in Brézis (1983), together with many other applications to ODE and PDE problems (see also Brézis (2011)).

11.7   Regularity of the Solutions to an ODS

11.7.1   *Regularity and boundary-value problems*

The concepts and results in Chapter 7 can be applied to the solutions to boundary-value problems. For example, let us set $I = [\alpha, \beta]$ and let us consider the problem

$$\begin{cases} y'' = a_0(x)\,y + b(x), & x \in [\alpha, \beta] \\ b_0 y(\alpha) + b_1 y'(\alpha) = h_1, & c_0 y(\beta) + c_1 y'(\beta) = h_2, \end{cases} \tag{11.31}$$

where $a_0, b \in C^0(I)$ and $b_i, c_i, h_i \in \mathbb{R}$, with $|b_0| + |b_1| > 0$ and $|c_0| + |c_1| > 0$. The following result holds:

If $m \geq 2$, $a_0, b \in C^m(I)$ and $\varphi : I \mapsto \mathbb{R}$ solves (11.31), then $\varphi \in C^{m+2}(I)$.

On the other hand, the results dealing with continuous and/or differentiable dependence with respect to parameters have a lot of possible extensions to infinite dimensions.

For instance, for every $u \in C^0(I)$, we can consider the problem

$$\begin{cases} y'' = f(x, y) + u(x), & x \in [\alpha, \beta] \\ b_0 y(\alpha) + b_1 y'(\alpha) = h_1, & c_0 y(\beta) + c_1 y'(\beta) = h_2, \end{cases} \tag{11.32}$$

where we assume that $f : I \times \mathbb{R} \mapsto \mathbb{R}$ is continuously differentiable and satisfies (11.24) and (11.25).

It is possible to prove that, for every $u \in C^0(I)$, there exists exactly one solution φ_u to (11.32). Consequently, the mapping $u \mapsto \varphi_u$ is well defined from $C^0(I)$ into $C^2(I)$.

The following result holds:

Theorem 11.18. *The mapping $u \mapsto \varphi_u$ is continuously differentiable. Its derivative at u in the direction v is the unique solution to the linear boundary-value problem*

$$\begin{cases} z'' = \dfrac{\partial f}{\partial y}(x, \varphi_u(x))\, z + v(x), & x \in [\alpha, \beta] \\ b_0 z(\alpha) + b_1 z'(\alpha) = c_0 z(\beta) + c_1 z'(\beta) = 0. \end{cases}$$

It can be easily understood that these regularity properties are very relevant when considering *control problems* for nonlinear ODSs. Recall the related results in Chapter 10; for more results, see for example Fattorini (1999); de Guzmán (1975); Li and Yong (1995); Luenberger (1969) and Levine (2005).

11.8   Stability and Attractivity of the Solutions

11.8.1   The case of a non-autonomous ODS

What has been done in Chapter 8 can be adapted and generalized to the case of a non-autonomous ODS

$$y' = f(t, y) \tag{11.33}$$

with $f : \mathbb{R} \times B(0; R) \mapsto \mathbb{R}^N$ of class C^1 (for example) and such that

$$f(t, 0) = 0 \quad \forall t \in \mathbb{R}.$$

Indeed, it makes sense to speak of the uniform stability and attractivity of the trivial solution φ_0.

More precisely, it will be said that φ_0 is *stable* if, for every $t_0 \in \mathbb{R}$ and every $\varepsilon > 0$, there exists $\delta > 0$ such that, if $y_0 \in \mathbb{R}^N$ and $|y_0| \leq \delta$, then necessarily $I(t_0, y_0) \supset [t_0, +\infty)$ and

$$|\varphi(t; t_0, y_0)| \leq \varepsilon \quad \forall t \geq t_0.$$

It will be said that φ_0 is *uniformly stable* if we have this property and the δ can be chosen independently of t_0.

Here, for every $(t_0, y_0) \in \mathbb{R} \times B(0; R)$, $I(t_0, y_0)$ is the interval of definition of the maximal solution to the CP corresponding to (t_0, y_0). Of course, $\varphi = \varphi(t; t_0, y_0)$ is the maximal solution to (11.33) expressed in terms of the initial conditions, that is, the so-called general maximal solution to (11.33).

Obviously, if the ODS is autonomous, stability and uniform stability are equivalent concepts. But this is not true in general; for example, it is easy to check that the trivial solution to the ODE

$$y' = -ty$$

is stable, but not uniformly stable.

In the case of the linear ODS

$$y' = A(t)y, \tag{11.34}$$

where $A : \mathbb{R} \mapsto \mathcal{L}(\mathbb{R}^N)$ is continuous, it is possible to determine the stability properties of the trivial solution by observing the behavior of a fundamental matrix.

Thus, if F is a fundamental matrix, the trivial solution to (11.34) is stable (resp. uniformly stable) if and only if, for every $t_0 \in \mathbb{R}$ there exists a constant $C(t_0) > 0$ such that $\|F(t)\|_S \leq C(t_0)$ for all $t \geq t_0$ (resp. if and only if there exist a constant $C > 0$ such that $\|F(t)F(t_0)^{-1}\|_S \leq C$ for all t_0 and t with $t \geq t_0$).

It is possible to extend to the framework of non-autonomous ODSs the stability results by comparison and at first approximation presented in Section 8.3.

It is also possible to adapt Liapunov's method and determine sufficient conditions for stability and for instability; for more details, see for example [Rouché and Mawhin (1973)].

11.8.2 Global stability, cyclic orbits, and Poincaré–Bendixson's theorem

The stability of the solutions to autonomous ODSs like (8.1) can be approached from another viewpoint. Thus, it makes sense to introduce the concept of *orbit* corresponding to a datum y_0 and analyze its behavior as $t \to +\infty$.

By definition, for any given $y_0 \in B(0; R)$, the associated orbit is the set

$$\gamma(y_0) = \{\psi(t; y_0) : t \in J(y_0)\},$$

where $J(y_0)$ is the interval of definition of the maximal solution to the CP

$$\begin{cases} y' = f(y) \\ y(0) = y_0. \end{cases} \tag{11.35}$$

Obviously, if the usual conditions of existence and uniqueness of solution are satisfied, for every $y_0 \in B(0; R)$ the orbit containing y_0 is unique. We can also write that

$$\gamma(y_0) = \gamma_+(y_0) \cup \gamma_-(y_0),$$

where $\gamma_+(y_0) = \{\psi(t; y_0) : t \in J(y_0), \ t \geq t_0\}$ and a similar definition holds for $\gamma_-(y_0)$.

The maximal solution to (11.35) satisfying $y(0) = y_0$ is constant if and only if $J(y_0) = \mathbb{R}$ and $\gamma(y_0)$ is degenerate, i.e. $\gamma(y_0) = \{y_0\}$. In turn, this is equivalent to having that y_0 is a *critical point* of f, that is, $f(y_0) = 0$.

On the other hand, the orbit $\gamma(y_0)$ is said to be *cyclic* if $J(y_0) = \mathbb{R}$ and there exists $T > 0$ such that

$$\psi(t + T; y_0) = \psi(t; y_0) \quad \forall t \in \mathbb{R}.$$

Obviously, $\gamma(y_0)$ is cyclic if and only if $\psi(\cdot; y_0)$ is periodic.

When the ODS is plane, that is, $N = 2$, the critical points and cyclic orbits play a special role. In this context, the most relevant result is the so-called *Poincaré–Bendixson Theorem*.

Before stating this result, let us give several definitions.

Definition 11.2. Let us consider the autonomous system (8.1), where $N = 2$. Let $y_0 \in B(0; R) \subset \mathbb{R}^2$ be given and let us denote by $\psi(\cdot; y_0)$ the maximal solution to the associated CP (11.35). It will be said that $z \in \mathbb{R}^2$ is a (positive) limit point associated to y_0 if there exist $t_1, t_2, \ldots$ with the $t_n \in J(y_0)$ such that

$$t_n \to \sup J(y_0) \quad \text{and} \quad \psi(t_n; y_0) \to z.$$

The set of limit points associated to y_0 will be henceforth denoted by $\Lambda_+(y_0)$. Finally, for any non-empty set $G \subset B(0; R)$, we will set

$$\Lambda_+(G) := \bigcup_{y_0 \in G} \Lambda_+(y_0).$$

Of course, if $y_0 \in B(0; R)$ and $\beta := \sup J(y_0) < +\infty$, we have

$$\Lambda_+(y_0) = \{z \in \mathbb{R}^2 : (\beta, z) \in T^+(\psi(\cdot; y_0))\},$$

where $T^+(\psi(\cdot; y_0))$ is the terminal to the right of $\psi(\cdot; y_0)$.

Definition 11.3. We call no-contact segment any closed segment $L \subset B(0; R)$ such that, at any $y \in L$, $f(y)$ is neither 0 nor parallel to L.

Definition 11.4. Let $y_0, y_* \in B(0; R)$ be given and let us assume that $\gamma(y_*)$ is cyclic. It will be said that $\gamma_+(y_0)$ approaches spirally $\gamma(y_*)$ if, for any $t \in \mathbb{R}$ and any no-contact-segment L whose interior contains $\psi(t; y_*)$, one has

$$L \cap \gamma_+(y_0) = \{\psi(t_n; y_0) : n \geq 1\},$$

for some t_n satisfying the following:

(1) $\psi(t_n; y_0) \to \psi(t; y_*)$ as $n \to +\infty$.
(2) For every n, $\psi(t_{n+1}; y_0)$ is located on L between $\psi(t_n; y_0)$ and $\psi(t_{n+2}; y_0)$.

The announced result is the following:

Theorem 11.19 (Poincaré–Bendixson). *Let $y_0 \in B(0; R)$ be given. Let us assume that there exists a compact set $K \subset B(0; R)$ such that $\gamma_+(y_0) \subset K$. Let us also assume that $\Lambda_+(y_0)$ does not contain any critical point. Then:*

(1) $\Lambda_+(y_0)$ *coincides with the set of points of a non-degenerate cyclic orbit.*
(2) *Either* $\gamma(y_0)$ *is a cyclic orbit (and* $\gamma_+(y_0) = \Lambda_+(y_0)$*), or* $\gamma_+(y_0)$ *approaches spirally* $\Lambda_+(y_0)$*. In the second case, it is usual to say that* $\Lambda_+(y_0)$ *is a limit cycle.*

In the proof of this result, it is used in an essential way that $N = 2$; more precisely, *Jordan's Closed Curve Theorem* is required. In fact, the result is false for $N \geq 3$; for details, see for example Hurewicz (1966).

Example 11.5. Let us consider the autonomous plane ODS

$$\begin{cases} y_1' = -2y_2 + y_1(4 - y_1^2 - y_2^2) \\ y_2' = 2y_1 + y_2(4 - y_1^2 - y_2^2). \end{cases} \quad (11.36)$$

This system possesses a unique critical point (the origin) and exactly one nontrivial cyclic orbit, given by

$$\gamma = \{(y_1, y_2) \in \mathbb{R}^2 : y_1^2 + y_2^2 = 4\},$$

that is a limit cycle. Actually, all orbits starting from a point $y_0 \neq 0$ converge spirally to γ.

To prove these assertions, it suffices to perform the change of variables

$$y_1 = \rho \cos\theta, \quad y_2 = \rho \sin\theta,$$

rewrite the system in the variables ρ and θ, compute explicitly the solutions, and apply Theorem 11.19. $\qquad \square$

11.9 Partial Differential Equations of the First Order

11.9.1 *Motivations and applications*

In this note, we will see how first-order PDEs can be used to model two phenomena with very different origins: the car traffic flux in a road and the behavior of an inviscid fluid.

11.9.1.1 *Traffic problems*

In these problems, the functions depend on two variables, x and t, respectively, related to the distance to origin on the road and the time instant. It will be assumed that (x, t) belongs to $\mathbb{R}_+ \times (0, T)$.

The relevant variables are the *car density* ρ, i.e. the number of vehicles per length unit and the *flux* q, that is, the number of vehicles at a road point per time unit. We will assume that ρ and q are regular functions in $\mathbb{R}_+ \times (0, T)$ and, for simplicity, we will denote their partial derivatives with subindices.

The number of cars must obey a classical conservation law. More precisely, for any spatial interval $[x_1, x_2]$, we must have

$$\frac{\mathrm{d}}{\mathrm{d}t} \int_{x_1}^{x_2} \rho(x, t) \,\mathrm{d}x = q(x_1, t) - q(x_2, t),$$

This identity says that the time variation of the total amount of cars in $[x_1, x_2]$ must coincide with the amount of incoming cars through $x = x_1$ minus the amount of outcoming vehicles through $x = x_2$. If q is regular enough, this can be written in the form

$$\frac{\mathrm{d}}{\mathrm{d}t} \int_{x_1}^{x_2} \rho(x, t) \,\mathrm{d}x + \int_{x_1}^{x_2} q_x(x, t) \,\mathrm{d}x = 0$$

and, since $[x_1, x_2]$ is arbitrary, we deduce that

$$\rho_t + q_x = 0.$$

In order to find a true PDE for ρ, we have to incorporate an additional (constitutive) law relating q to ρ.

We will only consider a very simple model, where $q = Q(\rho)$ and Q is a given (nonlinear) function. This means that, at any (x, t), the speed of a car depends essentially on the road congestion level at (x, t).

This leads to the PDE

$$\rho_t + Q'(\rho)\rho_x = 0, \quad (x, t) \in \mathbb{R}_+ \times (0, T). \tag{11.37}$$

A relatively usual function Q is given by

$$Q(\rho) = \begin{cases} \rho(1 - \rho), & \text{if } \rho \in (0, 1) \\ 0, & \text{otherwise.} \end{cases}$$

It is natural to complement (11.37) with boundary conditions (for $x = 0$) and initial conditions (for $t = 0$):

$$\begin{cases} \rho(0, t) = \rho_*(t), & t \in (0, T) \\ \rho(x, 0) = \rho_0(x), & x \in \mathbb{R}_+. \end{cases} \tag{11.38}$$

For more details, see for example Zachmanoglou and Thoe (1986).

11.9.1.2   *Inviscid fluids (incompressible or not)*

Let $D \subset \mathbb{R}^3$ be a non-empty bounded connected open set. It is supposed that the particles of an *ideal* fluid fill D during the time interval $(0, T)$.

In fluid mechanics, the *state* of a fluid is completely determined by its *mass density* ρ and the *velocity field* **u**. In both cases, we are considering functions of the variables $\mathbf{x} = (x_1, x_2, x_3)$ and t (ρ is a scalar function and $\mathbf{u} = (u_1, u_2, u_3)$ is a vector function).

The *conservation laws* of mass and linear momentum establish that ρ and $\mathbf{u}$ must satisfy the identities

$$\begin{cases} \rho_t + \nabla \cdot (\rho \mathbf{u}) = 0 \\ \rho(\mathbf{u}_t + (\mathbf{u} \cdot \nabla)\mathbf{u}) + \nabla p = 0, \end{cases} \tag{11.39}$$

where p is a new variable (the pressure). Here, we have used the notation

$$\nabla \cdot (\rho \mathbf{u}) := \sum_{i=1}^{3} \frac{\partial}{\partial x_i}(\rho u_i), \quad \nabla p := \left(\frac{\partial p}{\partial x_1}, \frac{\partial p}{\partial x_2}, \frac{\partial p}{\partial x_3} \right)$$

$$(\mathbf{u} \cdot \nabla)\mathbf{u} := (\mathbf{u} \cdot \nabla u_1, \mathbf{u} \cdot \nabla u_2, \mathbf{u} \cdot \nabla u_3) \quad \text{with} \quad \mathbf{u} \cdot \nabla u_i := \sum_{j=1}^{3} u_j \frac{\partial u_i}{\partial x_j}.$$

The equations in (11.39) are deduced as follows.

Assume that ρ and $\mathbf{u}$ are regular functions. Let $W \subset D$ be a fixed non-empty open set and let us assume that a set of fluid particles fills W at time $t = 0$. These particles describe trajectories $X = X(t, \mathbf{x})$, defined by

$$X_t = \mathbf{u}(X, t), \quad X|_{t=0} = \mathbf{x}.$$

At any time $t \in (0, T)$, the same particles initially located in W fill the set

$$W_t = \{ X(t, \mathbf{x}) : \mathbf{x} \in W \}.$$

Consequently, since the total amount of mass corresponding to these particles is constant, we must have

$$\frac{\mathrm{d}}{\mathrm{d}t} \int_{W_t} \rho(\mathbf{x}, t) \, \mathrm{d}\mathbf{x} = 0 \quad \forall t \in (0, T).$$

It can be proved that this can also be written in the form

$$\int_{W_t} (\rho_t + \nabla \cdot (\rho \mathbf{u})) (\mathbf{x}, t) \, \mathrm{d}\mathbf{x} = 0.$$

Therefore, taking into account that W is arbitrary, we get the first PDE in (11.39).

On the other hand, *Newton's Second Law* applied to the set of particles that were initially located inside W yields the identity

$$\frac{\mathrm{d}}{\mathrm{d}t} \int_{W_t} (\rho \mathbf{u})(\mathbf{x}, t) \, \mathrm{d}\mathbf{x} = \mathbf{F}(W_t) \quad \forall t \in (0, T),$$

where $\mathbf{F}(W_t)$ is the resultant of the forces applied at time t to these particles.

It can be proved that

$$\frac{\mathrm{d}}{\mathrm{d}t} \int_{W_t} (\rho \mathbf{u})(\mathbf{x}, t)\, \mathrm{d}\mathbf{x} = \int_{W_t} [\rho(\mathbf{u}_t + (\mathbf{u} \cdot \nabla)\mathbf{u})]\, (\mathbf{x}, t)\, \mathrm{d}\mathbf{x}.$$

On the other hand, it is said that the fluid is *ideal* if there exists $p = p(\mathbf{x}, t)$ (the pressure) such that

$$\mathbf{F}(W_t) = -\int_{W_t} \nabla p(\mathbf{x}, t)\, \mathrm{d}\mathbf{x}$$

for all W and all t. Thus, we deduce in that case that

$$\int_{W_t} (\rho(\mathbf{u}_t + (\mathbf{u} \cdot \nabla)\mathbf{u}) + \nabla p)\, (\mathbf{x}, t)\, \mathrm{d}\mathbf{x} = 0 \quad \forall t \in (0, T)$$

and, since W is arbitrary, the second PDE in (11.39) is also found.

In order to get a system with so many equations as unknowns, we need some additional law(s). Indeed, note that in (11.39) there are $N + 1$ scalar equations and $N + 2$ unknowns: $\rho, u_1 \ldots, u_N$ and p. Let us see how to proceed in two particular cases.

A first possibility is to impose *incompressibility*, that is,

$$\nabla \cdot \mathbf{u} := \sum_{i=1}^{3} \frac{\partial u_i}{\partial x_i} = 0.$$

This leads to the following so-called incompressible Euler equations:

$$\begin{cases} \rho_t + \nabla \cdot (\rho \mathbf{u}) = 0 \\ \rho(\mathbf{u}_t + (\mathbf{u} \cdot \nabla)\mathbf{u}) + \nabla p = 0 \\ \nabla \cdot \mathbf{u} = 0. \end{cases} \qquad (11.40)$$

Another completely different way to complete the information is to assume that the fluid is *compressible and barotropic*. This means that $p = P(\rho)$ for some given function P. We arrive in this way to the system

$$\begin{cases} \rho_t + \nabla \cdot (\rho \mathbf{u}) = 0 \\ \rho(\mathbf{u}_t + (\mathbf{u} \cdot \nabla)\mathbf{u}) + \nabla P(\rho) = 0. \end{cases} \qquad (11.41)$$

Once more, it is natural to complement these PDE systems with boundary-value conditions (on the lateral boundary $\partial D \times (0, T)$ or a part of it) and initial conditions (at $t = 0$) for ρ and $\mathbf{u}$. For example, for (11.40), the following conditions are appropriate:

$$\begin{cases} \mathbf{u} \cdot \mathbf{n} = U_n, & (\mathbf{x}, t) \in \partial D \times (0, T) \\ \rho = \rho_*, & (\mathbf{x}, t) \in \partial_- D \times (0, T) \\ \mathbf{u}(\mathbf{x}, 0) = \mathbf{u}_0(\mathbf{x}), & \rho(\mathbf{x}, 0) = \rho_0(\mathbf{x}), \quad x \in D. \end{cases} \qquad (11.42)$$

Here, $\mathbf{n} = \mathbf{n}(\mathbf{x})$ denotes the unit normal vector external to D. Accordingly, the first identity in (11.42) indicates that the fluid particles cannot cross the solid boundary ∂D; this is called the *slip boundary condition*.

Above, we have used $\partial_- D$ to indicate the part of ∂D where particles are supposed to *enter*, that is,

$$\partial_- D = \{\mathbf{x} \in \partial D : U_n(\mathbf{x}) < 0\}.$$

Thus, the second identity in (11.42) means that the mass density of incoming particles is prescribed.

For more details on the equations with origins in fluid mechanics, see for instance Chorin and Marsden (1990).

11.9.2 *The Cauchy problem for a nonlinear PDE*

In this note, we consider the CP

$$\text{(IVP-NL)} \qquad \begin{cases} F(x, u, Du) = 0 \\ u|_S = u_0, \end{cases}$$

where $F : U \subset \mathbb{R}^{2N+1} \mapsto \mathbb{R}$ is a given function (U is a non-empty open set), S is a hyper-surface of class C^1 in $\mathbb{R}^N$ given by

$$S = \{\, \xi(s_1, \ldots, s_{N-1}) : (s_1, \ldots, s_{N-1}) \in \mathcal{O} \,\}, \qquad (11.43)$$

where $\mathcal{O}$ is an open neighborhood of 0 in $\mathbb{R}^{N-1}$, $\xi \in C^1(\mathcal{O}; \mathbb{R}^N)$ and

$$\text{rank} \left[\frac{\partial \xi}{\partial s_1}(s) \mid \cdots \mid \frac{\partial \xi}{\partial s_{N-1}}(s) \right] = N - 1 \quad \forall s = (s_1, \ldots, s_{N-1}) \in \mathcal{O}$$

and $u_0 \in C^1(\mathcal{O})$.

We will see that, under these assumptions on the data, (IVP-NL) possesses a unique local solution (defined in a neighborhood of $a = \xi(0)$).

In other words, we will prove that there exist an open set $D \in \mathbb{R}^N$ with $a \in D$ and exactly one function $u : D \mapsto \mathbb{R}$ with $u \in C^1(D)$ such that

- $(x, u(x), Du(x)) \in U$ for all $x \in D$,
- $F(x, u(x), Du(x)) = 0$ for all $x \in D$ and
- $u(\xi(s)) = u_0(s)$ for all $s \in \mathcal{O}$ such that $\xi(s) \in D$.

The result is the following:

Theorem 11.20. *Assume that $F \in C^2(U)$. Let us set $a := \xi(0)$ and $u_a := u_0(0)$. Also, assume that $p_a \in \mathbb{R}^N$ is such that $(a, u_a, p_a) \in U$,*

$$
\begin{cases}
F(a, u_a, p_a) = 0 \\[2mm]
\displaystyle\sum_{i=1}^{N} \frac{\partial \xi_i}{\partial s_j}(0)\, p_{ai} = \frac{\partial u_0}{\partial s_j}(0), \quad 1 \leq j \leq N-1
\end{cases}
\tag{11.44}
$$

and

$$
\det\left(F_p(a, u_a, p_a) \,\Big|\, \frac{\partial \xi}{\partial s_1}(0) \,\Big|\cdots\Big|\, \frac{\partial \xi}{\partial s_{N-1}}(0) \right) \neq 0,
\tag{11.45}
$$

where the p_{ai} are the components of p_a and F_p denotes the vector whose components are the partial derivatives of F with respect to the N arguments p_i. Then, there exist an open set D containing a and a unique function $u : D \mapsto \mathbb{R}$ that solves the CP (IVP-NL) in D.

Remark 11.1. The hypothesis (11.44) indicates that the problem data are compatible. The hypothesis (11.45) is, as in Chapter 9, called a *transversality condition;* it says that the vector $F_p(a, u_a, p_a)$ is transversal to the tangent hyperplane to S at a and obviously generalizes the assumption (9.14), that is essential when the PDE is quasi-linear.   □

The proof of the existence of a solution to (IVP-NL) can be achieved by a method that generalizes the arguments in the proof of Theorem 9.1. Again, this is known as the *method of characteristics.*

Let us explain the main ideas.

Let u be a solution to (IVP-NL) of class C^2 and set $p := Du$. Then, after differentiating the identity $F(x, u(x), p(x)) \equiv 0$ with respect to x_i, we see that the ith component p_i of p satisfies

$$
\frac{\partial F}{\partial x_i} + \frac{\partial F}{\partial u} p_i + \sum_{j=1}^{N} \frac{\partial F}{\partial p_j} \frac{\partial p_i}{\partial x_j} = 0.
\tag{11.46}
$$

We have used here that

$$
\frac{\partial p_j}{\partial x_i} \equiv \frac{\partial p_i}{\partial x_j}
$$

for all i and j.

We can interpret (11.46) as a quasi-linear PDE of the first order satisfied by p_i.

Consequently, in accordance with the results in Chapter 9, it seems reasonable to construct y and the p_i starting from characteristic ODSs with the following structure:

$$\begin{cases} y' = F_p(y, v, q) \\ p_i' = -\dfrac{\partial F}{\partial x_i}(y, v, q) - \dfrac{\partial F}{\partial u}(y, v, q)\, p_i, \end{cases}$$

where v and q are, respectively, u and p written in terms of t (the independent variable in this ODS) and s (the variable used to parameterize S). On the other hand, it also seems reasonable to assume that

$$\frac{dv}{dt} = \sum_{i=1}^{N} \frac{\partial u}{\partial x_i}\frac{dy_i}{dt} = \sum_{i=1}^{N} q_i \frac{\partial F}{\partial p_i}.$$

All this leads in a rather natural way to the following characteristic ODS:

$$\begin{cases} y' = F_p(y, v, q) \\ v' = q \cdot F_p(y, v, q) \\ q' = -F_x(y, v, q) - F_u(y, v, q)\, q. \end{cases} \tag{11.47}$$

Here, F_x denotes the vector whose components are the partial derivatives of F with respect to its N first arguments x_i, and F_u denotes the partial derivative of F with respect to the $(N+1)$th argument u.

As before, we must complete (11.47) with appropriate initial conditions.

Thus, using (11.44), (11.45), and the Implicit Function Theorem, it is not difficult to prove that there exists a unique function $p_0 = p_0(s)$ of class C^1 satisfying

$$\begin{cases} F(\xi(s), u_0(s), p_0(s)) = 0 \\ \displaystyle\sum_{i=1}^{N} \frac{\partial \xi_i}{\partial s_j}(0)p_0(s) = \frac{\partial u_0}{\partial s_j}(s), \quad 1 \le j \le N-1 \end{cases}$$

in a neighborhood of the origin $\mathcal{O}' \subset \mathcal{O}$. We then impose

$$y(0) = \xi(s), \quad v(0) = u_0(s), \quad q(0) = p_0(s), \quad s \in \mathcal{O}'. \tag{11.48}$$

Accordingly, we can solve the CPs (11.47)–(11.48) for every s. This process furnishes $\bar{y}$, $\bar{v}$, and $\bar{q}$, defined and of class C^1 in a neighborhood of the origin in $\mathbb{R}^N$, with $\bar{y}(0,0) = a$.

Using these functions, it is easy to get a solution to (IVP-NL). Indeed, it suffices to invert $\bar{y}$ in a neighborhood of $(0,0)$ (this is possible in view of (11.45)), denote by ψ the local inverse of $\bar{y}$, and take

$$u(x) = \bar{v}(\psi(x)) \quad \text{in a neighborhood of } a.$$

The details of this construction and the proof of uniqueness can be found in Hartman (2002).

11.10   Control of Differential Systems

This section is devoted to complementing the information on control problems provided in Chapter 10.

First, we will recall several historical facts about the origins of control theory. Then, we will consider several examples of optimal control problems motivated by real-world phenomena. Finally, some related topics will be briefly mentioned.

11.10.1   *Historical data*

Let us recall some classical and well-known events and discoveries that have to some extent influenced the development of control theory.

We can start with a mention of the irrigation systems shaped in the ancient Mesopotamia, more than 2000 years B.C., and the Roman aqueducts, some of them built about 300 years B.C. In both cases, ingenious appropriate systems allowed the use of water at a previously regulated rate.

There were real controls, there were objectives and methods, and there were satisfactory results.

The works by Ch. Huygens[31] at the end of the 17th century on the *oscillations of the pendulum* are more modern examples of development

---

[31]Christiaan Huygens (1629–1695) was a Dutch mathematician, physicist, and astronomer. He made important contributions in optics and mechanics. He improved the design of telescopes and invented the pendulum clock, the most accurate timekeeper for almost 300 years. Apparently, he was the first to use mathematics to describe physical phenomena in a systematic way, even in the absence of observations. He was also the first to understand the difference between "model" and "parameters" (or data). Accordingly, he has been recognized as the first theoretical physicist and one of the founders of mathematical physics.

of control. The goal was to achieve a precise measurement of time and location, needed for navigation.

His results were later adapted to regulating the velocity of ancient windmills. There, the main mechanism is based on a system of balls rotating around an axis, with a velocity proportional to the velocity of the windmill blades. When the rotational velocity increases, the balls move away from the axis, acting on the wheels of the mill through appropriate mechanisms.

In turn, these ideas were adapted to the conception and operation of the *steam engine,* an invention that provoked the industrial revolution and consequently changed the world.

In this mechanism, when the velocity of the balls increases, one or several valves open to let the vapor escape. This makes the pressure diminish. When this happens, the velocity starts to go down, the balls become closer to the axis, and, as an overall result of the process, it is possible to keep the velocity close to a prescribed constant.

The steam regulating system was analyzed by G. Airy[32] and then, in 1868, by J.C. Maxwell,[33] who was even able to find some efficient control mechanisms.

This was the starting point of control engineering, which has been covering since then a lot of applications: transport systems; mechanisms for textile industry, paper production, petroleum, and steel industries; distribution systems in electrical plants; and then *amplifiers* in telephone systems; aircraft design; chemistry, etc.

A subsequent crucial step was given about 1950, when the scientific community realized that the models considered up to that moment were not accurate enough to describe the complexity of real-word phenomena.

---

[32] George Biddell Airy (1801–1892) was an English mathematician and astronomer. He contributed many results in science. As Royal Astronomer, we was responsible for establishing Greenwich as the location of the Prime Meridian.

[33] James Clerk Maxwell (1831–1879) was a Scottish mathematician. He introduced the equations that bear his name, a system of PDEs that explain that electricity, magnetism, and light are different manifestations of the same phenomenon. He is frequently viewed as the scientist who made the greatest impact on physics in the 19th century. Einstein described Maxwell's work as the "most profound and the most fruitful that physics has experienced since the time of Newton." In 1922, he was asked whether he had stood on Newton's shoulders; he replied: "No I did not. I stood on Maxwell's shoulders."

In particular, the contributions of R. Bellman[34] in the context of *dynamic programming*, R. Kalman[35] in *filtering techniques* and the algebraic approach to linear systems, and L. Pontryagin[36] with the *maximum principle* for nonlinear optimal control problems led to the foundations of modern control theory.

This was followed by an impressive progress in stochastic, robust, adaptive, etc., control methods in the 1970s and 1980s.

Nowadays, many of the control systems are computer-controlled and they consist of both digital and analog components. Thanks to the advances in informatics, the possibilities have grown exponentially.

For instance, we see that computer-aided design can be applied not only individually, to a prescribed control system, but also globally, to the optimization of the system structure, parameter identification, decision on the type of control, etc.

The rank of applications is huge and covers in practice most sciences, including economics, sociology, and politics and, of course, modern engineering.

---

[34]Richard Ernest Bellman (1920–1984) was an American mathematician. He introduced the *Dynamic Programming Principle* and the equation that bears his name, a necessary condition for optimality in discrete time optimal control problems. Bellman's ideas applied to optimal control problems governed by ODEs (and also classical variational problems) lead to the *Hamilton–Jacobi–Bellman equation,* a PDE whose solution is the *value function,* that is, the minimal cost as a function of initial data. If the Hamilton–Jacobi–Bellman equation is solved, the computed value function allows us to determine the optimal control. Bellman is also responsible for coining the expression "curse of dimensionality," to describe the problem caused by the exponential increase in computational work associated to adding dimensions to a working space. He also made contributions in biomathematics and founded the journal *Mathematical Biosciences.*
[35]Rudolf Emil Kalman (1930–2016) was a Hungarian-American electrical engineer and mathematician. He is known for his co-invention and development of the Kalman filter, a mathematical technique widely used, for instance, for navigation. The algorithm works by a two-phase process: a prediction phase, where estimates of the current state variables and their reliability are produced; and an update phase, relying on a weighted average, with the weight chosen in accordance with certainty. It has numerous technological applications: guidance, navigation, and control of vehicles; time series analysis in signal processing; robotics; etc. Kalman introduced the concepts of controllability and observability, eventually leading to control criteria and the Kalman decomposition.
[36]Lev Semenovich Pontryagin (1908–1988) was a Soviet mathematician. Although he was blind since 14, he was able to contribute with outstanding results in several fields of mathematics. In particular, he deduced what is called the *Maximum Principle* for optimal control problems, a fundamental result in this area. He also introduced the concept of *bang-bang control* and proved its optimality under certain circumstances and its usefulness.

Several examples are the systems of heating, ventilation, and air conditioning in big buildings; the *pH* control in chemical reactions; the computer-aided task optimization in automobile industries, nuclear security, defense, etc.; the control along unstable trajectories of the dynamics of aircrafts; the optimum design of space structures, optical reflectors of large dimensions, satellite communication systems, etc.; the control of *robots*, ranging from the most simple engines to the *bipeds* that simulate human locomotive abilities; the optimization of medical actions, ranging from the design of artificial organs to mechanisms for insulin supply or optimal therapy strategies in cancer, etc.

Recently, strategies related to *machine learning* or *deep learning* are being considered, specially in connection with optimal control problems involving many data.

More information can be found for instance in Norkin (2020) and Vrabie *et al.* (2013).

11.10.2 *Other more complex control problems*

In this section, we will mention some relatively involved optimal control problems, of interest for some applications. The available theoretical results are more difficult to establish than those in Chapter 10.

Many more details can be found for instance in Lee and Markus (1986) and Trélat (2005).

Let $T > 0$ be given and let α_i, a_i, b_i, B_i, μ, and k be positive constants. The following is an interesting control problem related to the evolution of HIV/AIDS virus, see Joshi (2002):

$$\begin{cases} \text{Minimize } J(v) := \dfrac{1}{2} \displaystyle\int_0^T \left(\alpha_1 |v_1|^2 + \alpha_2 |v_2|^2 \right) \mathrm{d}t - \int_0^T Y \, \mathrm{d}t \\ \text{Subject to } v = (v_1, v_2) \in \mathcal{U}_{\text{ad}}, \ (Y, V, v) \text{ satisfies } (11.50), \end{cases} \quad (11.49)$$

where

$$\mathcal{U}_{\text{ad}} := \{ v \in (L^2(0,T))^2 : a_i \le v_i \le b_i \ \text{ a.e.} \}$$

and

$$\begin{cases} Y' = s_1 - \dfrac{s_2 V}{B_1 + V} - \mu Y - kYV + v_1 Y, \quad t \in (0,T) \\ V' = \dfrac{g(1 - v_2)V}{B_2 + V} - cYV, \quad t \in (0,T) \\ Y(0) = Y_0, \quad V(0) = V_0. \end{cases} \quad (11.50)$$

Here, Y and V must be, respectively, interpreted as the densities of anti-bodies and virus units in the process.

On the other hand, v_1 and v_2 must be viewed as functions that describe the actions of an *immuno-reinforcing* and a *citotoxic* therapy, respectively. Of course, (v_1, v_2) is the control and (Y, V) is the state.

It can be proved that (11.49) possesses at least one solution $(\hat{v}_1, \hat{v}_2)$.

Furthermore, there exist functions $\hat{Z}$ and $\hat{W}$ such that $(\hat{v}_1, \hat{v}_2)$, the associated state $(\hat{Y}, \hat{V})$ and $(\hat{Z}, \hat{W})$ fulfill a coupled ODS of dimension 4, together with the equalities

$$\hat{v}_1 = \mathbb{P}_1\left(-\frac{1}{\alpha_1}\hat{Y}\hat{Z}\right), \quad \hat{v}_2 = \mathbb{P}_2\left(-\frac{g}{\alpha_2}\frac{\hat{V}\hat{W}}{B_2 + \hat{V}}\right),$$

where the $\mathbb{P}_i : \mathbb{R} \mapsto [a_i, b_i]$ are the usual orthogonal projectors.

A similar problem, taken from d'Onofrio *et al.* (2009), is found when we want to describe (and optimize) the evolution of a tumor submitted to a chemotherapy process:

$$\begin{cases} \text{Minimize } J(v) := x_1(T) + \alpha \int_0^T |v_1|^2 \, dt \\ \text{Subject to } v = (v_1, v_2) \in \mathcal{U}_{\text{ad}}, \ x \in \mathcal{X}_{\text{ad}}, \ (x, v) \text{ satisfies (11.52)}, \end{cases}$$
$$(11.51)$$

where

$$\mathcal{U}_{\text{ad}} := \{v \in (L^2(0, T))^2 : 0 \le v_i \le b_i \text{ a.e.}\},$$

$$\mathcal{X}_{\text{ad}} := \{x \in C^0([0, T])^4 : x_i \ge 0 \text{ for } 1 \le i \le 4, \ x_i \le B_i \text{ for } i = 2, 3\}$$

and

$$\begin{cases} x_1' = -ax_1 \log \dfrac{x_1}{x_2} - bx_1 v_2, \quad t \in (0, T) \\ x_2' = cx_1 - \mu x_2 - dx_1^{2/3} x_2 - cx_2 v_1 - fx_1 v_2, \quad t \in (0, T) \\ x_3' = v_1, \quad t \in (0, T) \\ x_i(0) = x_{i0}, \quad 1 \le i \le 4. \end{cases}$$
$$(11.52)$$

Again, it is assumed that we apply a double therapy, anti-angiogenic and citotoxic, respectively, determined by the controls v_1 and v_2.

The quantities $x_1(t)$, $x_2(t)$, and $x_3(t)$ must be, respectively, viewed as the values at time t of the densities of tumor cells, endothelial cells (with origin in a close blood vessel), and antibodies (furnished by the therapy).

The following is an apparently elementary problem with constraints in the control and the state:

$$\begin{cases} \text{Minimize } J(v) := \dfrac{1}{2} \displaystyle\int_0^1 |v|^2 \, dt \\ \text{Subject to } v \in \mathcal{U}_{\text{ad}}, \ x \in \mathcal{X}_{\text{ad}}, \ (x,v) \text{ satisfies } (11.54), \end{cases} \tag{11.53}$$

where

$$\mathcal{U}_{\text{ad}} := \{ v \in L^2(0,T) : a \le v_i \le b \ \text{ a.e. in } (0,T) \},$$

$$\mathcal{X}_{\text{ad}} := \{ x \in C^1([0,T]) : x'(0) = -x'(T) = 1, \ x(t) \le 1/9 \ \forall t \in [0,T] \}$$

and

$$\begin{cases} x'' = v, \ \ t \in (0,T) \\ x(0) = x(T) = 0. \end{cases} \tag{11.54}$$

In (11.53), $x = x(t)$ describes the trajectory of a particle propelled with force $v = v(t)$ and subject to some constraints during the whole time interval $(0,T)$.

Let us now consider a problem related to the production optimization in a factory, taken from Anita *et al.* (2011):

$$\begin{cases} \text{Minimize } J(v) := - \displaystyle\int_0^T e^{-at}(1-v)x \, dt \\ \text{Subject to } v \in \mathcal{U}_{\text{ad}}, \ (x,v) \text{ satisfies } (11.56), \end{cases} \tag{11.55}$$

where

$$\mathcal{U}_{\text{ad}} := \{ v \in L^2(0,T) : 0 \le v \le 1 \ \text{ a.e.} \}$$

and

$$\begin{cases} x' = \gamma v x, \ \ t \in (0,T) \\ x(0) = x_0. \end{cases} \tag{11.56}$$

In this problem, we accept that, at each time t, $x(t)$ is the production rate and $v(t)$ is the percentage dedicated to inversion. It is reasonable to assume, at least as a first step, that the production increase is proportional to the inversion. The goal is to maximize the consumption and, consequently, the gain.

It is usual to say that (11.55) is a *bilinear control problem,* in view of the way the control acts on the state in (11.56). This can also be seen, for instance, in (11.52).

The analysis of bilinear control problems is more complex. In practice, one is led to control-to-state mappings where the latter is less sensible to the action of the former.

Another bilinear optimal control problem, related to the behavior of a prey–predator eco-system and also taken from Anita *et al.* (2011), is the following:

$$\begin{cases} \text{Minimize } J(v) := x(T) + y(T) \\ \text{Subject to } v \in \mathcal{U}_{\text{ad}}, \ (x, y, v) \text{ satisfies (11.58)}, \end{cases} \tag{11.57}$$

where

$$\mathcal{U}_{\text{ad}} := \{v \in L^2(0, T) : 0 \le v \le 1 \ \text{a.e.}\}$$

and

$$\begin{cases} x' = r_1 x - \mu_1 xyv, \ t \in (0, T) \\ y' = -r_2 y + \mu_2 xyv, \ t \in (0, T) \\ x(0) = x_0, \ y(0) = y_0. \end{cases} \tag{11.58}$$

This time, $x = x(t)$ and $y = y(t)$ can be viewed as population densities of a prey and a predator species that interact along the time interval $(0, T)$. On the other hand, v must be interpreted as a *catalyst,* that is, an activating control for the interaction.

Bilinear controls are frequent in problems with origins in biology. A reason is that it is natural to modify the evolution of a set of individuals through an action whose effects are proportional to their number.

11.10.3 *Additional comments on control problems*

In view of the large amount of applications, the control of differential equations and systems has been the objective of many works, especially since the 1970s.

Lots of interesting questions that generalize, modify, or extend optimal control and controllability have been formulated (and solved in many cases).

Among them, let us mention briefly the following:

- Time optimal control problems.
  Here, the goal is to find a control that (for instance) minimizes the time of arrival of the state to a prescribed target. An example is as follows:

$$\begin{cases} \text{Minimize } T \\ \text{Subject to } T \geq 0, \ v \in \mathcal{U}_{\mathrm{ad}}(T), \ (11.59) \text{ is satisfied}, \ y(T) = y_d \,, \end{cases}$$

where $\mathcal{U}_{\mathrm{ad}}(T)$ is a non-empty closed convex set of $L^2(0,T)$ for every T, (11.59) is the (generalized) Cauchy problem

$$\begin{cases} y' = g(t,y) + v(t), \quad t \in (0,T) \\ y(0) = y_0 \end{cases} \tag{11.59}$$

and the function $g : \mathbb{R}^2 \mapsto \mathbb{R}$ and the real numbers y_0 and y_d are given. More details on time optimal control problems can be found in Hermes and LaSalle (1969), Lu *et al.* (2017) and Wang *et al.* (2018).

- Multi-objective optimal control problems.
  Sometimes it can be interesting to construct controls able to achieve several tasks simultaneously with a reasonable performance.
  For example, we may be interested in a control that furnishes small values of two independent cost functionals. In these cases, we have usually to look for equilibria.
  A particular bi-objective optimal control problem is the following:

$$\begin{cases} \text{Find a Pareto equilibrium for } J_1 \text{ and } J_2 \\ \text{Subject to } v \in \mathcal{U}_{\mathrm{ad}}, \ (11.60) \text{ is satisfied}, \end{cases}$$

where $\mathcal{U}_{\mathrm{ad}} \subset L^2(0,T)^2$ is non-empty closed and convex, (11.60) is the boundary-value problem

$$\begin{cases} y' = \begin{bmatrix} a_{11} & a_{12} \\ a_{21} & a_{22} \end{bmatrix} y + \begin{bmatrix} v_1(t) \\ v_2(t) \end{bmatrix}, \quad t \in (0,T) \\ y_1(0) = y_1(T), \quad y_2(0) = y_2(T) \end{cases} \tag{11.60}$$

and

$$J_i(v) = \frac{\alpha}{2} |y_i(T) - z^i|^2 + \frac{\mu}{2} \int_0^T |v_i(t)|^2 \, dt, \quad i = 1, 2.$$

Here, the $a_{ij}, z^i \in \mathbb{R}$ and $\alpha, \mu > 0$.

By a *Pareto equilibrium* we mean a control $\hat{v} \in \mathcal{U}_{\mathrm{ad}}$ such that there is no other $v \in \mathcal{U}_{\mathrm{ad}}$ satisfying

$$J_i(v) \leq J_i(\hat{v}) \quad \text{for} \quad i = 1, 2, \text{ with strict inequality for some } i.$$

For the analysis and computation of solutions to multi-objective optimal control problems, see for instance Vasile (2019).

References

H. Amann: *Ordinary Differential Equations: An Introduction to Nonlinear Analysis*. Walter de Gruyter, New York, 1990.

J.M. Amigó: La ecuación de Riccati, *Bol. Soc. Esp. Mat. Apl.*, 45 (2008), 147–170.

N. Andrei: *Nonlinear Conjugate Gradient Methods for Unconstrained Optimization*, Springer Optimization and Its Applications, 158. Springer, Cham, 2020.

S. Anita, V. Arnautu, V. Capasso: *An Introduction to Optimal Control Problems in Life Sciences and Economics. From Mathematical Models to Numerical Simulation with MatLab*, Modeling and Simulation in Science, Engineering and Technology. Birkhäuser/Springer, New York, 2011.

T. Archibald, C. Fraser, I. Grattan-Guinness: *The History of Differential Equations, 1670–1950*, Report No. 51/2004, Mathematisches Forschungsinstitut Oberwolfach, 2004.

V.I. Arnold: *Ordinary Differential Equations*, 2nd printing of the 1992 edition, Universitext. Springer-Verlag, Berlin, 2006.

V. Barbu: *Differential Equations*, Translated from the 1985 Romanian original by L. Nicolaescu, Springer Undergraduate Mathematics Series. Springer, Cham, 2016.

V. Berinde: *Iterative Approximation of Fixed Points*, 2nd edn., Lecture Notes in Mathematics 1912. Springer, Berlin, 2007.

S. Bittanti: The Riccati equation today, in "The Riccati and frontier culture in eighteenth-century Europe" (Castelfranco Veneto, 1990), Bibl. Nuncius Studi Testi V, Olschki, Florence, 1992, pp. 161–171.

F. Brauer, J.A. Nohel: *Qualitative Theory of Ordinary Differential Equations*. W.A. Benjamin Inc., New York, 1969.

M. Braun: *Differential Equations and Their Applications. An Introduction to Applied Mathematics*, 4th edn., Texts in Applied Mathematics, 11. Springer-Verlag, New York, 1993.

H. Brézis: *Analyse fonctionnelle*, Collection Mathématiques Appliquées pour la Maîtrise. Masson, Paris, 1983.

H. Brézis: *Opérateurs maximaux monotones et semi-groupes de contractions dans les espaces de Hilbert*, North-Holland Mathematics Studies, No. 5. North-Holland Publishing Co., Amsterdam-London; American Elsevier Publishing Co., Inc., New York, 1973.

H. Brézis: *Functional Analysis, Sobolev Spaces and Partial Differential Equations*, Universitext. Springer, New York, 2011.

A. Buică: Gronwall-type nonlinear integral inequalities, *Mathematica* 44 (67) (2002), no. 1, 19–23.

H. Cartan: *Formes différentielles. Applications élémentaires au calcul des variations et à la théorie des courbes et des surfaces*. Hermann, Paris, 1967.

C. Chicone: *Ordinary Differential Equations with Applications*, 2nd edn., Texts in Applied Mathematics, 34. Springer, New York, 2006.

C. Chicone, Y. Latushkin: *Evolution Semigroups in Dynamical Systems and Differential Equations*, Mathematical Surveys and Monographs, 70. American Mathematical Society, Providence, RI, 1999.

A.J. Chorin: *Lectures on Turbulence Theory*, Mathematics Lecture Series, No. 5. Publish or Perish, Inc., Boston, MA, 1975.

A.J. Chorin, J.E. Marsden: *A Mathematical Introduction to Fluid Mechanics*, 2nd edn., Texts in Applied Mathematics 4. Springer-Verlag, New York, 1990.

S.N. Chow, J.K. Hale: *Methods of Bifurcation Theory*, Grundlehren der Mathematischen Wissenschaften, 251. Springer-Verlag, New York–Berlin, 1982.

P.G. Ciarlet: *Introduction to Numerical Linear Algebra and Optimisation*, Cambridge Texts in Applied Mathematics. Cambridge University Press, Cambridge, 1989.

E.A. Coddington, N. Levinson: *Theory of Ordinary Differential Equations*, McGraw & Hill, New York, 1955.

C. Cordunneanu: *Principles of Differential and Integral Equations*. Chelsea P. Co., The Bronx, New York, 1971.

J.-M. Coron: *Control and Nonlinearity*, Mathematical Surveys and Monographs, 136. American Mathematical Society, Providence, RI, 2007.

J. Cronin: *Ordinary Differential Equations. Introduction and Qualitative Theory*, 3rd edn., Pure and Applied Mathematics (Boca Raton), 292. Chapman & Hall/CRC, Boca Raton, FL, 2008.

M. de Guzmán: *Ecuaciones diferenciales ordinarias. Teoría de estabilidad y control*. Ed. Alhambra, Madrid, 1975.

M. de Guzmán, I. Peral, M. Walias: *Problemas de ecuaciones diferenciales ordinarias*. Ed. Alhambra, Madrid, 1978.

T. Ding: *Approaches to the Qualitative Theory of Ordinary Differential Equations. Dynamical Systems and Nonlinear Oscillations*, Peking University Series in Mathematics, 3. World Scientific Publishing Co., Hackensack, 2007.

A.L. Dontchev: The Graves theorem revisited, *J. Convex Anal.*, 3 (1996), no. 1, 45–53.

A.L. Dontchev, R.T. Rockafellar: *Implicit Functions and Solution Mappings. A View from Variational Analysis*, 2nd edn., Springer Series in Operations Research and Financial Engineering. Springer, New York, 2014.

A. Dou: *Ecuaciones diferenciales ordinarias. Problema de Cauchy*, 2a edición, Editorial Dossat, S.A., Madrid, 1969.

A. d'Onofrio, U. Ledzewicz, H. Maurer, H. Schättler: On optimal delivery of combination therapy for tumors, *Math. Biosci.*, 222 (2009), no. 1, 13–26.

M.S.P. Eastham: *Theory of Ordinary Differential Equations*. Van Nostrand Reinhold Co., New York, 1965.

I. Ekeland, R. Témam: *Convex Analysis and Variational Problems*, Classics in Applied Mathematics, 28. Society for Industrial and Applied Mathematics (SIAM), Philadelphia, PA, 1999.

H.O. Fattorini: Infinite Dimensional Optimization and Control Theory, in *Encyclopedia of Mathematics and Its Applications*, Vol. 62: Cambridge University Press, Cambridge, 1999, pp. xvi–798.

D.T. Finkbeiner: *Introduction to Matrices and Linear Transformations II*, 2nd edn., W. H. Freeman and Co., San Francisco–London, 1966.

R. FitzHugh: *Mathematical Models of Excitation and Propagation in nerve*, Chapter 1 (in H.P. Schwan, ed. *Biological Engineering*. McGraw-Hill Book Co., New York, 1969, pp. 1–85.

J.R. Graef, J. Henderson, L. Kong, Lingju, X.S. Liu, Xueyan Sherry: *Ordinary Differential Equations and Boundary Value Problems, Vol. I, Advanced Ordinary Differential Equations*, Trends in Abstract and Applied Analysis, 7. World Scientific Publishing Co. Pte. Ltd., Hackensack, NJ, 2018a.

J.R. Graef, J. Henderson, L. Kong, Lingju, X.S. Liu, Xueyan Sherry: *Ordinary Differential Equations and Boundary Value Problems, Vol. II, Boundary Value Problems*, Trends in Abstract and Applied Analysis, 8. World Scientific Publishing Co. Pte. Ltd., Hackensack, NJ, 2018b.

P. Hartman: *Ordinary Differential Equations*, Corrected reprint of the second (1982) edition, Classics in Applied Mathematics, 38. Society for Industrial and Applied Mathematics (SIAM), Philadelphia, PA, 2002.

P. Henrici: *Discrete Variable Methods in Ordinary Differential Equations*. John Wiley & Sons, Inc., New York–London, 1962.

H. Hermes, J.P. LaSalle: *Functional Analysis and Time Optimal Control*, Mathematics in Science and Engineering, Vol. 56: Academic Press, New York-London, 1969.

M.R. Hestenes, E. Stiefel: Methods of Conjugate Gradients for Solving Linear Systems, *J. Res. Nat. Bur. Standards* 49 (1952), 409–436 (1953).

H. Hochstadt: *Differential Equations. A Modern Approach*, republication of the 1964 original, Dover Publications, Inc., New York, 1975.

S.-B. Hsu: *Ordinary Differential Equations with Applications*, Series on Applied Mathematics, 16, World Scientific Publishing Co., Hackensack, 2006.

W. Hurewicz: *Sobre Ecuaciones Diferenciales Ordinarias*. Rialp, Madrid, 1966.

F. John: *Partial Differential Equations*, reprint of the 4th edn., Applied Mathematical Sciences, 1. Springer-Verlag, New York, 1991.

H.R. Joshi: Optimal control of an HIV immunology model, *Optimal Control Appl. Methods* 23 (2002), no. 4, 199–213.

A. Kiseliov, M. Krasnov, G. Makarenko: *Problemas de Ecuaciones Diferenciales Ordinarias*.x Ed. Mir, Moscú, 1973.

A.N. Kolmogorov, A.P. Yushkevich: *Mathematics of the 19th Century. Function Theory According to Chebyshev, Ordinary Differential Equations, Calculus of Variations, Theory of Finite Differences.* Birkhäuser Verlag, Basel, 1998.

P.L. Korman: *Lectures on Differential Equations*, AMS/MAA Textbooks, 54. MAA Press, Providence, RI, 2019.

S.G. Krantz, H.R. Parks: *The Implicit Function Theorem. History, Theory, and Applications*, Modern Birkhäuser Classics. Birkhäuser/Springer, New York, 2013.

J.D. Lambert: *Numerical Methods for Ordinary Differential Systems. The Initial Value Problem*, John Wiley & Sons, Ltd., Chichester, 1991.

S. Lang: *Real and Functional Analysis*, 3rd edn., Graduate Texts in Mathematics, 142. Springer-Verlag, New York, 1993.

E.B. Lee, L. Markus: *Foundations of Optimal Control Theory,* 2nd ed., Robert E. Krieger Publishing Co., Inc., Melbourne, FL, 1986.

W.S. Levine: Fundamentals of dynamical systems, in *Handbook of Networked and Embedded Control Systems* Control Eng. Birkhäuser, Boston, MA, 2005, pp. 3–20.

X. Li, J. Yong: *Optimal Control Theory for Infinite Dimensional Systems.* Birkhäuser, Boston 1995.

L.A. Liusternik, V.J. Sobolev: *Elements of Functional Analysis,* Russian Monographs and Texts on Advanced Mathematics and Physics, Vol. 5: Hindustan Publishing Corp., Delhi, Gordon and Breach Publishers, Inc., New York, 1961.

X. Lu, L. Wang, Q. Yan: Computation of time optimal control problems governed by linear ordinary differential equations. *J. Sci. Comput.* 73 (2017), no. 1, 1–25.

D.G. Luenberger: *Optimization by Vector Space Methods.* John Wiley & Sons, Inc., New York–London–Sydney, 1969.

R.K. Miller, A.N. Michel: *Ordinary Differential Equations*, Academic Press, London, 1982.

A.K. Nandakumaran, P.S. Datti, R.K. George: *Ordinary Differential Equations. Principles and Applications,* Cambridge-IISc Series. Cambridge University Press, Cambridge, 2017.

J.E. Nápoles: De Leibniz a D'Alembert: Diez problemas que marcaron un siglo, Boletín Mate., Nueva Seri XV 2 (2008) no. 2, 130–161.

J.E. Nápoles, C. Negrón: La Historia de las ecuaciones diferenciales contada por sus libros de texto, Xixim Rev. Electrón. Mate., 2(2002) no. 2., http://www.uaq.mx/matematicas/redm/index2.html.

S. Nikitin: *Global Controllability and Stabilization of Nonlinear Systems,* Series on Advances in Mathematics for Applied Sciences, 20. World Scientific Publishing Co., Inc., River Edge, NJ, 1994

V.I. Norkin: *Generalized Gradients in Dynamic Optimization, Optimal Control and Machine Learning Problems*, Cybernet. Syst. Anal., 56 (2020), no. 2, 243–258.

H. Okamura: Sur l'unicité des solutions d'un système d'équations différentielles ordinaires, *Mem. Coll. Sci. Kyoto Imp. Univ. Ser. A*, 23 (1941), 225–231.

I.G. Petrovskii: *Ordinary Differential Equations*, Prentice Hall Inc., Englewood Cliffs, NJ, 1966.

L. Piccinini, G. Stampacchia, G.V. Vidossich: *Ordinary Differential Equations in* $\mathbb{R}^n$: *Problems and Methods*, Springer-Verlag, New York, 1984.

E. Polak: *Optimization. Algorithms and Consistent Approximations*, Applied Mathematical Sciences, 124. Springer-Verlag, New York, 1997.

M.R. Rao: *Ordinary Differential Equations. Theory and Applications*. E. Arnold Ltd., London, 1981.

A.P. Robertson, W. Robertson: *Topological Vector Spaces*, reprint of the 2nd edn., Cambridge Tracts in Mathematics, 53. Cambridge University Press, Cambridge-New York, 1980.

J.C. Robinson: *An Introduction to Ordinary Differential Equations*, Cambridge University Press, Cambridge, 2004.

N. Rouché, J. Mawhin: *Equations différentielles ordinaires*, Tomos 1 y 2, Ed. Masson, Paris, 1973.

W. Rudin: *Principles of Mathematical Analysis,* 3rd edn., International Series in Pure and Applied Mathematics. McGraw-Hill Book Co., New York–Auckland–Dusseldorf, 1976.

H. Rund: *The Hamilton-Jacobi Theory in the Calculus of Variations: Its Role in Mathematics and Physics*. D. Van Nostrand Co., Ltd., London–Toronto, Ont.–New York, 1966.

F. Simmons: *Ecuaciones Diferenciales*, McGraw & Hill, México, 1977.

D.R. Smith: *Variational Methods in Optimization*, Dover Publications, Mineola, NY, 1998.

I.N. Sneddon: *Elements of Partial Differential Equations*, MacGraw & Hill Book Co., Inc., New York, 1957.

E.D. Sontag: *Mathematical Control Theory. Deterministic Finite-Dimensional Systems*, Texts in Applied Mathematics, 6. Springer-Verlag, New York, 1990.

J. Speck: *Shock Formation in Small-Data Solutions to 3D Quasilinear Wave Equations*, Mathematical Surveys and Monographs, 214. American Mathematical Society, Providence, RI, 2016.

M. Spivak: *Cálculo en variedades*. Reverté, Barcelona, 1970.

E. Trélat: *Contrôle Optimal. Théorie & Applications*, Mathématiques Concrètes. Vuibert, Paris, 2005.

M. Vasile: *Multi-Objective Optimal Control: A Direct Approach*, Satellite dynamics and space missions, Springer INdAM Ser., 34. Springer, Cham, 2019, 257–289.

D. Vrabie, K.G. Vamvoudakis, F.L. Lewis: *Optimal Adaptive Control and Differential Games by Reinforcement Learning Principles*, IET Control Engineering Series, 81. Institution of Engineering and Technology (IET), London, 2013.

G. Wang, L. Wang, Y. Xu, Y. Zhang: *Time Optimal Control of Evolution Equations*, Progress in Nonlinear Differential Equations and their Applications, 92. Subseries in Control. Birkhäuser/Springer, Cham, 2018.

E.C. Zachmanoglou, D.W. Thoe: *Introduction to Partial Differential Equations And Applications*. Dover Publications, New York, 1986.

E. Zeidler: *Nonlinear Functional Analysis and its Applications. I. Fixed-Point Theorems*, Springer-Verlag, New York, 1986.

E. Zeidler: *Nonlinear Functional Analysis and Its Applications. IV. Applications to Mathematical Physics*, Springer-Verlag, New York, 1988.

Index

www.ingramcontent.com/pod-product-compliance
Lightning Source LLC
Chambersburg PA
CBHW052118230326
41598CB00080B/3848